PARTICLE or WAVE

PARTICLE or WAVE

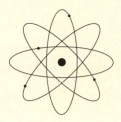

THE EVOLUTION OF THE
CONCEPT OF MATTER IN
MODERN PHYSICS

CHARIS ANASTOPOULOS

PRINCETON UNIVERSITY PRESS · PRINCETON AND OXFORD

Published by Princeton University Press, 41 William Street, Princeton,
New Jersey 08540

In the United Kingdom: Princeton University Press, 6 Oxford Street, Woodstock,
Oxfordshire OX20 1TW

ISBN-13: 978-0-691-13512-0

This book has been composed in Minion

Printed in the United States of America

To Maria-Elektra

CONTENTS

ILLUSTRATIONS

TABLES

ACKNOWLEDGMENTS

I would like to offer my greatest thanks to Ntina Savvidou, who urged me to start working on this book and for subsequent strong encouragement and advice. The perspective presented here has been strongly influenced by countless discussions on these topics. For encouragement and advice, I also thank Nikos Plakitsis. Also many thanks to Yannis Kouletsis for helpful comments on the manuscript at a critical stage, and to Yannis Savvidis for both reading the manuscript and help with an illustration. Many of the concepts and ideas presented here have been clarified in the course of discussions with many people: I would like to thank Chris Isham, Jonathan Halliwell, Bei-Lok Hu, Gerard 't Hooft, and Dimitris Ghikas in this regard. I also thank Utrecht University and the University of Patras for their hospitality during the time this book was being written. Special thanks to my editor, Ingrid Gnerlich, who trusted and supported this book even when it was at a stage of a rather digressive manuscript. Finally, I thank the anonymous readers of Princeton University Press for their comments, which were very important for the focus and balance of this work.

A NOTE ON TERMINOLOGY

S ome of the expressions and terms employed in this book diverge from the ones commonly encountered in the physics literature. This is unavoidable since the book is addressed to nonexperts. The purpose of this note is to identify all such expressions, explain the way they are used in the text, and relate them to the more standardized terms that appear in the physics bibliography.

Physicists have abandoned the use of the term *laws of motion* ever since the late nineteenth century, and perhaps even earlier. They prefer to employ the word "dynamics" in its place. Here I have decided to keep the term throughout (even though the notion of "dynamics" is fully explained). The first reason is convenience—it would be awkward to switch to a different word in the middle of the book, even though it refers to the same concept. The second reason is that the term "laws of motion" is slightly more general than "dynamics." It is a priori conceivable that the laws of motion in a physical system may be completely kinematical, even though this does not seem to be the case in physics. The reader should interpret "laws of motion" as "rules of motion," namely, as the rules that determine how a physical system evolves in time.

The *quantum phase* we refer to in chapters 6 and 9 is a term that is used rather loosely in contemporary quantum physics. It refers to different things according to the context. In the present context, it denotes the redundancy in the description of quantum states, which essentially corresponds to an anglelike physical variable. Its physical implications are, however, the same as described in the text. The *connection* of the quantum phase refers to a mathematical object by the same name that appears in the so-called theory of geometric quantization, and its physical function is essentially the one described in chapter 6. The notions of Bose and Fermi connection are also defined in this theory. I employed them in order to provide a mechanical representation of the spin-statistics relation.

The use of the expression *q-number* has been largely abandoned in contemporary physics talk. The preferred term is *operator*, which refers to the mathematical object that represents observables in quantum theory.

The words *state, configuration,* and *profile* convey largely the same meaning, namely, the most complete possible description of a physical system at a moment of time. Even though I occasionally use them interchangeably, I employ "profile" in reference to classical fields, "state" in reference to the quantum state, and "configuration" in a generic sense. The expression *irreducible system associated with a symmetry* is a translation of a concept, referred to in the mathematical literature as *irreducible unitary representation of a symmetry group* (in the context of quantum theory), or *transitive symplectic action of a symmetry group* (in classical mechanics).

While in the first two chapters of the book the word *force* denotes the specific physical concept that is employed in Newton's laws of motion, it is employed only as a colloquial term in later chapters (in the discussion of the microscopic "forces"). Force is not a fundamental concept in modern physics, but it remains as a descriptive term.

PARTICLE or WAVE

INTRODUCTION

Perhaps Universal History is nothing more than the history of a few metaphors.
　　　　　—Jorge Luis Borges, *The Sphere of Pascal*

I f we ask individuals from a Western society to ponder the meaning of the word "matter" (in the sense of material), they will probably come up with many different images. They may think about the atoms and all the tiny particles that have been discovered by modern science. Or perhaps they will think about the successes of technology in controlling and manipulating all different types of material processes and then consider how this affects their everyday life. They may also realize that they are themselves material beings and will inevitably remember the oldest of contrasts, "mind vs. matter," or perhaps "mind over matter." Or they may be quick to think in moral terms and focus on the distinction between the "person of spirit" and the "person of matter." One way or another, the word "matter" will have a meaning for them—perhaps not a unique or an unambiguous one, but they will definitely conclude that it refers to something so fundamental as to be really important.

We may next imagine asking the same question to people of other cultures and times. We will have to find an appropriate word in their language to translate "matter." It is rather remarkable—given the immense diversity of living and dead languages—that most of the time we will be able to find an adequate translation. The corresponding word may not have exactly the same connotations—the idea of atoms would hardly come into the mind of a person unfamiliar with modern science, or at least with a developed philosophical tradition. There may also exist fine differences of meaning between these words. Still, any person who understands the question will not fail to think of matter as the substance manipulated with one's hands, however primitive the technology may be. After all, humans are and have always been toolmakers and tool users. And neither will the person fail to grasp the distinction between matter and mind because this is akin to a distinction familiar to every human being, that between the things of the outside world that we perceive through our senses and the emotions and thoughts of our internal life.

It follows, then, that "matter" is an abstraction common to people of different times and cultures. It arises out of the one single thing that a modern person shares with a Greek of Plato's times, a tribesman of New Guinea, and a courtier of the Forbidden City: our common humanity in the way we sense, feel, act, and live our lives in the world. Matter is therefore a universal concept. A magician of old times would not hesitate to call it a "word of power"; a modern psychologist would perhaps prefer the word "archetype." Of course "matter" is not the only such word: time and space, emotion and reason, divine and profane, multiplicity and unity, flux and immobility, cause and effect, psyche and matter—all these are ideas that resonate in the thoughts and languages of all times, and their significance is explored in the people's philosophies, myths, religion, songs, or dreams. There exist differences of meaning, of course. One concept may be understood differently in different cultures. But no matter how important such differences may be, they do not affect the essence of the fundamental concepts. These concepts are perceived at a deep level of the human psyche, before any association to words, logical definitions, or social practices. At that level, they are the same everywhere because they refer to an experience of reality common to all people.

In modern societies, however, the study of matter is perceived to be a prerogative of the sciences, mainly physics. Indeed, the development of physics during the last couple of centuries has provided deep insight into the deepest structure and organization of matter. The success of the scientific method has been so overwhelming that by the end of the twentieth century many scientists did not hesitate to make an astonishing claim that it is only a matter of time before we construct a "theory of everything."

The scientific description of the world, however, does not grow in a vacuum. It is equally impossible to detach the concepts of science from ideas, perceptions, and feelings that arise in other fields of human culture. Scientific concepts may be presented in their final form detached, self-contained, and logically complete, but this is mostly an outward appearance, similar to the precise, clean-cut, and orderly performance given by a military unit during an inspection by their commander in chief. In reality, the concepts of science and the thoughts of scientists are immersed in the ocean of the society people live in, and their origins are blended with a myriad of other ideas, images, and thoughts. This is the reason that even an account of the scientific theories of matter cannot be restricted to the scientific theories themselves but has to seek their roots in other fields of human activity.

History also plays an important role. Hermann Weyl, one of the greatest mathematicians of the past century, has remarked that it is impossible to understand the character of modern mathematics unless it is considered in the perspective of its history over the last two thousand years. The same is also true for modern physical theories, and even more emphatically so. Mathematical concepts are abstract by nature, while the physical ones have a more earthly character. They are more closely related to the concrete experience of the senses and the way the world is perceived in different times and places. Indeed, mathematical purists are often exasperated with the ambiguous character of the physical concepts, which still seem to carry the "dirt" of their lowly origins.

If we follow Weyl's advice in the context of modern physics and attempt to trace back in time the origin of the modern physical concepts about matter, we see that they arise from two conflicting theories, whose seeds in art and myth are probably lost in the mists of prehistory. These theories were first explicitly stated and rationally analyzed in Greece during the fifth century B.C. The first of these theories is the *atomic theory*; it describes matter as consisting of discrete, indivisible pieces, which move in the void and through their motion create all things that come into our perception. Indeed, even the human soul consists of such particles, finer in their form than the ones of ordinary matter. The motion of these *atoms* is due to a necessity external to them, which determines precisely and unambiguously how and toward which direction each atom will move. Everything that we see in the world may be analyzed in terms of the distinctive properties of the atoms and their motion in the void; nothing else exists—all other images are simply figments of our imagination.

In opposition to the atomic theory stands the *theory of the elements*. Matter, it proposes, consists of elementary substances (the *elements*), which are distinguished from each other by means of their intrinsic qualities. The elements are pure and simple; they represent elementary qualities and cannot be analyzed in simpler components. Initially only four such substances were postulated: Earth, Water, Air, and Fire. All material things that appear to our senses arise from the elements being blended together, and what we perceive as creation and decay is nothing but the change of the analogies in the blend. This change is due to powers that are inherent in matter (originally denoted as love and strife), whose effect is to bring elements together or take them apart. Since the elements refer to qualities, and qualities are seemingly preserved when we divide matter in pieces, it is more natural in the theory of the elements to postulate that matter can be subdivided indefinitely.

These two theories not only differ in their details but also represent very different attitudes toward the world. The atomic theory has the ambition to explain all material phenomena through the motion of undivided entities in space. If the human intellect succeeded in understanding the specific rules that guide the atomic motions, it would be able to explain and describe every single process of nature. The atomic theory then lends itself to an image of predictability and control, a demystification of matter. The theory of the elements is the exact opposite. It places quality at the center of its explanations—and qualities, however much abstracted, are by their nature closer to sense perception than to logical calculation. Moreover, the theory of the elements attributes the cause of change to the intrinsic forces and powers of matter. It emphasizes the spontaneous character of matter, its mutability and aversion to control.

But as opposites flow to each other, the two conflicting perspectives were destined to a marriage. A third party was needed to bring them together, and this had appeared about fifty or sixty years earlier. A rather shadowy figure of antiquity, Pythagoras, had come up with an idea that may perhaps appear self-evident to a modern person. *Reality is structured upon symmetry*, and symmetry—the proper proportion—is something that subsists in number and figure, the subject matter of mathematics. The two conflicting theories for matter together with the realization of the mathematical character of reality were the germs of subsequent ideas about the structure of matter, to which we may trace the lineage of even the most complex theory of modern physics.

We follow somewhat the development of these theories in antiquity and the Middle Ages, but we shift our attention to the sixteenth century A.D. At the westernmost corner of the Eurasian continent, we encounter a young culture that had a deep fascination with number and shape; a love for the rediscovered ancient traditions mixed with a curiosity for the new worlds that lay open before them; an unprecedented strong inclination toward precision of observation and logical deduction; and a strong desire for the power bestowed by the unraveling of nature's secrets. That intellectual environment needed only one single spark before it caught fire. It took thirty-six years of inspired work by a single person to produce that spark. That person was Nicolas Copernicus, the Polish monk who resurrected the then obsolete idea that Earth and all planets move around the Sun. He expressed a deep conviction that the language of mathematics captures nothing less than the whole "System of the World." Copernicus's spark became fire, and the fire brought revolution—what we nowadays describe as the era of the Scientific Revolution was born.

It took almost a century and a half for the revolution to reach its apex in the work of Newton. His creation, unprecedented in scope, united "the things of the heavens and the things of the earth." They were all subject to the same laws, which were expressed unambiguously in a powerful mathematical language. The new theory kept the old atomic theory's vision of particles moving in the Void, but, influenced by a specific emphasis on precision and strict logical succession, the necessity that caused all motion was so strict, unyielding, and ever present as to be beyond anything that had ever been dreamed up before. The new theory was named mechanics. This name brings forward the image of matter as a machine, an object whose motion is governed by strict, unambiguous rules that apply separately to each of its parts. The image of the machine was so well attuned to the new theory that it slowly assumed its powers—in spite of the misgivings of its creator and loud protests from every direction.

One after the other, the physical phenomena succumbed to the lure of mechanical explanation and description. The machine moved eventually from the realm of science, and rational explanation into the realm of practice and everyday life. A new revolution—the industrial one—materialized, and it transformed not only people's perceptions of the world but the world itself. The new world brought wealth, comfort, and the promise to fulfill all material needs. However, progress came through the machine, and it had its price. It made people fear losing their humanity: many artists, philosophers, and scientists attempted to provide a different view of the world, but their efforts seemed at that time to be in vain.

Then came the twentieth century, bringing with it war and revolution. To the new generation, the comfort and security that progress had promised them appeared more and more like an illusion. Consequently, the perception of the world was radically transformed. In particular, the image of the machine was dealt two strong blows. They came from the same realm of thought from which the image had originally appeared: physical theory. The first blow was the work of a single person—Albert Einstein and his theory of relativity. The other—the more powerful one—was the theory of quanta, which was delivered through the efforts of a generation of physicists. All of a sudden the world looked different. Matter seemed to have been dematerialized, its mechanism dismantled and substituted by an uneasy abstraction, while the old but never forgotten ideas of inherent powers in matter returned to complement the atomic perspective. But the same moment that matter seemed more uncontrollable and uncertain, control over it increased rather than lessened, and technology moved in directions of greater promise and power, and of greater destruction.

The story above provides the background of the issues addressed in this book, namely, the images of matter in modern physical theories. The key principle of modern quantum theory is the remarkable assertion that the elemental and the atomic aspects of matter coexist. In technical language, this principle is referred to as the *field-particle* (or wave-particle) duality. I argue that the field description of matter is obtained from the development and refinement of basic insights of the theory of the elements, and in a similar way the atomic theory carries the conceptual germs of the particle description.

However, such identifications are neither simple nor straightforward. Even though a central thesis of this book is that the complex concepts of modern physics can trace their origins to simple and intuitive ideas, we cannot ignore the fact that modern physical theories weave these ideas in a distinctive and complex pattern that cannot be perceived unless one follows the twists of every single thread. If the atomic perspective of particles moving in the Void remains fundamental to modern physics, the three words "particle," "moving," and "Void" involve so many more associations and properties than their traditional counterparts that they appear substantially different. And whereas some scientists have seen in modern physics the rebirth of ideas about intrinsic powers of matter, the carrier of this idea is a mathematical object, the *quantum field*, which is so abstract in its character that its true essence, if such exists, remains hidden from our eyes.

For this reason, I found it necessary to expand on some of the intricate details of modern theories about the motion and structure of matter. Keeping the wave-particle duality and its implications in focus, I survey the state-of-the-art understanding of the structure of matter, namely, the theories of *quantum fields* and *elementary particles*. I place particular emphasis on identifying the limits of the present theories, and hence on determining the boundaries beyond which our present knowledge is not secure.

I hope I have been able to communicate the scope of this book and the main ideas that motivated it. The story narrated here is hardly exhaustive, and the choice of what to include and what to leave out reflects largely my personal tastes and preferences. My main guideline in the choice of the material was to look for the components that form the images of matter in modern scientific theories.

As far as the description of physical theories is concerned, I have tried to include all important concepts and ideas, and to present diverse and even conflicting attitudes and opinions. Still, the emphasis reflects inevitably my personal perspective. The balance of the presentation was another criterion in the choice of material. As the book deals primarily with the properties and structure of matter at the fundamental level, I have practically ignored specific fields of modern physics that deal with the larger-scale organization of matter—such as the theory of condensed matter or the physics of stars and galaxies. Clearly, I am solely responsible for any omissions, errors of fact or judgment, partiality of perspective, or mere ignorance. I can only ask for the reader's patience with any such inadequacies.

Finally, I wish the reader as much enjoyment in reading this book as the pleasure I have had in writing it.

1

FROM MYTH TO MACHINE

IMAGES OF MATTER FROM ANTIQUITY TO
NEWTONIAN MECHANICS

1.1 First Questions

It is a fact that the totality of experiences is so constituted as to permit
putting them in order by means of thinking—a fact which can only
leave us astonished, but which we shall never comprehend. The eternally
incomprehensible thing about the world is that it is comprehensible.
—Albert Einstein, *Physik und Realität*

Leaf through a dictionary or try to make one, and you will find that
every word covers and masks a well so bottomless that the questions you
toss into it arouse no more than an echo.
—Paul Valery, *Analects*

"Man by nature desires to know"—this is the opening line
of one of the most influential treatises in the history of
philosophy, Aristotle's *Metaphysics*. It is a statement so profound as to
be almost trivial. However, one should add that it is also in a person's
nature not to live alone—"unless he is an animal or a God." Whatever
people learn, they always have the desire to communicate it to their
fellows. Knowledge arises from a common effort, and it is common

effort that turns—with the passing of generations—mere observation into general fact, and general fact into concrete understanding. Every culture therefore develops a body of knowledge about the world. That knowledge is essentially of two types: one type refers to the external world, in which we live, act, and apprehend through our senses (to which we refer today as physics), and the other refers to the internal world of personal experience, emotion, and thought (which we speak of as psychology).

If the desire for knowledge constitutes a basic feature of humanity and is therefore present in all cultures, the description of the world varies extensively from culture to culture. Most civilizations of the past have been content with a description coined in mythological terms. They do not phrase their observations or their theories about the world in a direct definite manner, but they prefer to rely on the language of myth, allegory, and metaphor.

It is a noteworthy achievement for any society to go beyond the indirect descriptions of the world and attempt a description in terms of analysis and logical argument. The first step in this "rational" approach to the description of the world is the realization that the rules of thoughts and language, which are processes of humanity's internal world, have a correspondence with the phenomena of the external world. Disciplined thinking and precise language may discover connections between the events of the external world, even connections that remain hidden from direct experience.

In other words, language and thought satisfy certain rules of universal validity that are innate to human beings. These rules have something in common with the ones that govern the physical processes of the external world. People eventually realize that there is a basic principle, which transcends the dichotomy between the external and internal things and suggests the existence of an underlying unity. The Greeks of ancient times, loving discourse and argumentation, denoted this principle as *Logos*, a word translated invariably as speech, word, or rationality. A derivative word is logic, which was originally defined as the study of the rules of language in their relation to the things of the world.

This distinction between the "rational" and the "mythical" description of the world is important, but not absolute or oppositional. The mythical perspective underlies, supports, and nourishes all rational manifestations of the human intellect. After all, the myth and the metaphor are the language of dreams, of the unconscious. Modern psychology has repeatedly demonstrated the importance of the human unconscious in the development of the conscious experience, even the intense and disciplined intellectual activity required in the formulation of modern scientific theories.

I explained in the introduction that the subject of this book, matter, is a deep and fundamental concept that has been used in all cultures to denote the fundamental substance of the external world: the things we act upon with our hands and shape into a work of art or technology, but also the things that act upon us by stimulating our senses or by threatening to damage our bodies. I shall focus on the structure and properties of matter, especially as they have been understood by modern physics. However, the remarkably successful modern concepts about matter have their origin in much older ideas that had emerged in the study of nature. By modern standards, these ideas are perhaps vague and imprecise. But, like seeds buried in the ground, they waited for suitable conditions to grow and bring fruit. Such ideas incorporated insights about the world that were deep and powerful. They evoked mental images that were stronger than the initial phrasing or context of the idea. For this reason, they transcended the bounds of time and space and affected the thinking of people very different in culture to the ones who first perceived them. Deep ideas achieve, very literally, mythical status in human history.

What we call the "rational" approach for the description of nature is based in a very important ability of the human intellect: the ability to transfer familiar concepts, ideas, and descriptions into a new context, completely unrelated to the initial one. We call this process metaphor (a transfer). Metaphors make possible the description of things that lie beyond our immediate experience through a comparison with things that are well known and understood. *The things we see are a sign for the things we do not see,* states an ancient proverb. Indeed, the things we know can lead us—after strenuous efforts of will and intellect—to concepts that refer to a world beneath and beyond anything we perceive through our senses.

Language is steeped in metaphor. The metaphor is not simply a feature of fancy or funny expressions. Abstract words incorporate a metaphor in their meaning. For example, the word "physics" comes from the word "physis," usually translated as nature. This word denoted originally vegetation, growth, and change. Hence, when we discuss physical knowledge there is the subtle implication that we need to find out how things change and grow, and nature, physis itself, is this process of growth and change. Similarly, the word "theory" is related to vision; a theory is then an attempt of the human intellect to achieve as clear an understanding of things as the one provided by the most fundamental of all senses, vision. The best theories are the ones that allow us to visualize concepts as though they are concrete things or processes that take place before our eyes.

1.2 The Two Paradigms

> It is evident, then . . . that all men seem to seek the causes named in the Physics, and that we cannot name any beyond these; but they seek these vaguely; and though in a sense they have all been described before, in a sense they have not been described at all. For the earliest philosophy is, on all subjects, like one who lisps, since it is young and in its beginnings.
> —Aristotle, *Metaphysics*

It is often possible to explain the physical phenomena in terms of familiar objects or processes. We observe, for instance, the clouds, large white or gray bulks moving in the sky. We also notice that clouds bring rain, and that rain clouds often disappear afterwards. Since rain is water, clouds must also be made of water. But how does water stay up in the sky, when at all other times it falls to earth? To answer this question, we come up with a similar occasion: water evaporates when it boils within a pot. It is then natural to assume that clouds consist of evaporated water.

It requires only a bit more effort to realize that rain falls when the cloud cools sufficiently for the vapor to become liquid and fall to earth.

The explanation above demanded quite a bit of imagination from the person who first conceived it, but it involved only a simple observation of everyday phenomena. But when one asks a deeper question, like "what do things consist of?" the answer is far beyond immediate experience. Any answer needs the creation of new concepts. These will necessarily be phrased in terms of familiar things; otherwise we would explain something unknown in terms of something also unknown, which is a self-defeating procedure. The process of metaphor comes to our rescue here, always the first step for an attempt to understand the world that goes beyond immediate experience.

One possible answer to the question "what do things consist of?" is that all things arise out of water. This was put forward by the person traditionally credited as the first of the Western philosophers, the Greek Thales from the city of Miletus, who lived during the seventh century B.C. This is the first recorded instance of an explanation of the constituents of the world, supported with logical argumentation and put forward in a nonmythical language.[1] The answer is not as naive as it looks at first sight. Thales started from the observation that water seems to appear in all earthly phenomena, and he generalized this fact to the larger context of the fundamental principles of matter. Clearly, the word "water" was used metaphorically in the latter context. Thales postulated a fundamental ingredient of the world, a primordial state of existence for matter that underlies all things that appear to us. This fundamental substance shared with water some important properties: flow, mutability, and the ability to act as source of life. Ordinary water itself is not the fundamental ingredient. Water was considered in most mythologies, at almost all times, as the source of all existence. In Greek mythology, for example, it was represented by the Ocean, an ancient, formless, and hidden deity that had not been born from anything and constituted "the origin of all Beings."[2]

Thales's attempt to explain the basic nature of the world was not very far off the mark, even though it may seem a mere semirational restatement of mythological tradition. Any theory about the fundamental constituents of things has at its roots one out of two alternatives—two different metaphors that generalize different observations of everyday life.

The first such description may arise from the observation of ordinary sand. We may imagine, for example, a beach. It is a large expanse of sand that looks like a continuum, much like our experience of the world through the senses is continuous. However, if we take a handful of sand and look closely, we notice that the sand consists of grains, that is, small, discrete things. We cannot easily split these grains into smaller ones, the way we can separate a handful of sand into two smaller lumps.

We next transfer the image of the sand to the level of the fundamental character of matter. We arrive at the statement that things consist of small, fundamental bodies, which cannot be cut into smaller pieces. This is the "atomic theory," the first recorded instance of which is credited to Leukippus, a rather shadowy figure of early Greek philosophy. (We should note that a different concept of "atoms" also appeared at about the same time in India.[3]) The development of the atomic theory as an overriding theoretical model with the ambition to account for all experience is due to Democritus of Abdera, a great traveler and prodigious writer, who lived in the fifth century B.C. In modern language, the atomic theory is translated into the statement that all things consist of fundamental particles.

On the other hand, we may follow Thales and consider water, or rather any fluid, as our basic material. We readily observe that many different objects can be created when fluids are mixed in different proportions. By transferring this simple observation to the level of the deeper structure of the world, we arrive at a different conception: material bodies arise from the mixing of different types of fundamental fluids. The property of fluids we concentrate on is not fluidity, but *continuity* up to the tiniest of scales; that is, the fact that we can split a drop of water into smaller drops without ever reaching an end to this process—or at least this is how it appears to the senses unaided by any modern instruments. The fundamental elements, like the primordial water, are simple; they cannot be analyzed into something more elementary.

In the theory of the elements, a physical body owes its characteristics to the mixing proportions of the fundamental fluids that are involved in its formation. Such a description in terms of "fluids" has appeared in many civilizations,[4] but historically the most influential for the formation of science was the version that appeared in ancient Greece. Its form was sufficiently sophisticated that it could qualify as a theory for all material things. Its highest point was the theory of Empedocles from southern Italy. Empedocles identified four fundamental fluids, or rather *elements*, which he conveniently described as Earth, Air, Water, and Fire. I emphasize again that these names are metaphorical. The element called Earth derives its name from its defining property of solidity; Water, from fluidity; Air, from its gaseous nature; and Fire, from its relation to heat and its mutability. Common earth, water, fire, and air are not themselves the fundamental elements. The names assigned to the fundamental elements describe some of their characteristic properties. In Empedocles' words,

> Through earth we perceive Earth, through water Water,
> Air through air and from the fire hidden Fire[5] . . .

The main distinction between the atomic theory and the theory of the elements lies in the fact that an element is characterized by a fundamental *quality*. Qualities always refer to our sense experience. For instance, Earth incorporates the notion of solidity, which is something we experience by touch. In contrast, the atomic theory describes all these qualities in terms of the shapes and the motions of atoms, which are both geometric concepts. Even though geometry refers to our sense experience of shapes in material objects, it is largely autonomous from the senses as a discipline. It may be described in terms of abstract concepts and relations that do not refer to any datum of experience. This makes the geometric description far more objective than any description based on qualities. Democritus stated that qualities are derived things, images of the mind:

> [S]weetness and bitterness are just human conventions.
> All that really exists is atoms and the Void.[6]

The majority of physicists during the late nineteenth and early twentieth century came to repeat that statement almost word for word.

The atomic theory necessarily assumes that the fundamental particles exist and move within some "container." This container is not material itself. Otherwise, it would also consist of atoms. Being void of matter, this container was referred to as the Void. The atomist worldview, therefore, needs two concepts to explain our world of experience: the fundamental material particles and the Void. The Void is not nothingness. It is empty *space*: each of the points of space is characterized by the *potentiality* that at some moment of time a particle may appear there. One of the first lessons of ancient philosophy was the realization that potentiality is very different from nothingness.

The second basic feature of the atomic theory is the postulate of a law that governs the motion of the particles. The atomic motion cannot be without order. Democritus postulated the existence of an overriding, absolute necessity governing the motion of these particles. Indeed, this necessity is the *cause* of the atomic motion.

Democritus's "successor" as promoter of an atomic theory, the philosopher Epicurus, had a different attitude: his motives were different. Democritus had been first and foremost a physicist, a student of nature. Epicurus was mostly a moral philosopher and he cared more about the behavior of man and his status in the world. He realized that if the atomic theory is correct, then man also consists of such particles and of nothing else. Hence, man's motions and choices were fully determined by the iron necessity that guided the motion of atoms. But Epicurus believed in man's freedom of will. This could only be maintained by somehow making the necessity of Democritus more flexible. Therefore, he postulated a *random* component in the motion of the atoms[7] to escape the clutches of a stern necessity. Epicurus's dilemma was an early instance of the conflict between the determinism of the laws of matter and human beings' free will. We shall see that this debate was kindled afresh during the time of the scientific revolution. It was a central dilemma to the philosophers of eighteenth-century Europe, the age of Enlightenment. It still rages, stirred by the emergence of quantum theory in the twentieth century.

According to the atomic theory, the properties of the objects in our immediate experience can be explained in terms of atoms and the laws that govern their motion. However, the objects of our immediate experience are different from each other (a rock is different from a flower), even though they all consist of atoms. The reason for this difference is that atoms differ in certain fundamental parameters. In various forms of atomic theory (to an extent even in modern atomic theory), such parameters are the size, shape, and weight of the atoms. However, the diversity of macroscopic properties is also due to the multiplicity of the different possible ways through which these basic particles combine during their incessant motion in the Void.[8]

The atomic theory was the first recorded *materialist* theory in history—not only did it seek to explain the physical phenomena, but also the psychic ones fell under the scope of its explanation. Life and soul are also, Democritus thought, manifestations of atoms. The atoms responsible are lighter and finer than those of ordinary matter, but they still move in the Void and endure the rule of an iron necessity.

The question of how the things of our experience form and degenerate also arises in the theory of fundamental elements. Empedocles identified two basic principles that bring the fundamental fluids together or take them apart. The names he gave to those principles were rather explicit: love and strife, the principle of unity and the principle of discord. Love and strife are, in a sense, the psyche of matter, the origin of all its motions and changes, the principles that make matter always shifting in its appearance. Later generations gave different names and descriptions to these principles: they were called forces, energies, or even the universal psyche. Still, in all their different forms and descriptions, these powers were always thought to be *inherent in matter*. Matter does not exist without them, because continual motion and change are the defining features of the physical world.

The most sophisticated account of the intrinsic powers of matter in antiquity was Aristotle's. Aristotle moved beyond the love and strife polarity. He started from the realization that all things in the world are in a continuous flux. One thing may appear different at different times, much like clay may be worked in many different shapes. Things by nature, then, have an intrinsic potentiality for change (*potentia*). But what chooses what of all possible behaviors will be manifested in a physical body? It is certainly conceivable that an external agent may act on the body and enforce a specific behavior; for example, an artist may bring a lump of clay in the form of a statue. However, if no external agent is present, the physical body will behave in a way that is appropriate to its own nature and its natural "place" in the order of things. There exists, therefore, a power intrinsic in matter that guides every single body toward the realization of a behavior that is according to its own nature. This power is to be considered the principle of motion, change, and, as Aristotle argued, life. This principle he called the *entelechy*, namely, *what has its objective inside itself.*

The postulate of elemental powers inherent in matter enjoyed a long life in the history of ideas. It eventually led to the concept of force, the central point of Newtonian mechanics. Its appeal was strong because it guaranteed that matter is alive, impossible to bind and control. It also guaranteed freedom of will, viewing people as part of a living and dynamic Universe. Indeed, the theory of elements does not care too much about the notion of absolute necessity. Elements are defined by their qualities, which have a less objective character, and they are less amenable to precise rules. For this reason, the idea of powers inherent in matter was usually affirmed in opposition to the materialist worldview of the atomic theory.

A crucial observation was that elements always combine through fixed proportions in the formation of the physical bodies of our experience. There is therefore a natural tendency to associate the elemental description of nature with quantitative procedures. A physical body is distinguished from another solely because of the different proportion of the fundamental elements it contains. The understanding of matter is thus strongly related to the study of proportions; therefore it has a mathematical component.

1.3 Images of Synthesis

What is the thing that always exists and is never born, and what is the thing that always changes and never is? The first may be perceived by thought and reasoning, because it is unchanged, the latter by sense without reasoning, because it is continuously born and dying, never existing truly.

—Plato, *Timaeus*

In the previous section I sketched the two fundamental theories about the structure of matter, the atomic theory and the theory of the elements and of powers inherent in matter. The latter dominated the explanations of the physical world during antiquity. The atomic theory—in spite of the enormous respect granted to its founder—did not attract many followers among philosophers or the educated public. The explanation of the totality of experience in terms of atoms that move incessantly and chaotically was widely perceived to be incompatible with the highest ideal of ancient philosophy: the existence of a world ordered according to the plan of a perfect harmony that encompassed human beings and their place in the Universe. The human psyche with its wealth of emotions, experiences, and thoughts was simply too vast to be adequately explained by the atomic theory.

The physical sciences in antiquity were mostly qualitative, and the few exceptions, such as astronomy and optics, serve only to highlight this rule. This proved a great obstacle for the atomic theory, which could unravel its potential only within a tradition that emphasized the quantitative study of the world, like that of modern science. On the other hand, a theory based on fundamental qualities—like the theory of the elements—could be adapted to fit different contexts with relative ease. It proved remarkably successful in providing adequate qualitative explanations for many physical phenomena. Moreover, the theory of the elements did not make any extravagant claims about explaining the totality of experience—including the human psyche. For this reason, it could be readily translated into the language of different philosophical systems. But in no science and in no philosophy did the theory of the elements play such a crucial role as in the discipline of alchemy.

Alchemy appeared in the later years of antiquity, during the third and fourth centuries A.D., but its roots are to be found in the earliest civilizations of the Middle East. It housed under the same roof ideas from philosophy and practical chemistry, but also from astrology and magic. The name "alchemy" is the Arabic rendition for the Greek word chemistry—whose root is probably Egyptian—and testifies to the fact that the main body of this tradition entered western Europe via the Arabs.

The alchemists' attitude toward the world is immortalized in the never-ending quest for the philosopher's stone. The philosopher's stone was viewed as the purest form of matter, the control of which would allow in particular the transformation of base matter into gold. It was the primordial substance, the source of unity behind the incessant transformations of ordinary matter of everyday experience, the essence that "unites together the powers of things superior and things inferior."[9] The philosopher's stone was often conceived as a spiritual element, distinct from the four fundamental elements and superior to them because of its spiritual nature. Its transformation powers were so great that even with a small grain of that substance "one would feel no more as a man but akin to a spirit."

The alchemists emphasized the living and transformative powers of matter, its inherent ability for change and motion. The qualities of the fundamental elements were thought to refer to the realities of a world beyond immediate perception and to be strongly connected to the internal life of human beings. Alchemists postulated a unity of the deepest parts of a human's internal life and the basic properties of matter—external to humans—so that obtaining control over matter would allow a deep and fundamental control of human nature. The process of turning base matter into gold was often viewed as analogous to the purification of the human soul, and its growth into a state that would allow it to apprehend the inner character of reality.

Alchemy affected modern science in many different ways. Alchemists performed a huge number of experiments and bequeathed a significant amount of empirical data to their successors, physicists and chemists. But more important was the fact that the practice of alchemy provided the crucial psychological motives for both individuals and societies to focus on scientific research. The key point was the alchemists' emphasis on *control* of either human nature or the natural world. The motivation for their research was not only intellectual curiosity or the search for the cosmic balance. Alchemists sought to unlock the secrets of the universe in an expectation of a reward in terms of control and personal empowerment.

In most societies, more people appreciate promises of empowerment and material reward than promises of pure knowledge or personal growth. For this reason, the popularized ideas of the alchemists always attracted large audiences. The larger the audience, the easier it is for an idea to survive. Publicity makes it easier to attract the crucial thing that all thinkers of all times are in need of: economic assistance. Adequate sponsorship can guarantee both the physical survival of a person and the intellectual survival of that person's ideas. The promise of material reward is something that can be most easily understood by the funding agencies of all times, be it an inquisitive Renaissance prince or a modern research council. Modern science would simply be impossible if the societies were not convinced they had something concrete to gain by funding research.

This is far from saying that the pure desire to understand the world is missing from modern science. However, intellectual curiosity has not been by itself sufficient to sustain science as a social phenomenon in other cultures. In the absence of sponsorship, which is usually motivated by the need for technological gains, science would be practiced by only a small, scattered community of dedicated individuals, and it would have never taken up its role of today, that of a major force shaping modern society and transforming all aspects of everyday life.

However, the alchemical tradition did not create any new theories about the fundamental constituents of nature. It mostly propagated older physical and philosophical ideas, albeit dressed in an unrecognizable garb.[10]

Figure 1.1 **An alchemical drawing.** This drawing represents the unification of spirit and matter by the alchemists. Matter is represented by the square, which rests in the circle of the four elements and includes the triangle of spirit.

Most remarkable among all theories of antiquity is a theory of Plato, which was developed in his dialogue *Timaeus*. Plato's theory involved a creative combination of the two fundamental ideas about the structure of matter and exerted great influence on thinkers of later times, most importantly, thinkers of the period that witnessed the birth of modern science. In *Timaeus* Plato provides an imaginative, semimythological account of the reasoning of the creator in the building of the world. What is of interest to our discussion is Plato's description of the formation of the elements.

The building blocks of matter are, according to Plato, triangles of specific, regular shapes, which come in different sizes. Plato was not interested in the origin of these triangles because he only sought an economical way to describe the formation of the elements. "If somebody finds something more beautiful in their constitution [than the basic triangles], he will not be deemed an enemy but a great friend," he wrote.[11] The triangles combine among themselves, sharing some of their edges, and form geometric structures with the shape of regular polyhedra. Each of these polyhedra corresponds to and defines a different element (Earth, Water, Air, and Fire).

There exist four regular polyhedra that may be formed from identical triangles: the tetrahedron, the octahedron, the cube, and the

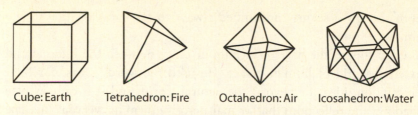

Cube: Earth   Tetrahedron: Fire   Octahedron: Air   Icosahedron: Water

Figure 1.2 **Plato's polyhedra and the formation of elements.**

icosahedron. The tetrahedron is the lightest (as it is constructed from the smallest number of identical triangles) and the sharpest (as its angles are the most acute). It corresponds to Fire because Fire is characterized by the qualities of lightness and sharpness. The most balanced and "solid" polyhedron is the cube; it naturally corresponds to Earth. Following a similar argument, the octahedron is said to correspond to Air and the icosahedron to Water. Plato also introduces another element, which corresponds to the dodecahedron (which is not made of triangles because its side is a pentagon). This fifth element (the *ether*) was deemed responsible for the constitution of the heavenly bodies.

Plato's theory combines the atomic perspective with the theory of the elements. Polyhedra are formed by fundamental triangles, which are in essence discrete objects. At the fundamental level, then, the theory is closer to atomism. However, the notion of the elemental fluids is preserved. Their qualitative differences are not due to the differences in the properties of the building blocks, but they arise because of the different geometric structures that define them. Thus a difference of structure or symmetry—*ultimately a mathematical difference*—is taken as the source of all qualitative distinctions between the elements. This possibility is the most spectacular characteristic of this theory. It is the main reason for the popularity Plato's theory enjoyed through the ages, in spite of its eventual inability to adequately describe physical phenomena. In fact, the aspiration to reduce qualitative distinctions to quantity appears repeatedly in the history of science. It brings spectacular progress, even though it seems condemned always to fall short of its ultimate aim.

Plato sought the basic principles of unity between the internal and external world of human beings. He used the word "ideas" for their description. Ideas lie beyond the natural world, and it is only as pale shadows and reflections that we may perceive them in everyday life. The human soul partly belongs in the word of ideas, and this explains a person's rational ability to grasp the inner workings of the world in which he or she lives. The science of mathematics is a concrete example of secure knowledge, which allows people to enter the world of ideas and see deeply into the roots of all things. For this reason, mathematics provides the key for the understanding of nature.

Plato's emphasis on mathematical knowledge for the description of the material world was inspired by an older tradition from Pythagoras of Samos and his disciples,[12] namely, the teaching that "all things we know have number; for without number we can neither think nor know."[13] Harmony of form, the most cherished ideal of the ancient world, could be expressed in the language of proportionate analogy—essentially ratios of numbers. And if the world is an ordered place—a cosmos—it should also be subject to those fundamental laws of order that are revealed to our intellect through mathematics.

Plato's theory of solids is based on geometry, which was still a young discipline in his days and exerted huge fascination to all who understood it, with its elegance of form and clarity of concept. It has never ceased to do so, even after the excitement of novelty wore off. The geometric ideal of reality has seduced and still seduces many a philosopher and scientist with its power and beauty, as though—Plato would say—it strikes chords deeply hidden in the human soul.

Plato's works with the emphasis on the mathematical structure of reality were transmitted during the fifteenth century A.D. to Renaissance Italy, and they played a significant role in the development of modern science. Because, if there is a thing that places modern physics apart from all previous efforts to describe physical phenomena, it is the fact that it is fundamentally *mathematical physics*. The language that describes the world is mathematics, and physical concepts are represented by mathematical objects.

The modern scientific theories about the constituents of matter are much more elegant and accurate than the simple picture of fundamental particles or fundamental fluids, from which they gradually emerged. Modern atomic theory, with its relation to chemistry, as well as the basic theory of modern physics, quantum field theory, contain a mixture of these two fundamental notions, which are composed on the basis of a powerful and elegant mathematical formalism. However, it is not the mathematical formalism itself that creates the physical theories. It is the determination of the basic *physical* concepts that renders any theory as a viable and comprehensive picture of reality. "Mathematics," Sir Francis Bacon emphasized, "should only give definiteness to natural philosophy, not generate or give it birth."[14]

1.4 The Roots of the New Era

The men of experiment are like the ant, they only collect and use; the reasoners resemble spiders, who make cobwebs out of their own substance. But the bee takes a middle course: it gathers its material from the flowers of the garden and of the field, but transforms and digests it by a power of its own. Not unlike this is the true business of philosophy.
—Sir Francis Bacon, *The New Organon*

The mathematical description of nature is not an exclusive feature of modern science. After all, the idea of mathematical physics is attributed to the Pythagoreans, and traces of it appear in many civilizations of the past. Moreover, any engineer of the prescientific era would readily agree that the use of mathematics allows one to predict and explain the behavior of matter in many cases of practical interest. This was the case in architecture, the building of bridges, and, after the introduction of gunpowder in fourteenth-century Europe, the making of guns. Hence, the emphasis on mathematics is not the only distinguishing feature of modern physics. Rather, the birth of modern science is due to different tendencies and attitudes that came together in western Europe during the sixteenth and seventeenth centuries. Together they shaped a rather special worldview, which proved remarkably successful in the description of nature.

A dominant feature of modern physics is *empiricism*. Empiricism is largely a creation of the British, and its roots seem to be deeply planted in the British cultural tradition. The thirteenth century saw the beginnings of an intellectual tradition that placed stronger emphasis on the knowledge coming from experience than was common in the Western philosophy of the time. Roger Bacon and John Duns Scotus were the most famous of the medieval philosophers, who expressed this attitude in an eloquent fashion.

This is an appropriate point to comment on a widespread popular belief about the history of science, namely, that that the dominant intellectual attitude before modern science was the study of the world only through logical arguments and without recourse to the results of experience. This belief is unjustified by the historical facts: it was true neither in antiquity nor in the Middle Ages. If anything, Aristotle in the fourth century B.C. explicitly stated that knowledge is based on observation and experience, and that "basic concepts are formed when a sufficient number of facts from experience come together."[15] These concepts lead the intellect to state certain principles, which lead logically to conclusions about the world surrounding us. Detailed observation and its relation to theories was the main vehicle for the study of the world in all ages, and probably in all cultures that had an interest in doing so. After all, this procedure needs only a minimum of common sense.

It is true, though, that intellectuals in western Europe in the twelfth to fourteenth centuries had been very enthusiastic about the sudden import of the philosophical writings of the ancients (twelfth century A.D.) and were deeply impressed by the achievements that rigorous and precise reasoning could achieve. This enthusiasm brought about a period of great creativity, as the young societies of western Europe assimilated the old knowledge and interpreted it in a novel way, creating an intellectual tradition summarily described as Scholastic. However, young societies—like young people—are often excessively impressionable; Scholastic philosophers tended to throw caution away in their expectations of what theoretical reasoning could achieve unaided by experience. They placed more emphasis on theoretical deductions from logical principles than any common sense (or good sense) would suggest. It was against this attitude that the forefathers of empiricism protested.

The widespread belief I commented upon was created from the intellectual polemics of a different era, the age of Enlightenment—mostly the early eighteenth century A.D. An intellectual battle raged during these times about the relative merit of the ancients (meaning the interpretation of ancient philosophy by the Scholastic tradition) and the moderns (meaning the supporters of the worldview associated with modern science). Part of the argument concerned the validity and universality of the conclusions of early modern science and of the empirical method. This conflict had strong political overtones because it touched upon issues such as the relation of church and state and the divine right of the monarch. In any such conflict, the truth is the first to suffer because none of the combatants stays away from pamphleteering. The moderns won this war, thus opening the door to many of the ideologies of the modern Western world. As history is usually written from the victor's side, the slander against the ancients has persisted until modern times.

Modern empiricism is distinguished from the more traditional "common sense" empiricism by the former's overwhelming emphasis on the notion of *experiment*. The experiment is not, by itself, a different type of observation. The emphasis lies on phenomena that are *repeatable* and created in conditions such *that all relevant external parameters can be controlled*. What constitutes an external parameter depends on the system under study. For example, if we study the physical properties of a gas, the relevant external parameters are the gas's temperature, its pressure, or the size of the vessel that contains it. The studied phenomena must also possess a quantitative component, so that the observations can be written down in terms of numbers rather than verbal descriptions. In modern science, it is not enough to state that a gas seems to expand when heated.

One must find the exact law that determines the exact value of the expanded volume for each recorded value of the gas's temperature.

The emphasis on the precision of the experimental method is not a strictly scientific or intellectual achievement; rather, it is a social phenomenon. Besides technological advancement, it presupposes a society with a developed and sophisticated *trade culture*. To see this, one should recall that the birth of modern science in western Europe (sixteenth–seventeenth century A.D.) coincides with the era of exploration, which was spurred by the desire of European states to dominate intercontinental trade.

The European societies of that time had an active merchant class, and they had appreciated ever since medieval times that meticulous bookkeeping was fundamental for the health and growth of any commercial enterprise. There is an obvious similarity between keeping the accounts of a commercial enterprise and keeping the accounts of an extended series of experimental outcomes. This is not only because the skills necessary for these activities are the same (basic literacy and high skill in numeracy). More important is the logic that the widespread application of commercial bookkeeping impresses upon a society. It familiarizes people with the benefits of a detailed and controlled record of proceedings. The success of this method in an important field of human endeavor like finance creates the confidence that such an attitude might bear fruit in other fields. One of these fields is the study of nature.

Of course, the possibility of carrying out accurate experiments presupposes the existence of rather advanced technology, in the form of accurate clocks, weight scales, methods of insulation, and so on. The technology of sixteenth-century Europe had not been inconceivable for many cultures of the past. However, the political fragmentation of sixteenth- and seventeenth-century Europe resulted in this technology becoming quite widespread. A person with sufficient funds could access this technology, even if that person did not live in the largest urban or political centers of the time. This implied that experiments could be carried out not only in the vicinity of palaces, when sponsored by benevolent rulers, but also in the quietude of a country house and away from inquisitive or curious eyes.

Of crucial importance was also the invention of printing and its rapid assimilation as a part of everyday life in sixteenth-century Europe, which made books much cheaper and easier to produce. It thus promoted the circulation of interesting discoveries to a large readership in a wide geographical area and within a relatively short amount of time. This greatly facilitated the exchange of ideas.

Modern historical research has identified other social factors that played an important role in the creation of modern science. The existence of universities and colleges with relative autonomy and independence from the local authorities and the development of a legal system in which awards and penalties were distributed according to logical reasoning from some basic laws are two examples that are often cited in this regard. Nonetheless, the cultural and technological developments of early modern Europe, important as they were, would not have been sufficient by themselves to create modern science. The reason is that "the book of nature cannot be grasped unless one knows the characters in which it is written. . . . These characters are triangles, circles and other geometric figures, without which man cannot grasp a word of the language and wanders in vain through a dark labyrinth."[16] If mathematics had not been sufficiently developed, all efforts to create a new science would have faltered and perhaps died out. To appreciate this fact we need to recall some elements from the history of mathematics.

1.5 Mathematics and the World

SIMPLICIUS: . . . But I still say, with Aristotle, that in matters of nature one need not always require a mathematical demonstration.
SAGREDO: Granted, where none is to be had; but when there is one at hand, why do you not wish to use it?
—Galileo Galilei, *Dialogue Concerning the Two Chief World Systems*

How can it be that mathematics, being after all a product of human thought, which is independent of experience, is so admirably appropriate

to the objects of reality? Is human reason, then, without experience, merely by taking thought, able to fathom the properties of real things?
—Albert Einstein, *Sidelights in Relativity*

The very first steps in the development of mathematics are probably rooted in ages for which we have no historical record. Hence, we do not know the place and time that two of the arguably most fundamental discoveries of mathematical thought took place, namely, that there is no limit in how large a number one may conceive, and that numbers can be used fruitfully for the study of geometrical shapes.

In the first civilizations of the Middle East, we find many examples of sophisticated mathematical thinking. Intellectual curiosity apart, there were very practical reasons. For example, "it was an important task for the rulers of Mesopotamia to dig canals and to maintain them, because canals were not only necessary for irrigation but also useful for the transport of goods and armies. The rulers or high government officials must have ordered Babylonian mathematicians to calculate the number of workers and days necessary for the building of a canal, and to calculate the total expenses of wages of the workers."[17] Besides engineering, architecture and astronomy (or astrology) also provided motives for the development of mathematical techniques. The Babylonians and the Egyptians developed practical number systems, pioneered geometry and astronomy, and created methods for the solution of complicated equations.

However, the mathematical knowledge was at this stage almost purely empirical. The idea of *proving* a mathematical proposition from a number of basic principles did not exist. Mathematical facts were obtained from generalization of experience. As an example, let us consider the statement that the area of a field with the shape of a rectangle is the product of the lengths of its sides. To verify this statement, we compare a measure of area, like the number of plants that grow in the field, with a measure of length, like the number of wooden stakes needed to make a fence around it. The statement above is true in many different circumstances. It is true irrespective of the field's location, the field's soil, and whether the field was planted with wheat or barley. There exists a universality and simplicity in mathematical statements, which suggests a hidden underlying structure.

The identification of this underlying structure was the novelty brought by Greek mathematics: the whole world of mathematical statements is described concretely and compactly in terms of an *axiomatic system*—that is, one starts with the definitions of the mathematical objects under study together with some basic principles, the truth of which is intuitively apparent (axioms). One then uses some commonly accepted rules of inference—often known as rules of logic—and reaches conclusions about the properties of mathematical objects. These conclusions are the *theorems*, literally "the things we see." One has, therefore, to follow strict rules before ascertaining any proposition as valid in a satisfactory way. The rigor and precision of mathematics has always served as a model for the description of the world. We shall see that the modern concept of the laws of nature was built around the demand that the changes of matter should take place according to precise and unambiguous rules similar to the ones encountered in mathematical practice.

The axiomatic method in Greek mathematics was applied almost only in geometry. The most famous compendium is Euclid's *Elements*, which was written in Alexandria during the fourth century B.C. Euclid derives all geometry from five basic axioms, together with the introduction of elementary (non-analyzable) concepts such as a point, a line, and a plane. There was substantial development of other fields of mathematics such as arithmetic (the study of numbers) and algebra, but these did not lie within a unified axiomatic framework.

The other branches of mathematics—apart from geometry—developed more slowly, and it was within a timescale of many centuries that significant new insights were achieved. The Middle Ages witnessed the gradual development of *algebraic* thinking. While contributions to this development came from many cultures (mathematical ideas are easily diffused), we should distinguish those of the Arabs; indeed, the name "algebra" comes from the Arabic word "al-Jabr," which means "reunion." Unlike geometry, which deals with the properties of shapes, and arithmetic, which deals with the properties of numbers, algebra in its essence involves the study of properties, relations, and rules. As such, it introduces a level of abstraction that allows mathematical reasoning to address problems not on a case-by-case basis, but in their most general context. This abstraction turned out to be crucial for the birth of mathematical physics.

In particular, algebraic thinking led to the development of an *abstract formalism* for the study of numbers. This formalism involves the introduction of variable quantities (*variables*, in short). For example, one may employ the letter "x" as a variable. It is an *abstract symbol*, and it represents a generic number without specifying which particular one. The introduction of Arabic numerals was crucial for this development because previously one expressed numbers using letters of the alphabet; Arabic numerals left the letters free for other uses.

If, in addition to the letters that represent variables, one introduces symbols to represent arithmetical operations (like addition or multiplication) and relations (like equality), one can write mathematical statements solely in terms of the abstract symbols. We cannot attribute this ingenious discovery to one person alone.[18] It was a collective achievement, which took place over the course of generations. The result of this process was that relations between numbers were expressed solely in terms of symbols. For instance, the string of symbols $(x + 1)^2 = x^2 + 2x + 1$ represents a statement that holds for any possible number that can be substituted into the place of x.

Symbolic notation leads naturally to the introduction of the most fundamental notion of modern mathematics, that of the *function*. The function concept is so important that we cannot even imagine how mathematics would have developed in its absence.[19]

A function is a relation between two mathematical objects, and it can be expressed as a relation between the abstract symbols that represent them. More precisely, a function states a *rule* by which one can go from a given number represented by x to some other number represented by y. So when we write the equation $y = x^2$, we mean that the number y is obtained from the number x through the operation of multiplying x with itself, for all possible values of x. The expression above is very simple, and one can easily state it explicitly instead of using symbols. However, the abstract notation facilitates the writing of much more complicated relations between mathematical objects; these relations would be extremely time consuming (not to mention paper consuming) when written explicitly.

The introduction of the function did not only achieve convenience in arithmetical notation. It also brought a revolution in the study of geometry because it provided a completely new way of representing geometrical objects. To see this, we consider a problem of some relevance to the engineers of the sixteenth century: the determination of a cannonball's trajectory. In this case, one wants to determine the vertical distance of the ball from the ground as the horizontal distance from the cannon increases. The rule that does this amounts essentially to the definition of a function that specifies the cannonball's vertical distance for every value of its horizontal distance. The knowledge of this function fully determines the cannonball's orbit. However, an orbit is nothing but a line in space. We thus recognize the remarkable fact that we can use a function to represent a geometrical object. This is not an isolated coincidence; we can find a similar description for points, surfaces, and so on.

For instance, one needs three numbers to determine a point in space uniquely. These numbers measure the point's distance from three straight lines, which intersect vertically at one point. These lines are the *coordinate* axes, and the point of intersection is the *origin* of the coordinates. If one employs numbers to represent points, one may describe the whole of Euclid's classical geometry with the language of functions. This is a huge achievement, connected with the name of the French philosopher René Descartes.[20]

Descartes' work is a turning point in the history of ideas. His thought constitutes a bridge from ancient to modern mathematics, and from the Scholastic philosophical tradition to the modern Western one. Few people in the history of human thought share such a pedigree of achievement. Unusually for such a brilliant philosopher and mathematician, Descartes started his adult life as a soldier. After a rather conventional education in a Jesuit college, at the age of twenty-two he entered a military academy in the Dutch city of Breda. But even as a soldier, Descartes remained deeply

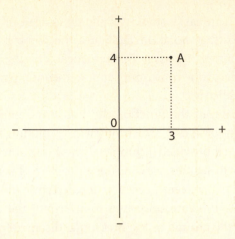

Figure 1.3 **Coordinate axes.** The coordinate axes on the plane are a pair of
vertical axes, which intersect at the point O, the origin of the coordinate system. We project a point A of the plane to both axes, and we measure the
lengths of the projected points from the origin. These lengths are either positive or negative according to whether the projections lie on the positive or negative part of the axes. The pair of numbers we obtain specifies uniquely the
point A: it provides its coordinates with respect to the system of axes. In order
to specify a point on space, rather than on a plane, we set a coordinate system
involving three axes, vertical to one another, which meet at a single point.

fascinated by the mathematical sciences, to which he had been introduced
during his student days.

Descartes stayed only one year in the army. Leaving Breda, he traveled to Copenhagen to meet a friend of his. His mind during the trip
was almost obsessively centered on issues of mathematics and philosophical method. On one extremely cold day, Descartes was warming
up deep in thought near a fireplace when, perhaps mesmerized by the
fire, he had a vision. Three dreams followed the same night. This experience created in him the belief that he had been selected for the task of
unfolding all branches of science under the auspices of a new mathematical method. Descartes' ambition and the power of his vision are
reflected in the provisional title of his most important work: *Plan of a*

Universal Science Capable of Raising Our Nature to Its Highest Degree of Perfection.

With Descartes' mathematical method, a new field of mathematics came into completion: *analytic geometry*. Not only did it achieve a unification of geometry and algebra, but it also provided a method to describe geometrical objects of high irregularity and complexity, which were treated with great difficulty, if at all, by the traditional methods.

It is evident that the immediate application of analytic geometry to physics would be in the study of particle motion. One may idealize a particle by a point moving in space. Its motion is described by a function that specifies its position at each moment of time. The position of a particle is determined by the knowledge of three coordinates, so the necessary function should provide a rule giving three numbers for each instant of time. Consequently, any law that governs the motion of particles can be written in *the form of equations* about the *functions of time* that represent the particle's motion.

The French philosopher Henri Bergson has strongly emphasized the importance of the point above: "Modern science must be defined primarily by its tendency to treat time as an independent variable."[21] This tendency allows an efficient and effective description of motion that takes into account the properties of physical objects at any moment of time. It therefore provides a priceless tool for the description of nature because "the nature of things is the principle of their motion and change and since we study Nature we must first understand what motion is— for if we do not know this, neither do we understand what Nature is."[22]

At the time of the Scientific Revolution, the most pertinent and concrete problem of physical philosophy was related to astronomy. The exact observation of the planetary motions had already been achieved in antiquity, and it was probably due to the Babylonians. In ancient times, the dominant idea was that Earth was the center of the world: in this view the Sun, the Moon, and the other planets revolved around Earth. The theory that Earth and the planets revolve around the Sun was known, but observation seemed to rule it out. For, if Earth revolved around the Sun, one should be able to notice a substantial change in the relative position of the stars during the year. This had never been observed, because Earth's

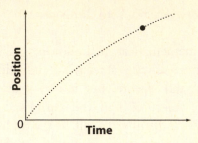

Figure 1.4 **Particle motion and its graphical representation.** A function is a rule that takes one number, the variable, into another, its value. A pair of numbers is represented by a point on the plane. As the variable of the function changes, the point on the plane moves and forms a curve, which provides a graphical representation of the function. A particular example of such function is the position of a particle as it changes with time. Here time is the variable and the position coordinate the function's value.

distance from the Sun is much smaller than its distance from even the nearest of stars. The difference in scale of these distances is so huge that the naked eye cannot distinguish any change in the position of the stars as a result of Earth's motion. The explanation that the stars were too far away was considered as an unjustified hypothesis with no concrete evidence to support it. Moreover, this discrepancy of distances had no place in any balanced and harmonious plan of the Universe: it seemed too ugly and disproportionate to be true. The assumption of a geocentric system was the simplest solution that fitted the observed facts. This, however, implied a very complicated description of the planetary motions. The observed orbits around Earth were not simple, and they could be described only through an extremely complex system that involved synthesis of motions within circles of different lengths and origins (known as *epicycles*).

With the coming of the sixteenth century, a large and detailed body of astronomical observation had been accumulated. Nikolas Koppernigk or Copernicus was the first to propose that the Sun was indeed the center of the planetary system. Copernicus was a Polish monk who had been educated in Renaissance Italy. His study of the classics brought

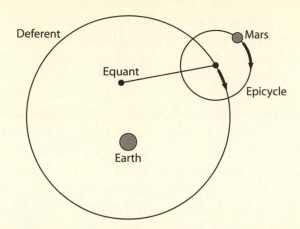

Figure 1.5 **Epicycles.** The ancient astronomer Hipparchus had perfected the model of the epicycles, which described the motion of the planets, while keeping Earth at the center of the Universe. A planet revolves about a small circle, an epicycle, which in turn revolves about a larger circle, the deferent. Earth may or may not lie at the center of the deferent—in the latter case the deferent is called eccentric. The equant is a noncentral point about which the epicycle moves at a constant angular rate

him into contact with an ancient theory by Aristarchus of Samos that the Sun is indeed the center of the Universe. He soon realized that such an assumption would lead to a dramatic simplification of the geometrical descriptions of planetary motions by rendering all complications due to the epicycles unnecessary. It took Copernicus "four times nine years" before he published his full-fledged ideas (admittedly through a tortuous path of argumentation) in an epoch-making book titled *On the Revolution of the Planetary Spheres.*

Copernicus's most important legacy to future generations was his belief that the successful geometrical description he had discovered was not just an abstract tool for astronomical calculations but provided a physical account of the way the world really worked. Number and shape, he thought, do contain the essence of things. The reason why the mathematical description of Earth revolving around the Sun provides a simple account of the celestial phenomena is that Earth *actually* rotates around the Sun.[23] In modern times, accustomed as we are to the basic ideas of the sciences, this step may seem trivial. But in Copernicus's time and culture it was nothing short of a revolution. His theory not only went against the habits of thought of his time, but it also suggested a new way of making theories about nature: one should rely primarily on mathematics because mathematics is in tune with the deepest structure of the world. Indeed, the tradition of modern physics begins with Copernicus.

Supporting the Copernican thesis involved a large amount of work and a lot of courage. The leading academics of his day (and of later days) fought strongly against his idea. In fact, some of them had no scruples in bringing the persecuting force of the Inquisition to bolster their arguments.

One generation after Copernicus, however, his theory was still a source of inspiration. Johannes Kepler, an astronomer who, being near-sighted, could not observe the stars himself, achieved much more than simply providing additional support for the Copernican system. In two books, *The New Astronomy* and *On World Harmony*, both remarkably charming in spite of their technical nature, Kepler presented three laws that governed the motion of the planets around the Sun.

- Planets move in elliptic orbits, with the Sun lying at one focal point of the ellipse.
- The line connecting a planet to the Sun spans an area proportional to the duration of the planet's motion.
- The square of the period of the orbit of a planet around the Sun is proportional to the cube of the distance of the planet from the Sun.

Kepler's laws were empirical: they constituted a triumph of meticulous bookkeeping, dedication, and careful observation that heralded the dawn of the new science. This, of course, had to be combined with Kepler's ingenious problem solving and the courage to put forward his results. The laws he discovered were not simple or intuitively apparent. For this reason, the few people who accepted the Copernican system tried to explain them in terms of simpler principles.[24]

The Copernican system found its most eloquent and convincing champion in Galileo Galilei, a remarkable mind whose work exerted a

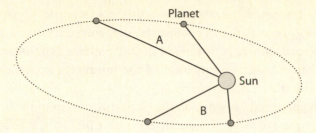

***Figure 1.6* Keplerian orbit of a planet around the Sun.** The planet moves in an elliptic orbit around the Sun, which is located in one of the two focal point of the ellipse (not in its center). The area connecting the planet to the Sun spans equal areas in equal times. Hence, the area of the triangular region A equals that of the region B, if the times are equal.

major influence on all his contemporaries. Galileo became a professor of mathematics in the University of Pisa before his twenty-fifth year of age. Three years later he became a professor in the most influential university of Padua. In Padua he had the opportunity to employ the newly discovered telescope in the observation of the skies. He turned into a passionate advocate of the Copernican system, as he realized that his more accurate astronomical observations provided it with strong support.

Galileo realized that the consequences of the heliocentric system were not restricted to astronomy. Copernicus had brought mathematics to the forefront as the key for unlocking the "secrets of nature." The Copernican system did not simply introduce a new astronomy, it proposed a new attitude for the physical sciences. For this reason, it had to fight its way against the established physical theories. In Galileo's time the dominant philosophy revolved around a particular interpretation of the works of Aristotle. Hence, Galileo saw it as his task to attack basic premises of this Aristotelian physics—and this involved both the execution of carefully designed experiments and mathematical work. He had to tread carefully, though. He was a *mere* professor of mathematics, a discipline of lower prestige (not to mention salary) than physical philosophy. His colleagues would not view his new ideas very kindly. More than that, the new ideas about the world had necessarily theological implications, ones that ran against the contemporary trend toward a literal interpretation of scripture. At times of intense religious strife and passions, such disagreements implied the intervention of the office of the Inquisition.

Galileo published his work on the Copernican system in the form of a dialogue, the *Dialogue Concerning the Two Chief World Systems*. This book was published late in his life, at a time when political circumstances enabled him to get a license for publication. It presented a debate between three persons: Salviati, Sagredo, and Simplicio. Salviati was a mouthpiece for Galileo, putting forward Copernican and Platonic ideas, Simplicio supported the established Aristotelian physics, and Sagredo was supposed to keep a middle road. Intelligent and witty, this dialogue is still attractive to modern eyes. At the end the reader is left to believe—after a masterful development of argument and counterargument—that only a fool would fail to be convinced by the Copernican system or would disagree with the idea of the mathematical character of matter.

The rest of the story is very well known. It has increased in stature to become a legend of the modern world, an archetype of the conflict between free inquiry and established authority. The book was a success, but Galileo's enemies managed to ban it a few months later. The Inquisition summoned Galileo to Rome, in spite of his old age and infirmity (he was almost seventy). He was treated humanely there, but the Inquisition made the error Galileo had always feared. They branded the heliocentric system as heretical and insisted that Galileo—himself a faithful Catholic—should recant. He did so and spent his remaining years in comfortable imprisonment, from which he managed to smuggle a manuscript for publication. A few years before his death, the *Discourses and Disputations Concerning Two New Sciences* was published. This book was a compendium of Galileo's lifetime work. It did not demonstrate the artistic skill of his previous dialogue, but it provided the germs of the new science. It contained the fruits of many years of experimentation, but more important, it provided the first fully mathematical description of physical phenomena. This publication signaled the birth of modern mathematical physics.

Nevertheless, Galileo's last book fell short of his ambition. He was old, tired, and almost totally blind by the time he started writing it. He felt the shadow of time over his life, and he persisted on his ambition to build a completed worldview to rival the established one. Some of his explanations were therefore rushed. He left gaps behind, and these led him astray on the most crucial issue: the explanation of motion. He did sense the right direction, but destiny did not reserve for him the honor of founding the new physics. The world would have to wait another fifty years for that.

1.6 The Metaphor of the Machine

> God is to the created creatures as an inventor is to his machine.
> —G. W. Leibniz, *Monadology*

The age of Copernicus and Galileo—the sixteenth and seventeenth centuries—was a remarkably active period in the study of nature. Old ideas were rediscovered and tailored to adapt to new facts. The Platonic emphasis of the mathematical description of matter and the ancient atomic theory were placed alongside the heliocentric system of Copernicus and Kepler and were enriched by the growth of experimental thinking. This process led to laying the foundations of modern science by Galileo and Descartes, and eventually to the brilliant synthesis of Isaac Newton.

One central thread underlay these developments, a thread so strong that even people who felt uneasy about the new sciences found difficult to refute it. It was felt that the world is governed by a strict necessity that is logical in nature, and that the causes of nature may be understood up to their finest detail through the application of careful and rigorous thinking.

In a sense, the most extreme form of this attitude is expressed in a statement by Gottfried Wilhelm Leibniz that everything in this word happens for a reason. A sufficiently large intellect should be able to state explicitly the *logical* reason for which each and every thing happened as it happened and not otherwise, deriving it as a mathematical theorem from some initial postulate. In this sense, it is logically impossible for our world to have been any different from what it is: we may imagine that different possibilities exist, but this is only due to our ignorance of the deepest causes. Our world is logically complete and unique, and for this reason it is the *best* one possible. One cannot fail to be impressed by the power of this vision: it brings the idea of the rational character of the world to its logical limits. On the other hand, it is totally horrifying. There is no place for freedom in such a world: everything, even our deepest thoughts and desires, our loves and our sacrifices, our choices and our commitments, has been fixed by causes external to us and outside our influence. Human beings are nothing but puppets in a cosmic process of logical deduction.

Not all thinkers of this time shared the extremity of this viewpoint. Even Leibniz was too deep a philosopher to trap himself in the ultimate logical consequences of this perception. He did not think that the logical completeness of the world should be reflected in the properties of matter. The material Universe is only a part of a greater whole; for this reason, the full knowledge of the necessity that governs the world is a sole prerogative of God. It was, however, only a small step to transfer the absolute necessity—which had been perceived a long time before Leibniz—from the realm of metaphysics to the level of the physical processes and phenomena. The natural world was then visualized as a *machine,* consisting of interconnecting parts that act upon each other according to predetermined and unchanging rules.

The mechanistic worldview is most aptly summarized in Descartes' philosophical investigations. Descartes believed that the analytic geometry he had developed provided the key to the workings of nature. He had realized that geometry may describe the motion of bodies, and he found no reason to assume that there exists anything other than that in the world. A thing is material, he conceived, only as far as it is a body, namely, if it occupies a place in space. "There is but one kind of matter in the whole universe," *Descartes wrote,* "and this we know only by its being extended. All properties we distinctly perceive to belong to it are reducible to its capacity of being divided and moved according to its parts . . . all variation of it, or diversity of form depend on motion."[25] Hence, there is no need to introduce any qualities for the description of the material world—all that is needed is the knowledge of the geometrical properties of matter.

In spite of some similarities of his theory with that of the atomists, Descartes emphatically rejected one of their key concepts. While he would accept that matter could structure itself in a form of particles, he did not accept the idea of empty space. The defining property of matter was its extension. That "which constitutes the space occupied by a body, is exactly the same as that which constitutes the body," Descartes wrote. Hence, there cannot exist space separate of body since all spatial extension is by definition body. What we perceive as motion is nothing but the effect of matter pushing other matter, or, in a more precise language, "the transfer of one piece of matter or of one body, from the neighbourhood of those bodies immediately contiguous to it and considered at rest, into the neighbourhood of others."

It was imperative for Descartes to find the laws that determine the motion of matter because these laws would provide the description of any phenomenon in the natural world. These laws of motion would have to be rigorous and predetermined. Otherwise, chaos would reign in the world. Descartes took mathematical reasoning as a prototype for the laws of motion: one proposition leads to another according to precise and unambiguous logical rules. When he looked for an example that manifested this behavior in the physical world, he was quick to find one: mechanical motion. In any well-designed machine, the motion is transferred from one part to another unambiguously.* In the machine, Descartes had found his metaphor for the basic principle of nature.

Anything in the material universe is a machine: there exists an unbending necessity that governs the motion of matter, which should be expressed by precise rules, similar to the ones that machines are

---

*The analogy of mechanical motion to logical reasoning suggests the possibility of building machines whose motion mirrors logical operations. As such, it forms the basis of modern computer science. This possibility has been fully exploited in the course of the twentieth century. However, its seeds can be found in the thinkers during the time of the Scientific Revolution, most notably in the works of Leibniz.

designed to follow. "The only difference . . . between machines and physical objects is that, while the function of the machines is mostly through processes that can be perceived by our senses . . . physical processes almost always depend on parts so small that they are not accessible to our senses." Animate matter is no different: plants, animals, and the human body are also machines, perhaps of higher complexity but machines nonetheless. Life is nothing but a derivative property of matter in motion. However, this is not the case for the human soul. In contradistinction to the material world of iron necessity, there exists the world of spirit, of soul, of intellect, which is free in its essence. Humans live at the crossroads of these two worlds, free because of their soul, but chained to matter's iron necessity through their body.

Descartes' philosophy is radical: it splits reality into two worlds, separate in their essence and character, which come grudgingly into contact through the human experience. However, this split was necessary from the moment he believed matter to be subject to an iron necessity. In his belief, Descartes reflected the intellectual climate of his era. This specific conception of the laws of nature is a unique and special creation of the societies of western Europe at the times of the Scientific Revolution.

Other cultures of the past—Greeks and Romans, for example—had also talked about laws of nature, but what they considered as such was the power that bound things together in a well-ordered and balanced world. One may say that they visualized the world in analogy to a building—a temple—and the laws of nature were its architectural plans. In this sense, the place and properties of one part had to be thought of, and explained, in terms of the harmony of the whole construct and vice versa: the harmony of the whole was brought upon by the combination of the properties of the individual components.

The mechanistic description was born out of a different metaphor for nature. The world is a machine, and the laws of nature are similar to the engineer's blueprints that describe the possible motions and connections of the machine parts. The properties of each part are thought of in terms of the part's function in the working of the machine. The explanation for the behavior of concrete things is to be sought primarily in their specific interconnections at each moment of time and at each point of space, and not necessarily in the harmony of the whole construct.

More specific toward the construction of the scientific worldview was Galileo's contribution. One can learn more, he said, about nature by abandoning, temporarily at least, the quest to find the true essence and reason for existence of things—why they appear in the architectural plan the way they do—and focus on *how* things happen, on the specific laws that accompany the manifestations of phenomena. In other words, one should seek *descriptions* of events rather than their true essence or their place in the grand scheme of things. A very successful attitude this was, and quite healthy if not taken too far. In Galileo's words, the idea of the laws of nature as an engineer's blueprint is firmly established.

Descartes' worldview had a tremendous effect on the subsequent development of both physics and philosophy. This was not at all obvious at the time. His theories faced hostile criticism, both among the established order that was faithful to the older "Scholastic" view of nature and among radical philosophers who could not accept the idea of a sharp split between a mechanistic matter and an immaterial free soul. Descartes is among the thinkers who have been most vilified and whose ideas have been most aggressively attacked in history, up to and including the twenty-first century. The reason is that he brings to their logical conclusion some of the basic ideas that characterize the "scientific" worldview. If one accepts the fact that things in the world take place according to strict and unbending laws that govern the behavior of matter, it is difficult to avoid the conclusion that material things are essentially machines. Much ink has been spent on a refutation of the word "machine," but this is often a game of words. The fact remains that the word "machine" or "automaton" provides the most convenient and accurate metaphor for the structure of things that this view of nature entails. One has then either to follow Descartes and place the internal life of human beings in another realm, district in essence from the material world, or to consider human experience as a mere epiphenomenon, a trivial consequence of the fundamental law of matter, with no intrinsic value in the order of things. Both positions are problematic and difficult to hold.

It is important to emphasize that the necessity in the mechanical description is quite distinct from the necessity of older philosophical theories that postulated a world in which every single phenomenon is predetermined by a specific cause—like the theories of Leibniz or of several ancient philosophical schools. The first difference is that necessity in the other theories referred not only to the material Universe, but to the totality of the world, including soul, spirit, and intellect. Second, the necessity in those theories could be perceived only if one had a full and comprehensive knowledge of the totality of the world. Such knowledge was, according to Leibniz, possible only for God. For human beings, restricted as they are to a small part of space and endowed with a small lifespan, it would be impossible to discern the true causes behind the things of immediate experience.

On the other hand, the mechanical necessity is *local in space*. The causes of a physical phenomenon can be found exclusively in the neighborhood of that phenomenon, the same way that the motion of a lever in a machine can be explained in terms of another lever adjacent to it, and that in terms of yet another adjacent lever, and so on. Hence, even a finite being would be able to comprehend the full chain of cause and effect (or at least a large part of it) that is involved in the realization of any physical phenomenon.

The mechanistic worldview was therefore very ambitious about the scope of human knowledge. For this reason, it turned out to be remarkably successful. If the things of the external world are machines, it is possible to distance the human psyche from them. Knowledge becomes something very concrete that has nothing to do with one's personal growth, as almost all philosophies of the past claimed. There is no need to relate a person's inner life, his or her moral development, religious practices, or mystical experiences, with the study of nature. One's intellectual ability is sufficient for the observation and understanding of the machine that is the Universe. Humans may then be (or pretend to be) impartial observers of the workings of the world, they may pretend to forget the deepest parts of their psyche in their study of nature, because these are simply excess baggage as far as humans' understanding of the world is concerned. It is possible to set down precise and strict rules about proper reasoning. The mechanistic worldview places human intellect as an unmoving point of reference for the study of the Universe. Anything that does not touch upon a person's internal world can be described in an impartial, objective, even cold, manner. This way an idealized sketch (almost a caricature) of a new human type is born, that of the *scientist*.

The great Newtonian synthesis, which signals the birth of modern physics, follows Descartes' spirit closely. This is quite remarkable, since Newton himself was critical of Descartes' ideas and considered his own work to provide a refutation of the mechanistic worldview. However, the products of scientific endeavor do not always satisfy their author's expectations, not even the expectations of the scientific community that bred this author. They have a life and almost a will of their own.

The development of Newtonian physics led toward a mechanistic worldview that brought many of Descartes' intuitions into flesh. How this took place, I shall narrate in the following chapter. The mechanistic worldview has been seriously challenged in physics only in the twentieth century, with the advent of quantum theory. However, it made a spectacular comeback in the later years of the twentieth century due to the development of the science of "reasoning machines"—computers. Mechanicism still provides a basic reference point for modern science, even though it lacks support at the level of fundamental physics. The least we can say about the incisiveness of Descartes' perception is that not only did he express the spirit of his times, but he was a prophet of an era that was yet to come, an era that reached its climax at least two centuries after him.

1.7 Newton's Achievement

The incomparable Author having at length been prevailed upon to appear in publick, has in this Treatise given a most notable instance of the extent of the powers of Mind. And has at once shewn what are the principles of natural philosophy, and so far derived from them their

consequences, that he seems to have exhausted the argument and left
little to be done by those that shall succeed him.
—Edmond Halley, review of *Principia Mathematicae*

I described in the previous sections how Descartes and Galileo had laid
the groundwork for a new science that would account for the motion of
matter, and I noted that they had fallen short of bringing their construction to completion. They had left many open problems, the explanation of
the planetary motion being the most important. These gaps allowed them
to catch but a momentary glimpse of the grand architectural plan that
would unify all separate developments in astronomy, mathematics, and
natural philosophy. What was missing was the unity of vision afforded by
a simple theory that would enable a systematic study of all problems of
nature from very few selected principles. History reserved this achievement for Isaac Newton.

Newton "combined in one person the experimenter, the theorist, the
mechanic and not least the artist in exposition. . . . Nature to him was
an open book." That was Einstein's assessment, written more than two
centuries after Newton's death. Indeed, Newton was a man of many
gifts. He combined an exceptional aptitude for mechanical construction with a deep intuition into the workings of mathematical theory,
talents most fitting for the task science faced in his day. He was a lone-
some figure. His solitary upbringing together with the complexities of a
mind of genius isolated him from close personal contact with other
people. For this reason, he devoted all his energies and all his passion to
study. He moved to Cambridge at the age of seventeen, where he came
into contact with the works of Galileo, Descartes, and the atomic theory
recently resurrected by Pierre Gassendi. He was self-taught in mathematics, but his discoveries by the age of twenty-three had made him
perhaps the leading mathematician in Europe. However, he was not
immediately acknowledged. He had a deep reservation toward publishing his results, partly because of a fear of exposing himself and partly
because of a streak of intellectual stinginess, both common aspects of
character in people of an intellectual bent who live a secluded life.

Newton very early in his life devoted his attention to the problem of describing and explaining the motions of matter. This was a topic of excitement in the circle of natural philosophers at the time. The realization that the natural state of matter is motion was still novel. Today, educated as we are from a tender age in the basic ideas of modern science, it is difficult to realize how remarkable this realization actually is. Our sense experience suggests that motion is a fickle thing; one needs to power things up in order to keep them moving. This suggests that matter by nature tends to find a resting place. In western Europe, in particular, this idea had dominated physical thinking since the rediscovery of Aristotle's thought. It was only through Galileo that this idea of static matter was laid to rest. In Newton's times, confusion was still reigning about the origin and character of motion. There were many conflicting opinions and perspectives, and the first crucial step in the search for a solution was to decide what to keep and what to leave behind.

Newton's first and most important decision concerned the treatment of time and space. The discussion about the nature of time and of space goes back to the earliest moments of philosophy and even earlier; it had not abated by Newton's time and has not abated now. Without a theory about space and time, any physical theory is lame because all physical phenomena refer to them, directly or indirectly. Newton—like most of his contemporaries—faced the dilemma of dealing with space as something external, a vessel within which things move (like in atomic theory), or that space has no existence by itself, it only signifies a relation between material bodies. A similar dilemma was present about time: is time something external, outside material things, or is it nothing but a measure of change of things?

Newton decided to consider space and time as fixed, absolute, and unchanged structures that subsist in the background of any physical phenomenon. This was his first and foremost assumption. In his words, "Absolute, true, and mathematical time, of itself, and from its own nature flows without regard to anything external. . . . Absolute space, in its own nature, without regard to anything external, remains always similar and immovable."

The introduction of absolute time and space offered Newton a dramatic simplification in the description of motion. Since time and space are absolute, they form a background structure, to which any physical phenomenon must inevitably refer. The existence of a background makes it possible to compare and relate immediately motions of different bodies in different parts of the Universe, something that would otherwise be extremely difficult.

The motion of any physical body is completely specified by the knowledge of the *functions* that determine its place in absolute space in terms of absolute time (see sec. 1.5). Consequently, any law of motion takes the form of a rule that governs the construction of these functions. Such a rule is likely to be very complicated because it would need to take into account the complex actions of the bodies upon each other. However, it is possible that the discovery of proper physical concepts would make the form of this rule reasonably simple—at least that was the hope entertained by Newton and his contemporaries.

We do not know exactly when Newton came to the discovery of his laws of motion because he persistently refrained from publishing his results. However, by 1684 he had realized that he could no longer not keep his silence. A friend of his, the astronomer Edmond Halley, visited him and mentioned a chance conversation he had heard in a coffeehouse. Robert Hooke, whom Newton detested, had claimed that he could derive the laws of Kepler from basic principles of motion. Newton was spurred into action. He had conceived much earlier how to do this, but he had to hurry if he were to get the appropriate credit. He entered into a frenzy of work that lasted two years. He clarified his ideas, developed them into a coherent, logical system, and applied this system to as many concrete physical situations as possible. Halley helped him tremendously in publishing his work. The result was Newton's masterpiece, a treatise titled *Mathematical Principles of Natural Philosophy* (usually referred to by a part of its Latin title as *Principia Mathematicae*). It was an immediate success, the ultimate synthesis—it was perceived—of all developments in natural philosophy during the last two hundred years.

The central idea in Newton's theory was that bodies act one upon the other by causing direct change to their *accelerations*. The acceleration is the rate of change of velocity, and velocity is the rate of change of the position (that is, of the numbers that determine the position as coordinates). The mathematical implementation of this law was the statement that the acceleration of the particle is proportional to a physical quantity—the external *force*. This statement carries the name of Newton's *second law of motion*.

The force is the means by which a body acts upon another. The word had itself a long history because it referred to an old concept. However, Newton used it in a new and creative way. In older philosophical usage, the word "force" denoted yet another conception for a power inherent in matter. For example, in Aristotle's philosophy, force was identified with the intrinsic potentiality of things to behave in many different ways. However, the same word was also used to denote the more conventional concept of pushing and pulling, the action of human muscle on physical bodies. Later philosophers interpreted this concept differently but always saw in it a principle that causes change and motion. Force was supposed to be an immaterial agent, and for this reason it was easy to associate it with "occult" or spiritual qualities. In fact, this was the dominant tendency, especially in relation to the motions of stars. It was Kepler who first conceived that forces should be subject to precise mathematical treatment, thus taking the crucial step toward their demystification.

Galileo, Descartes, and several physicists of the seventeenth century—most notably the Dutch Christiaan Huygens—employed a quantitative description of force in the study of motion. Galileo, in particular, had stopped just short of stating Newton's second law. However, the incorporation of force into the overriding framework of Newton's theory provided it with more precise features: force is exerted from one physical body onto another, and it comes in pairs of action and reaction. This last statement is known as *Newton's third law*:

> Whatever draws or presses another is as much drawn and pressed by that other. If you press a stone with your finger, the finger is also pressed by the stone. If a horse draws a stone tied to the rope, the horse (if I may so say) will be equally drawn back towards the stone.

Newton defined force as "an action exerted upon a body, in order to change its state, either at rest, or of moving uniformly forward in a right line."[26] This definition implies that in absence of forces, a body continues its natural state of motion, which is a motion with constant speed in a given direction. This statement, first made by Descartes, carries the name of Newton's first law of motion.

Still, the definition of force above is hardly adequate. It simply describes the role of force in Newton's theory, and it does not provide much information outside this specific context. To understand "force," one needs to refer to metaphor and intuition: for example, force could be viewed as a generalization of the action of human muscle on physical bodies. The fact that a fundamental concept of the theory is not defined independently signifies the change of perspective afforded by modern science. One does not seek an absolute definition of a theory's fundamental concepts but is mostly interested in the *mathematical description* of the physical phenomena.

However, physical bodies resist the change in their state of motion that is caused by force. This resistance is called *inertia*. A strong resistance implies that the physical body is accelerated less violently when a force acts on it. The inertia is then determined by the constant of proportionality between acceleration and force. This constant is known as the *inertial mass*, and it is an intrinsic and unchangeable characteristic of a physical body. Intuitively, it seems evident that a heavy body will resist external action more effectively than a light one. Therefore, the inertial mass is a measure of the body's *quantity of matter*.

Together with the quantity of matter, Newton also defined the *quantity of motion*. This is the physical magnitude that is obtained by the product of a body's mass to its velocity; modern physicists prefer to call this magnitude *momentum*.[27] It measures how much a body's motion may affect its surroundings. In bowling, for example, a ball will be more effective in bringing down the pins if it is very heavy or if it moves very fast. This means that the ability of a moving body to cause damage—or simply change—depends both on its mass and on its velocity. The suitable quantity is their product; the potential for damage increases when either mass or velocity increases. With this definition, Newton's second law of motion states that the external force equals the rate of change of the quantity of motion.

The three laws of motion provided the foundations of Newton's system. His greatest ambition was to find an explanation for the motion of the planets. Kepler, we should recall, had discovered that planets move in elliptic orbits. Since the motion was not in a straight line, a force must be exerted on the planets—most probably by the Sun.

This had been evident to Kepler, who had suggested that "the planets are magnets and are driven around the Sun by magnetic force."[28] One would therefore have to identify the type of force that produces elliptic motions in agreement with Kepler's laws. This was not a difficult task. Hooke and others had probably found the solution earlier, but the derivation was much more convincing within the full framework of Newtonian theory. The force between the Sun and one planet had a remarkably simple form. It was

- attractive,
- with a direction along the line that connects the Sun and the planet,
- with a magnitude proportional to the *inverse square* of their distance.

Newton then took a further step. Galileo had suggested that the celestial phenomena are of the same character as the ones on Earth. Since the language of mathematics is unique, and since mathematics describes the physical world in its totality, there should be no distinction between the heavenly spheres and Earth. Galileo had seen the surface of the Moon with his telescope: it was rocky, and rock is abundant also on Earth; it is hardly an ethereal state of matter. It was then reasonable to assume that the attractive force between the Sun and the planets had to appear on earthly phenomena. This could be nothing but the all-too-familiar pull of physical bodies toward the center of Earth: gravity.

In addition to the above, Galileo had also pioneered the experimental study of gravity. His work led Newton to the realization that the gravitational pull on a body is proportional to the body's inertial mass—hence to its quantity of matter. He eventually postulated the law of *Universal Gravitation*: all bodies are attracted to each other through a force

- whose direction is along the line connecting these bodies, and
- whose magnitude is proportional to each of the inertial masses of the bodies and inversely proportional to their distance.

The law of Universal Gravitation and the three laws of motion provide a single unifying description of the astronomical phenomena of the sky and of the mechanical ones of Earth. Because of Newton's theory, the distinction between celestial and earthly phenomena, present in many philosophies of the past or in practices like astrology, was irrevocably shattered. The unity of the mathematical description guarantees that the laws that govern the motion of matter are the same everywhere.

Descartes had been the strongest champion of the belief that physical concepts should be explained in terms of pure mathematics and geometry, and he had attempted a unified theory of celestial and earthly phenomena that employed only geometrical concepts. He postulated the existence of ethereal vortices as the main cause of planetary motion: bodies fall to Earth because they enter a vortex, in the center of which Earth is situated. The fall of the bodies toward Earth is not very different from the motion of a feather in a bathtub full of water after one removes the plug.

The success of Newton's theory dealt a strong blow to the Cartesian-Platonic desire to reduce all material phenomena to pure geometry: force is not a geometric concept. However, the blow proved not to be lethal. Future developments brought this desire back into the forefront: it is a siren that lures many a physicist even today, and my guess is that it will always continue to do so if only because of its intrinsic elegance. Nonetheless, the Newtonian theory forced a distinction that proved very useful in physical theories, even in contemporary ones: concepts and constructions that are purely geometrical, like positions and velocities, are to be distinguished from concepts or mathematical objects that are not so, like the force or the inertial mass. The former came to be called *kinematical* (pertaining to motion) and the latter *dynamical* (pertaining to force).

Another very important consequence of Newton's laws of motion is that they were *deterministic*. In the words of a great mathematician of a later century, Pierre-Simon de Laplace,

Basic Kinematical Concepts

Time
Positions of physical bodies
Velocities of physical bodies

Fundamental Quantities

Quantity of matter (mass)
Quantity of motion (momentum)

Basic Principles

The principle of inertia
Quantity of motion equals mass times velocity

The Concept of Force

Three Laws of Motion

1st Law	In absence of forces bodies move at a straight line with constant velocity
2nd Law	The force acted on a body equals the rate of change of the body's quantity of motion
3rd Law	Forces come in pairs of action and reaction

Figure 1.7 **The structure of Newton's theory.**

We may regard the present state of the Universe as the effect of its past and cause of its future. An intellect which at any given moment knew all the forces that animate nature and the mutual positions of the beings that compose it, if this intellect were vast enough to submit the data to analysis, could condense into a single formula the movement of the greatest bodies of the Universe and that of the lightest atom: for such an intellect nothing could be uncertain; and the future just like the past would be present before his eyes.[29]

Newtonian theory encapsulates a very rigid necessity. It provides, on one hand, an excellent example of scientific theory in that, with sufficient empirical information, one can determine everything about the physical system, its past and the future. On the other hand, it seems to imply that the initial conditions of the Universe—its state at the moment of creation—allow an absolute determination of every single event that has happened or will ever happen. There is slightly more freedom here than in the vision of a unique, logically consistent world. At least the choice of the initial conditions is free. However, the initial conditions of the Universe do not refer to humans and their immediate experience; they have been set up a long time in the past. Humans—or at least their bodies—are in the Newtonian Universe slaves to necessity.

Still, the introduction of the force seemed at the time to remove the specter of Cartesianism. The force concept carried within it traces of its past as a power inherent in matter, and it suggested a description of nature very different from that of a soulless predetermined mechanism. It is true that the transmission of motion in machines and the action of one machine part on another (like a cog to a lever) could be described in terms of Newtonian forces. However, the law of Universal Gravitation refers to an interaction that takes place instantaneously, *without physical contact* between the interacting bodies, unlike what is the case in actual machines. The introduction of the gravitational force moved the new physics away from the mechanistic ideal; it was even perceived as a conclusive refutation of mechanicism.

The nature of force was definitely the most contested point of Newton's theory. Its vagueness made it an object of mystery. For some people it was almost an occult element; it reminded them too much of the souls that older theories attributed to the planets, and it seemed completely at odds with the spirit of the new science. Even to people who did not support a mechanist viewpoint, or did not admit to doing so, the idea of action at a distance entailed by Newtonian gravity was troublesome. As Descartes had eloquently stated, force acting from a distance is tantamount to "making material particles truly divine, as they could be aware without intermediation to what happens to places far away from them."[30]

Newton believed that gravitational force could be explained in terms of more profound processes and concepts. "It is inconceivable," he wrote, "that inanimate brute matter, should without the mediation of something else, which is not material operate upon and affect other matter without mutual contact."[31] It was easy to assume an intermediary object that transferred the gravitational forces by means of strict mechanical contact. However, Newton refrained from making any hypotheses he could not prove. "Hypotheses," he exclaimed, "have no place in experimental philosophy." We have too little information, and we may only conjecture about, his thoughts on this issue. We should remark, however, that Newton was a fervent practitioner of alchemy and a voracious reader of alchemical literature. Perhaps he saw the elusiveness of the concept of force as a sign that matter was not simply a mechanism but had inherent powers of life and transformation, similar to the ones of the alchemical tradition. Perhaps he had the broad vision that the discovery of the roots of force—especially if they are of a nonmaterial origin—would lead to a unification of mathematical science with the deepest principles of alchemy. If he had such a vision, he never followed it openly.[32] A few years after the publication of the *Principia,* his creative activity came to a sudden and inexplicable halt. He left his theory practically unchanged until the day of his death.

The concept of force persisted as the sore spot in Newton's synthesis. On one hand, it provided the desired disagreement with the mechanistic worldview, but on the other, it could be accepted only with the deepest reserve. As usual in science, a sore spot leads eventually to progress: the effort to find a more acceptable description of force provided the next great breakthrough in physics. The introduction of the field concept eventually led to the description of forces in terms of direct contact between material objects. This breakthrough placed the Newtonian theory firmly within the mechanistic worldview: a true *mechanics,* namely, a theory about machines and their motions.

2

PROGRESS!

FROM NEWTONIAN MECHANICS TO NINETEENTH-CENTURY PHYSICS

2.1 Newton's Successors

If matter acts, if it moves through mechanisms yet unknown to us, if
motion is inherent in matter, if, in a nutshell, it may produce, preserve,
keep in balance, through the infinite fields of space, everything our
vision encounters, the united and immutable nature of which fills
us with awe, where is the need to find an external cause, since this active
principle is founding Nature itself, which is nothing but matter in
motion?

—Le Marquis de Sade, *La philosophie dans le boudoir*

The mechanics of Newton aroused immense fascination in Europe's
intellectual circles. It provided a comprehensive theory about the
workings of nature, which started from specific, basic principles and led
with the rigor of mathematical reasoning to results that were in remarkable agreement with the experiment. This was a vindication of the new
way of thinking: rationality guided by experiments sufficed to uncover
the secrets of nature. Many intellectuals were more ambitious. They
viewed Newton's laws of motion as prototypes for every human activity.

Humans are part of nature, they thought, so reason and effort will lead to other laws of nature, in biology, in human relations, in government, in religion, everywhere.

Newton would have been shocked had he lived to see the implications drawn from his work. Conservative, religious, and shy in his personal life, he was a world apart from the political revolutionaries, the antireligious pamphleteers, and the refined philosophes of the Parisian society, who were inspired from his work. However, the new era—the age of Enlightenment—needed a prophet, and Newton provided the most suitable figure for that role. His work was a symbol, and, as in most legends, symbols are more important than facts.

Few people could really follow Newton's work—the *Principia* hardly made easy reading. Most people dislike mathematical reasoning, even if they may swear by it, and this was as true then as it is now. The educated audience based its opinion on popular accounts. One of the most influential ones, the *Elements of Newton's Philosophy,* had been written by none other than Voltaire. Few people saw Newton as a proponent of a deterministic universe with no place for free will, or paused over his anxiety to avoid the dreadful consequences of his theory. For most thinkers in the century of "Light," Voltaire the most prominent among them, Newton was a liberator. He had liberated human thought from the grasp of authority and metaphysics and turned humanity toward its true mother, the natural world.

Ironic: a theory that cannot help but nullify a person's free will is celebrated as the champion of human freedom! However, for the philosophers of the Enlightenment the political battle against privileges in the state and the Church was more important than intellectual rigor. For this purpose, the repetition of a catchphrase was much more useful than philosophical consistency.

Newton had brought an epoch to a close: the heroic age of science was gone. His mechanics provided a compass for all further study of the natural world. The following generations of scientists and philosophers possessed something that Galileo and Descartes had lacked, namely, a detailed and comprehensive framework of concepts and techniques they could always fall back upon, a pillar to support and direct the development of new ideas. Perhaps this was not as obvious then as it is now. Who could have foreseen at the onset of the eighteenth century that they possessed a germ of concepts and ideas that would lead toward a deeper understanding of the phenomena of heat and electricity, of light and magnetism, of burning and life?

Still, the Newtonian theory was contested. The concept of force united many physical phenomena, but it divided the scientific community. The criticism came from two directions. On one side were the people who believed that Newton had given too much ground to the Cartesian conception of matter as things that extend in space and move according to mechanical laws. In opposition, they claimed that force is the ultimate component of the physical world. Material bodies are secondary objects: they appear to be extended and impenetrable, but in reality, they are nothing but the focal points of the activity of forces. After all, the only thing we can learn from experiments about the tiny material bodies that supposedly form matter is the forces they exert: we never observe the fundamental particles themselves. This school of thought was named *dynamism*. Its adherents included Gottfried Wilhelm Leibniz, the polymath philosopher, who had entered into a controversy with Newton about the discovery of the calculus; Roger Joseph Boskovitch, a Jesuit priest and a highly intuitive scientist, whose theories involved a synthesis of the ideas of Newton and Leibniz;[1] and Immanuel Kant, who, with Descartes, was the most influential of all modern philosophers.

The arguments of the dynamists were partly physical, partly metaphysical, and partly theological. However, their central idea is remarkably simple. They affirmed that *things exist only as long as they act*, in contrast to the underlying principle of the Newtonian (and Cartesian) theory that *the existence of things is prior to action*. As Leibniz explained:

There is something more to bodily things than extension, namely the force of Nature that God has given to everything. This force is not a mere faculty or ability . . . it is equipped with a striving or effort such that the force will have its full effect unless it is blocked by a contrary striving. . . . We should not attribute this force to God's miraculous action; it is clear that he has put it in the bodies themselves—indeed that it constitutes the inmost nature of bodies. For what makes a substance a substance is that it acts. Mere extension does not make something a substance; indeed extension presupposes a substance, one that exerts effort and resistance; extension is merely the continuation or spreading out of that substance.[2]

The dynamists' emphasis on the primacy of action had enjoyed a long history in philosophical thought, and it attracted many scientists to it. However, its results were rather poor: dynamism could not take the form of a concrete physical theory. After the Newtonian theory was extended and clarified in the course of the nineteenth century, dynamism was deemed untenable.

The other criticism to Newtonian theory came from the mechanicists. Matter, they thought, should be described only in terms of motion or of concepts that relate to motion. Perhaps we can make an exception to the concept of mass, which is after all a measure of the quantity of matter. But forces—gravity in particular—were inconceivable. If we cannot dispose of the concept of force as a convenient mathematical fiction, then we should attempt to explain forces in terms of simple mechanical contact and motion. This approach turned out to be more successful in physical theory. Even though the development of the field concept was initially viewed as a blow to the mechanistic description of nature, mechanicism managed to turn the tables by employing the field concept for its own benefit. By the end of the nineteenth century, mechanicism seemed to have scored an unchallengeable triumph. However, history turned the tables once more: the quantum theory of the twentieth century dealt an almost lethal blow to mechanicism, while it resurrected many of dynamism's ideas.

The conflict between mechanicism and dynamism was partly a new facet of the ancient conflict between the atomic theory and the theory of elements. For example, Leibniz supported dynamism because he wanted modern science to preserve the concept of entelechy, Aristotle's version of the powers inherent in matter. On the other hand, the mechanistic ideal (after Newton's theory) was an elaboration of the idea that everything may be explained in terms of atoms moving in the Void,

Newton's absolute space being a direct descendent of the Void of ancient atomic philosophy. Indeed, when Newton and most of his contemporaries thought about the basic constituents of matter, they thought in terms of atoms. However, this fact did not affect the use of elements in the description of the physical phenomena that fell outside the scope of Newtonian mechanics. For most of the eighteenth century, the favorite explanation of burning, heat, and electricity involved elementary fluids, which were the carriers of specific qualities. Eventually, burning and heat entered the realm of mechanistic explanation in terms of atoms. In electricity, however, the fundamental qualities survived longer—in fact, up to the present day.

2.2 The Atoms of Chemistry

> Our lamps brenning both night and day
> To bringe about our craft, if that we may.
> Our fourneys eek of calcinacioum
> And of watres albifinacioum . . ."
> —Chaucer, "The Canon's Yeoman's Tale"

Chemistry is the discipline that deals with the transformations between different forms of matter. As a practical science, it is one of the oldest, since a rudimentary knowledge of chemistry is necessary for the extraction and working of metals. For this reason, a substantial body of chemical knowledge was accumulated during antiquity and the Middle Ages. However, the understanding of chemical processes was gravely deficient: the practitioners completely ignored the participation of the gases in chemical phenomena. They therefore missed an important factor in the balance of their mathematical formulas, and as a result, they failed to find rules of general validity that would form the backbone of a precise and quantitative chemical science.

Chemists looked for fundamental principles that could provide some sort of logical coherence in their discipline. The most successful such principle was the *phlogiston,* first postulated by Georg Stahl in the mid-seventeenth century. The phlogiston was supposed to be the substance that lies at the root of combustion phenomena. Materials that are rich in phlogiston burn. In this process, the phlogiston escapes to the environment and leaves behind a formless heap of ashes. Phlogiston as a concept had marked similarities to the ancient element of fire. There was a crucial difference, though. The phlogiston was stripped of all mystical connotations of the alchemical tradition, and it was supposed to behave in many respects like ordinary matter.

The phlogiston theory dominated the thought of most eighteenth-century chemists. Its demise started with the realization that combustion phenomena are related to the presence of a specific gaseous substance, which could be isolated from other substances and studied separately. Joseph Priestley first isolated this gas: we now know it as *oxygen*. Priestley was part of a remarkable generation of British chemists—including Joseph Black and Henry Cavendish—who emphasized the need for precise quantitative experiments that would also take into account the contribution of the gases in the chemical processes. However, Priestley was conservative: he refrained from abandoning the phlogiston theory. He was in fact one of its main proponents.

The discovery of oxygen, together with the increasing concentration of experimental data, led to the realization that one would have to make too complicated assumptions in order to save the appearances for the phlogiston theory. The latter's demise signaled the birth of modern chemistry. It was delivered by a man of genius, Anton-Laurent Lavoisier.

Lavoisier, initially a geologist, was a member of the French Academy of the Sciences. Unlike his British colleagues, he had the opportunity to interact with mathematically minded physical philosophers that filled the ranks of the academy. The latter were as a rule dismissive of the whole discipline of chemistry because it had not developed as a quantitative science, and because it employed elusive concepts like that of the phlogiston. Nonetheless, the mystery of the chemical phenomena strongly attracted Lavoisier, and the criticism of his fellows bred in him the resolve to bring chemical theory out of its unsatisfactory state. The most paradoxical aspect of the phlogiston theory drew his attention: when one burns a metallic substance, the resulting ashes are always heavier than the original substance. However, the phlogiston was supposed to *leave* metals during combustion. The ashes left after burning should therefore be lighter than the original metal. Lavoisier analyzed this phenomenon in his first experiment. He came to believe that one should abandon the phlogiston theory, and he went on to plan "an immense series of experiments" that would lead him to a better understanding. His experiments involved the study of the gases liberated from or combined with metallic substances, and they eventually covered almost all aspects of chemical phenomena. It took Lavoisier twenty years to complete his plans, and he would have achieved more had he not been beheaded in the guillotine during the Terror years of the French Revolution—allegedly because of the professionally motivated hatred of a revolutionary "hero."

Lavoisier emphasized precision above all in his experiments—quite appropriately, one might say, for a person who earned his living and funded his research through his work as an accountant in a financial company. Lavoisier's work created a revolution, and it led to the recognition of chemistry as a science of the same level of rigor as Newtonian physics. His experimental methods strongly suggested that the quantity of matter (as represented in physics by the inertial mass) is conserved in chemical reactions. He therefore viewed the principle of conservation of matter as the basic principle of the new science.

"We must accept as an axiom," Lavoisier wrote, "that nothing is created in all process of art and nature. The same quantity of matter exists before and after the experiment."[3] Clearly, this principle had important philosophical implications. If the total content of matter in the world is constant, then matter can be neither created nor destroyed. The world has always had the same amount of matter, hence there was no reason to assume that it had ever been created. The conservation of mass suggested the idea of an eternal universe, with no beginning and no end. This idea had been a central point of the philosophy of materialism from the ancient times until the Enlightenment, and now it seemed to have acquired the status of a basic principle of science, a principle that could be verified in chemical experiments.

Lavoisier left an immense legacy to his successors. His discoveries were gradually systematized, and the various material substances were placed under a classification scheme. The first distinction in that scheme was one between elementary and composite substances. In Lavoisier's words, elementary substances are the ones "into which we are capable, by any means to reduce bodies by decomposition." Any substance that is not elementary is composite, and in principle it can be created by chemical interaction of elementary substances. Water, as Lavoisier had found, consisted of oxygen and another elemental gas, named *hydrogen*. The word *element* is henceforward reserved in chemistry for these elementary substances.

The elements—whether fluid, solid, or gaseous—combine through chemical reactions according to *fixed analogies of mass,* irrespective of any circumstances. It takes sixteen mass units of oxygen and two mass units of hydrogen to make eighteen mass units of water, irrespective of whether the reaction takes place at the top of a mountain or deep in a subterranean laboratory, in the heat of the summer or in the cold of winter, and irrespective of whether hydrogen was prepared by such and such method or through another, more efficient one.

The new developments in chemistry brought forward the issue of its relation to physics. How could one explain chemical phenomena using the physical concepts of Newtonian mechanics? Can we account for the properties of the elements and discover the law that governs their combinations?

An answer to these questions was provided by the modern atomic theory, which was pioneered by John Dalton. His was an ambitious attempt to unify the concepts of Newtonian mechanics with the new discoveries of chemistry. Dalton's atoms were very similar to billiard balls, and he considered chemical reactions to arise between mechanical interactions of the atoms. His theory proved to be remarkably successful, even though no definite proof for the existence of atoms was found before the twentieth century. For many scientists of the nineteenth century, the atomic theory remained nothing but an intelligent hypothesis that seemed to provide a good description of chemical phenomena. One should not be so hasty as to assume that atoms—endowed with the properties the chemists associated to them—exist in the real world.

The starting point in the modern atomic theory is a distinction between pure substances and mixtures. Salted water is a mixture of water and salt, and one may separate its components by nonchemical processes—for instance, by letting the water evaporate. *Pure substances are not necessarily elementary.* Water is a pure substance, but not an elementary one because it consists of two gases. Mixtures contain arbitrary proportions of their constituents, while in a pure substance the proportions of its constituents are always the same.

A pure substance consists of tiny particles, for which the word "molecules" (Latin for "piece") was employed. All *molecules* pertaining to one specific substance are identical. However, molecules are not themselves the elementary ingredients of matter. They consist of smaller particles, of different attributes and properties. These constituent particles are truly elementary. They cannot be further subdivided. Moreover, an atom of one type cannot turn into an atom of another type. Only these constituent particles deserve the name of *atoms* (indivisible). The molecules of each substance derive their identity from the specific recipe that determines the numbers and types of atoms used in their formation. Any molecule of water, for instance, consists of two atoms of hydrogen and one of oxygen. If it contained a different combination of atoms, say two oxygen atoms, it would simply not be a molecule of water.

What distinguishes elementary substances from composite ones is that their molecules consist of only one type of atom. This means that there exist as many types of atoms as there are elementary substances. Hence, a chemical reaction is a *mechanical* phenomenon by which atoms of different types combine to form molecules, or molecules break down to atoms, which may recombine to form other molecules. The fixed proportions of mass in chemical reactions follow from the fact that the properties of pure substances are fully determined by their molecules, while the molecules are themselves determined by the type and number of atoms that constitute them.

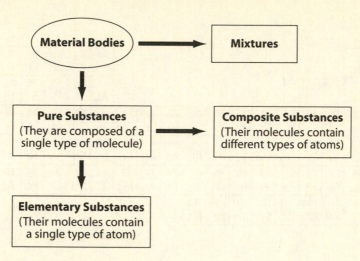

Figure 2.1 **Different types of substance.**

Two numbers distinguish between atoms of different types: the atomic number, which is always an integer, and the mass number, which denotes the *weight* of each type of atom. Since atoms had not been isolated, the mass number did not measure the atomic mass in an absolute sense, but rather the ratio of its mass to that for hydrogen, which was the lightest of the elements. One could estimate the mass numbers from the characteristic proportions involved in the chemical combinations of elements. Additional numbers appeared in the refinement of the theory, when it had to explain larger numbers of experimental data. One such example was *valence*, which is roughly the number of "grappling hooks" on each atom by which it can grasp or be grasped by neighboring atoms in the formation of molecules.

The importance of the atomic number was realized through the work of Dimitri Ivanovitch Mendeleev, a professor of chemistry in the University of St. Petersburg. Mendeleev's work culminated in one of the most celebrated discoveries of nineteenth-century science: that of the periodic system of elements."[4] Like most chemists of his time, Mendeleev was both intrigued and disturbed by the very large number—more than

1 H												2 He

Main table (atomic number / symbol):

1 H																	2 He
3 Li	4 Be											5 B	6 C	7 N	8 O	9 F	10 Ne
11 Na	12 Mg											13 Al	14 Si	15 P	16 S	17 Cl	18 Ar
19 K	20 Ca	21 Sc	22 Ti	23 V	24 Cr	25 Mn	26 Fe	27 Co	28 Ni	29 Cu	30 Zn	31 Ga	32 Ge	33 As	34 Se	35 Br	36 Kr
37 Rb	38 Sr	39 Y	40 Zr	41 Nb	42 Mo	43 Tc	44 Ru	45 Rh	46 Pd	47 Ag	48 Cd	49 In	50 Sn	51 Sb	52 Te	53 I	54 Xe
55 Cs	56 Ba	57-71	72 Hf	73 Ta	74 W	75 Re	76 Os	77 Ir	78 Pt	79 Au	80 Hg	81 Tl	82 Pb	83 Bi	84 Po	85 At	96 Rn
87 Fr	88 Ra	89-103	104 Rf	105 Db	106 Sg	107 Bh	108 Hs	109 Mt									

57 La	58 Ce	59 Pr	60 Nd	61 Pm	62 Sm	63 Eu	64 Gd	65 Tb	66 Dy	67 Ho	68 Er	69 Tm	70 Yb	71 Lu
89 Ac	90 Th	91 Pa	92 U	93 Np	94 Pu	95 Am	96 Cm	97 Bk	98 Cf	99 Es	100 Fm	101 Md	102 No	103 Lr

Figure 2.2 **The periodic table.** The chemical elements are placed in a sequence according to a number that characterizes them. Elements with similar chemical properties are placed in the same column. One identifies a period of 8 in the first half of the table and a period of 18 in the second half. The explanation of this periodicity was made possible only with the modern quantum theory of the atom.

fifty—of elementary substances that were known in his day. Surely, nature cannot be that uneconomical: there must be a law that governs the chemical properties of the elements. Mendeleev realized that these properties exhibit a kind of periodicity. Heavier elements share the same chemical properties with certain lighter ones, in a way that followed a kind of pattern. The elements could be put in order in terms of an index number, so that the lightest element (hydrogen) had number one, the next lightest element had number two, and so on; this number is the *atomic number*, and it serves to make the periodicity in the properties of the elements more apparent. The atomic number typically increased with the mass number, but not in a proportional way, so the relation between these two numbers remained a mystery for a long time.

Mendeleev's discovery is more conveniently described through the *periodic table*. In this table, the elements are placed in increasing order according to their atomic number and in columns according to their chemical properties. For example, the element indexed 3 has the same properties as the one indexed 11; the one indexed 4, the same as the one indexed 12, and so on. This suggests that the classification of elements involves a period of eight, at least as far as the lightest elements are concerned. For heavier elements, the rule turned out to be more complicated, but periodicity was still present. Mendeleev and the following generations of chemists lacked an explanation of this periodicity because they had no way of knowing the physical meaning of the atomic number. Nonetheless, the periodic table was of immense value because it suggested where one should look for new elements, in the gaps of unfilled values of the atomic number. Moreover, it hinted that there must be a deep law hidden beneath the atomic structure.

2.3 Energy: A First Encounter

> Energy is nature's efficacious motion; we call it efficacious because
> it contains the source of its motion in itself.
> —St. John of Damascus, *De fide Orthodoxa*

In the previous section, I described the shift in chemistry toward an atomic description of the elements. Similar developments took place in the study of the thermal phenomena. During the eighteenth century and the first part of the nineteenth, the dominant theory for heat involved an elementary fluid, the caloric, which was in some aspects similar to the phlogiston. Even Lavoisier, who had discredited the phlogiston theory, considered the caloric as essential for the explanation of thermal phenomena related to chemistry. The caloric was an *exquisite elastic fluid, which produces heat,*[5] abundant in hot bodies and scarce in cold ones.

When a cold and a hot body come into contact, the fluid moves from the latter to the former until equilibrium is established and both bodies are equally warm: the density of the caloric fluid in them is identical. This density corresponds to the physical magnitude we know as *temperature,* which can be measured by suitably designed instruments—the thermometers that are so familiar to us.

The only rival to the theory of the caloric fluid was the *kinetic theory*. This theory did not assume a new type of substance to explain the thermal phenomena but viewed heat as a consequence of the continuous random movement of the fundamental particles that compose ordinary matter. The kinetic theory had the support of Newton's authority behind it, but it was considered inadequate to explain all thermal phenomena.[6] The caloric theory, on the other hand, was preferred by some of the most eminent mathematical physicists of the early nineteenth century, such as Pierre-Simon de Laplace or Simeon-Denis Poisson.[7] If heat is a kind of matter, they thought, it would fit naturally within a mechanical description of the world. They proved doubly wrong. Not only was the caloric rapidly abandoned in favor of the kinetic theory, but also kinetic theory provided mechanistic science with its greatest triumph.

To understand the reasons for the success of kinetic theory, we must fall back to the early days of Newton's theory and trace a remarkable consequence of his laws of motion. A physical body in motion that is not acted upon by any forces is characterized by a physical magnitude (other than momentum), whose value does not change in time. Leibniz saw this magnitude as the true representative of the quantity of motion, the *living force*, as he named it.[8] A body with abundance of living force may resist more effectively any external agents that try to obstruct its motion—it has more options as to how it will move. In this sense, living force is the standard coinage for the mechanical universe; the more of it a body possesses, the more things it can do.

However, the living force is not conserved in all physical processes, in particular ones that involve the action of frictional forces. For this reason, it could not give rise to a law of universal validity. The crucial step in this direction was taken after the invention of a new type of machine, one that seemed to convert heat into mechanical motion: the steam engine.

"The steam engine exploits our mines, it moves our ships, it digs our ports and rivers, it moltens iron, it processes wood, it mills wheat, it spins our clothes, it carries the heaviest weights." This is a phrase in the preface of a small essay titled "Thoughts on the Moving Force of Heat," written by Sadi Carnot, a young engineer in the French artillery. The theory of the steam engine, he realized, was still in its infancy, and all efforts toward its improvement were random and lacking in physical insight. What are the rules, Carnot wondered, that govern the change of heat into the motive power of the steam engine? His analysis of the properties of the steam engine was remarkably prescient, and it was based on a simple principle—the caloric, he thought, is preserved in all processes, whether they are thermal or mechanical. He was wrong, but only barely so. Something was indeed conserved, but it was not the caloric. One should invert the argument and look more closely at the mechanical part of the steam engine. It took fifteen years for this fact to sink in; Carnot's work was all but ignored in his day.

Following Newton's work, mechanics had become the strongest pillar of the natural sciences. It was the only discipline characterized by precise and rigorous mathematical reasoning. However, in the first decades of the nineteenth century, competitors started making their appearance. Chemistry, as we saw, became increasingly quantitative, and a large body of information was slowly being gathered in other fields of research: the study of heat, biology, electricity, and magnetism. Each seemed to have a language and rules of its own. The new developments fascinated scientists, but they also created unease. How can one speak about the unity of nature if the mechanic has one set of rules, the chemist another, the engineer another, the biologist another?

Many people thought that there must be a single principle that will allow us to speak of all these phenomena in the same language. This principle should describe how the powers and forces of one type are translated into the ones of another: "besides the 54 known chemical elements there is in the physical world one agent only, and this is called Kraft [force]. It may appear, according to circumstances, as motion, chemical affinity, cohesion, electricity, light and magnetism; and from any one of these forms it can be transformed into any of the others."[9] Such thoughts were becoming more and more widespread after the 1830s, and many people contributed in ideas and experiments. It is difficult to pinpoint exactly the first person who stated that principle. Still, we must distinguish three people, for their contributions were crucial.

First among them was Julius Robert Mayer. Mayer was a medical doctor, and for this reason, his research was motivated by the study of living organisms. He had been a member of an expedition to the tropics when, in the treatment of a patient, he realized that the blood in his veins was brighter there than at home. This implied, he thought, that the metabolism consumes smaller quantities of blood oxygen in the tropics because the body needs less effort in a hot climate to preserve its temperature. Mayer followed this observation to its logical conclusion that there exists a relation between heat, chemical phenomena, and mechanical motion. It took a lot of effort and many experiments to make his initial hunch more concrete. He realized that heat is always converted into mechanical motion through a specific proportion. One was therefore entitled to state that heat and motion are simply different aspects of something universal, a force that resides in all matter and simply changes its appearance from one situation into another. This force has to be conserved, he thought, "if it disappears it must appear active in some other way."

However, Mayer—like Carnot before him—was ignored by the physics community. His intuition was deep and penetrating, but he was an amateur in physics, he had little knowledge of mathematics to support his claims, and his arguments were viewed as "too philosophical" by most physicists. His primacy in establishing the relation between mechanical work and heat was not recognized even after his ideas had been verified, and the disappointment sunk Mayer into a deep depression that lasted many years.

Five years after Mayer's first publication, his principle was at last presented in a way that would be irrefutable by even the most stubborn skeptic. The experimental work of James Prescott Joule provided the first step. Joule reached exactly the same conclusions as Mayer about the interrelation between heat and mechanical force. Unlike Mayer, whose work he had not known, he had the opportunity to catch the ears of his colleagues. "Wherever living force is apparently destroyed," Joule wrote, "whether by percussion, friction, or any other similar means, an exact equivalent of heat is restored. The converse of this proposition is all true . . . all three therefore—namely heat, living force and attraction through space . . . are mutually convertible into one another."

Almost immediately, Joule's experimental demonstration was followed by the theoretical work of Hermann von Helmholz. Helmholz placed the idea that all forces were manifestations of a single physical magnitude into a solid mathematical footing. He still used the word "force" for this magnitude in the beginning, but the name that eventually stuck to it was *energy*. The word is Greek for action, and it had acquired by that time a rich history of two thousand years in philosophical discourse. It had been used to denote an inherent power in beings that was the cause of change, motion, and life. This word was very appropriate for the principle that was thought to house every single physical phenomenon under one roof.

In contrast to his predecessors, Helmholz started his discussion from Newtonian mechanics. He proved that the conservation of energy was a mathematical generalization of the conservation of the living force in mechanics, and he argued that all phenomena of heat, electricity, magnetism, and even biology could be brought under the principle of energy conservation. The works of Joule and Helmholz complemented each other, and having theory and experiment at its side, the principle of energy conservation was firmly entrenched as a fundamental—perhaps the most fundamental—law of nature.

People who were unhappy with the mechanistic worldview—Mayer and Helmholz among them—saw the new concept of energy with its universal significance as the strongest of allies. Mechanics should not be the foundation of all physical sciences. It deals only with matter in motion, and the real world contains more than that. The activity of matter is equally real. "Energy is as real and eternal as matter," William Thomson and Peter Guthrie Tait wrote in one of the most important nineteenth-century surveys of physics.[10] Energy, the measure of this activity, is very different from the abstract forces of Newton's theory. It is conserved, as much as the quantity of matter is. Therefore, activity and matter should be placed on equal footing as fundamental principles of the world. A new science unifying all aspects of the physical world was envisioned. *Energism*, as it would be appropriate to call it, would emphasize the study of activity rather than that of inert matter.

However, energism fell short of its ultimate aim. It was killed by one of its children, a theory that built its success based on the energy principle—the kinetic theory we referred to earlier. The first successful formulation of kinetic theory was due to James Clerk Maxwell, a person we shall encounter again in our discussion of electromagnetism. Kinetic theory dealt with the physical properties of gaseous substances. Its first assumption is that gases consist of fundamental particles (the molecules of chemistry perhaps?) that move according to mechanical laws. In their motion, they collide among themselves and with the walls of their container. Molecules are extremely small, so there is an immense number of them even in the smallest of containers. Their motion is deterministic as Newton's laws specify, but because they are so many, it is impossible to keep track of them all. For practical purposes, the molecules' motion can be considered random. One then needs to use statistics for their study. It turns out that the temperature is proportional to the average speed of the molecules' random motion: *if the molecules move fast, the body is hot; if they do not, it is cold*. What is more important, the flow of heat between two bodies is nothing but a flow of the *energy* associated with the random motion of their constituents. In other words, heat is not a different form of energy; it is simply energy of mechanical motion. This was a tremendous realization: all forms of energy were nothing but different expressions of mechanical energy. The new horizons that had been opened by the universality of the energy principle collapsed back to the mechanical ideal that everything in the world could be explained in terms of fundamental particles and their motions.

The elaboration of kinetic theory led to the conclusion that thermal phenomena can be almost completely reduced to the concepts of

Newtonian mechanics,[11] when these laws are applied at the level of molecular motion. The theory that makes this reduction explicit is *statistical mechanics.* This constitutes one of the most elegant theories of modern physics. It was developed through the efforts of two of the greatest physicists of the nineteenth century, the Austrian Ludwig Boltzmann and the American Josiah Willard Gibbs. A description of statistical mechanics, however, would require a huge digression from the main topic of this book, which deals primarily with the structure and motion of matter at the fundamental level, and for this reason we shall not enter into further details.

2.4 Light

Where there is much light, the shadows are deepest.
—Johann Wolfgang von Goethe,
Wilhelm Meister's Apprenticeship

We saw in the last two chapters the prominence gained by the atomic theory in the thought of nineteenth-century scientists. It provided the basis of concrete physical theories that described the phenomena of heat and combustion. However, there were fields of research that were not adequately fertilized by the creative use of the atomic principle. Most important among them was the science of light, optics.

Optics is, together with astronomy, the oldest of the physical sciences. Already in antiquity, it was known that light rays propagate in straight lines. This implied that the science of light could be expressed in the language of geometry. Indeed, the development of geometry was soon followed by the birth of *geometric optics*, which described the properties of mirrors and lenses. Mirrors *reflect* light, and the law that guided this reflection could be summarized in a simple geometric statement: the angle between the incoming light ray and the reflected ray can be split in half by a line vertical to the mirror. On the other hand, the properties of lenses were explained in terms of the phenomenon of *refraction*: light rays change their direction when they pass through a transparent material. This is the reason, for instance, that a stick half-immersed in water appears crooked to our eyes.

Both reflection and refraction could be accounted for by two contrasting theories: According to the first, light consists of small particles; according to the second, it corresponds to waves formed in a continuous medium. It is rather easy to explain reflection and refraction in the particle theory of light. If one directs small bodies toward a polished surface, they will be bounce back in a way compatible with the basic law of optical reflection. Refraction is also explained by Newton's laws. Light particles move in a straight line when no force is exerted upon them. When they enter a material, their direction of motion changes because the material exerts forces on them. It is similar to a bullet changing its direction after it passes through a thin wooden plank.

The explanation in terms of waves is more elusive. Reflection is not a problem. One often observes the formation of waves in a cistern or in a bathtub, and it is easy to see that these waves are reflected on the walls. Refraction is rather difficult to visualize in terms of waves—strictly speaking, it can be fully accounted for only by the development of a mathematical language that described the propagation of waves.

The two conflicting theories for light were formulated during the era of the Scientific Revolution. Descartes, Hooke, and Huyghens advocated the wave perspective. Their theories differed in their details and in their explanatory power. The common underlying idea was that light is a manifestation of particles, which interact strongly with each other, thus transmitting the motion to *all directions* of space (and this is why one speaks of a wave). Huyghens, for example, visualized these particles as springy spheres. To understand his conception, one may think of a floor full of tennis balls. One then kicks one of them. If the balls are very densely packed, the disturbance will propagate along the room with the location of the kicked ball at its center.

The particle theory of light was pioneered by Gassendi within the context of his atomistic philosophy, and it was later developed into a coherent scheme by Newton. In distinction to the wave theory of light, in which the motion of the fundamental particles is transmitted to all directions, in Newton's theory the light particles only follow straight lines, which correspond to the lines employed in the description of geometric optics.[12]

Newton's work provided a turning point in the history of optics. He had devoted much time and effort to the study of light. He was the first to realize that sunlight may be analyzed into many colors (the ones that appear in the rainbow). His work, which had appeared before the *Principia*, had been the cause of the bitterest controversy. It was the first time that Newton had to support a theory of his in public, and the experience was almost traumatic. Newton became an advocate of the particle theory for light, even though he admitted that with the present state of knowledge this theory was nothing but a hypothesis. Eighteen years after the *Principia*, he published his treatise on the science of *Opticks*. That book was another masterpiece, in a vein very different from the *Principia*. In the *Opticks*, Newton manifested his skill as an experimenter, rather than as a mathematical physicist. The amount of empirical information he had amassed, and the skillful buildup of experiment upon experiment toward an elaboration and testing of his hypotheses were simply astonishing. For many intellectuals of the eighteenth century, this was the quintessential example of modern science as it emphasized the importance of careful experimentation toward the understanding of nature.

One of the most influential mathematicians of the eighteenth century, the Swiss Leonard Euler, pushed forward the wave theory of light, and, as we shall see in the next section, he identified the mathematical structures that are necessary for its description. Still, Newton's prestige was so great that the majority of physicists accepted the corpuscular theory of light. As a result, the wave theory was gradually removed to the sidelines. When a young British scientist, Thomas Young, tried to revive it in 1802, he faced so hostile a criticism that he abandoned optics altogether and he moved into a very different field of research, the decipherment of the newly discovered hieroglyphic script.

Young was an amateur physicist. He had been trained as a doctor, and he had an aptitude for languages, but his heart lay in natural philosophy. His ideas on optics were not very different from those of the seventeenth-century adherents of the wave theory. He believed that light was caused from vibrations in an underlying medium—the ether. The key difference is that he proposed an experiment that could distinguish between the particle and the wave description of light.

Young's experiment relies on the phenomenon of *interference*. To explain interference, we must first describe a related phenomenon (known since Newton's time) that appears in the propagation of waves. Suppose we have a solid wall, in which we drill a small hole. We then direct a wave toward one side of the wall. An observer at the other side will see a wave emerging from the hole and propagating from it in a symmetric way. This effect is very easy to visualize; one may even perform a similar experiment in a bathtub. Could we use this effect to distinguish between the particle and wave theories of light? Suppose we direct light toward a curtain with a hole in it. If we place a screen behind the curtain, we observe an extended bright band directly behind the hole. It is impossible to determine whether this is due to light behaving as a wave, or because the light particles are scattered from the forces exerted on them by the curtain.

But what happens when we open a second hole in the curtain? If light consists of particles, we would simply observe one thin light band behind each of the two holes. If the holes are close to each other, perhaps these bands will merge into a single one. However, if light is a wave, each hole will act as a source of a new wave. These two waves will merge, thus creating a new kind of wave, which will presumably have a very different behavior from that of a beam of particles. Indeed, the experiment demonstrated that the screen behind the holes shows a succession of dark and bright spots, something that could only be compatible with the wave description of light. For Young this phenomenon, interference, was conclusive—there was no way particles could produce the periodic pattern on the screen.

Young failed to convince many people. The bulk of the scientific community in Britain—respecting the Newtonian tradition—preferred the particle theory of light, and Young was unable to provide a mathematical justification of his experimental results. But the main problem was that Young could be arrogant and even insulting in the course of scientific

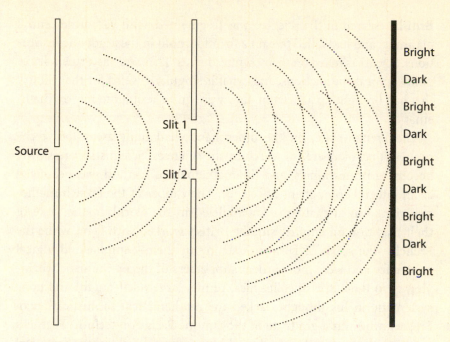

Figure 2.3 **Interference of light.** Interference is one of the most important demonstrations of the wave nature of light. A hole on a curtain acts as a source from which light propagates symmetrically. The wave then encounters a curtain with two slits, which behave like two light sources. Two waves with different profiles are formed, and the combined wave is obtained by the addition of the profiles. The profile of the combined wave can be monitored in a screen placed behind the two slits. What we observe is a periodic succession of bright and dark spots.

argument. This did not earn him many friends among his colleagues. Young's theory was consequently ignored—when it was not ridiculed.

Conclusive support for the wave theory of light came fifteen years later from across the Channel. Augustine Fresnel, a civil engineer who had been working on the construction of roads in Napoleonic France, provided what was missing from Young's conception: a precise mathematical description. Young and Fresnel are typical examples of their compatriots' attitude toward science during the early years of the nineteenth century.

British research in the sciences was fragmented, with very little central planning, and had failed to gain a firm foothold in the academic institutions. Scientists relied on their own resources. They were as a rule wary of philosophical speculations, and, what is singularly weird for the country that had brought up a Newton, they were quite hesitant to rely on mathematics.

The French state, on the other hand, had built great "polytechniques" that pioneered research in the sciences. Scientists were slowly becoming professionals, and the mathematical method was perceived as an integral component of science. The caliber of the French mathematicians was at that time unparalleled in the world. In the wake of their tradition on empiricism, the British worked with facts, while the French—the spirit of Descartes strong in them—emphasized logical coherence and were more ardent supporters of the mechanistic worldview than their British colleagues. Young, largely self-taught and nonsystematic in his thought, lacked the mathematical sophistication of Fresnel, who was a graduate of the famous Paris Polytechnic. Fresnel's mathematical competence allowed him to write explicit mathematical laws that described the propagation of waves, and from them to predict the behavior of light in the two-slit experiment. Perhaps he had not convinced the bulk of the scientific community by the time of his death—at only thirty-nine years of age—but he had provided a body of work that all future researchers could not help but rely upon. It was simply a matter of time before the wave nature of light became universally accepted.

2.5 From Flow to Field

> You could not step twice into the same river; for other waters are
> ever flowing on to you.
> —Heraclitus

The success of the wave theory of light presented a problem both for the Newtonian theory and for the mathematics of the day. Mechanics, traditionally conceived, described particles, that is, discrete entities that move in space. To describe light waves, which are continuous objects by nature, it would have to be upgraded. It was necessary to construct a comprehensive framework that would translate Newton's laws into the physics of continuous media and, following that, to develop a set of mathematical techniques that would allow one to make detailed calculations. Once the need was perceived, a solution was only a matter of time and effort. Already since the eighteenth century, one branch of mathematical physics had been developed whose techniques and ideas were ideally suited for the description of wave phenomena. Suitably modified, it could be applied to the wave theory of light. This branch was the theory of *fluid mechanics.*

Fluid mechanics—as the word mechanics indicates—studies the *flow* of fluids, and not their intrinsic properties or qualities, which may be possessed by a fluid similar to the caloric. But how can we describe flow in the language of mechanics? How can we translate a river, ever changing but always the same, into the language of equations? Fluid mechanics provided an answer in accordance with the principles of Newtonian physics: we should try to describe every physical magnitude that we can measure, and from them construct a chart that describes the water's flow.

It turns out that there exist two different strategies for the construction of such charts. In the first one, we fix our attention on the motion of specific material points of the river's waters. If one drops a leaf in the river, the leaf will follow the motion of the points of the fluid it first came in contact with. Hence, if we spread many leaves at different points of the water, we will obtain a very thorough mapping of the river's flow. The trajectory followed by a single leaf corresponds to the motion of a small element of the fluid. It is therefore called a *flow line* for the fluid. If we determine all flow lines, we have completely specified the characteristics of the river's flow. This approach was pioneered by the French mathematical physicist Joseph-Louis Lagrange, and for this reason it carries his name.

The alternative method involves less effort: we do not have to chase leaves along the river's course. The price we pay is that we get a more static

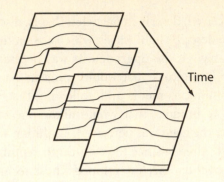

Time

Figure 2.4 The Euler picture. The evolution of a physical system in time is likened to a succession of snapshots of the system's profile.

description. We focus our attention on *fixed* points in space (rather than fixed material points of the fluid), and then we study how the behavior of the fluid at these points changes in time. For this purpose, we select points along the river's bank with a full view across. We then place a camera at each of these selected points. At any chosen moment of time, one activates all cameras instantaneously. Each camera takes a snapshot of the waters. Given enough pictures, we can determine many properties of the river's flow, as for instance the height of the water's surface at a given point. More importantly, if we take two pictures very close in time to each other, we can determine the velocity of the water at a moment of time, which is the quantity that really interests us.

The method above carries the name of the Swiss mathematician Leonard Euler. It allows us to construct *profiles* of the whole river, namely, to draw a map that describes the velocity of the waters at each point of space. The time evolution of the fluid appears then as a succession of different maps, one for each moment of time—much like cinema provides the illusion of motion by the rapid succession of instantaneous snapshots. This cinematographic representation of motion and change captures an essential structural feature of Newtonian mechanics, namely, that one can describe physical systems in terms of profiles that change in time. Euler's method not only proved successful in the mechanics of fluids, but also provided basic ideas that proved remarkably persistent. These concepts were eventually incorporated into the mathematical formalism of modern theories, such as electromagnetism or quantum mechanics.

In the Euler picture, all physical information is contained in the profile of velocities, that is, in the specification of the fluid's velocity at each point. This object is called the *velocity field*; the word "field" (originally the French *champs*) was chosen to designate the profile or the map. A geographical term like "field" is particularly relevant for this concept. A field is, in general, a specification of a physical magnitude at each point of space. This magnitude does not have to be a single number. For velocity, we need to specify its value in the three directions of space—therefore we need *three* numbers. In modern field theories, physical magnitudes are represented by more general mathematical objects: for instance, by a purely geometric object like a point of a sphere.

The concept of the field as a profile or a map contains as a special case the notion of a wave. A wave is a field whose profiles repeat a given pattern *in space and in time*. In other words, wave profiles exhibit a periodicity. To see this, we consider a vantage point that allows us to photograph a large segment of the river at once. We see peaks and crests of the waters in such a picture. The height of a peak is known as the *amplitude* of the wave; typically the larger the amplitude, the more energy the wave carries. For example, one may think of sea waves: the higher the wave, the more damage it causes when it bursts on the beach.

The distance between two successive peaks is called the *wavelength*. Similarly, one defines the *period* of a wave: if the field has exactly the same profile at successive time instants, the period is the time interval between these instants.[13] It is often convenient to describe waves in terms of their *frequency*, which is defined as the inverse of its period. The frequency, therefore, measures the number of wave peaks at a specific time interval—say, a second.

Whether a field exhibits wave behavior or not depends on its law of motion. By law of motion in this context, we mean the set of rules that

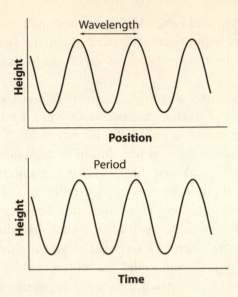

Figure 2.5 **Wave profiles, wavelength, and period.** The first drawing is essentially a snapshot that describes the profile of a wave at a fixed moment of time. It may refer, for instance, to the amplitude of a vibrating string's points. The distance between two successive peaks is the wavelength of the oscillation. The second plot demonstrates the amplitude of a given point of the string as it changes in time. The distance between two successive peaks now refers to a time interval; it is known as the wave's period.

takes the field profile at one moment of time into another profile at a next moment of time. For real fluids like water, the law of motion is extremely complicated. This is to be expected. Real fluids are collections of a huge number of particles, which interact with each other through complex forces. The description in terms of the velocity field is only an approximation, which is only meaningful at scales much larger than the distance between molecules. In this specific context, the field is only a mathematical idealization, which simplifies our description of physical phenomena. However, this does not imply that fundamental fields cannot exist. Nature may well have an attribute, which behaves continuously all the way down to the smallest scale. It turned out that fundamental fields are not only plausible but also apparently indispensable to contemporary physics.

2.6 Electricity and Magnetism

We therefore are entitled to use language fitted to deal with electrification as a quantity as well as quality.
—James Clerk Maxwell, *A Treatise on Electricity and Magnetism*

Imagine that we could bring a great mind of the early days of science—Newton, for example—back to life and familiarize him with all the understanding about the works of nature that has been achieved by modern science. What would be the discovery that would most surprise and intrigue him? It would not be, I think, the theory of relativity or of quanta. Remarkable as they are, their description of the world would not have been completely alien to the author of the *Principia*. However, Newton would not have foreseen, even in his most perceptive flashes of intuition, that the key principles leading to a unification of so disparate fields of natural science as chemistry, biology, optics, and mechanics were to arise from one of the humblest corners of physical philosophy of his time, the study of the electric and the magnetic phenomena. In reality, modern physics is unintelligible unless seen through the light of electromagnetism.

The study of electricity and magnetism has its roots in antiquity. The word "electric" refers to amber, which develops the ability to attract small pieces of wood after it has been rubbed on a woollen cloth. There existed speculations about the origins of electricity, but little substantial progress throughout the Middle Ages or even during the era of the Scientific Revolution. The breakthrough came with the advent of Newton's theory: attraction could be described in terms of forces, the effects of which could be measured in the laboratory. The discovery of a precise law for the electric force is attributed to Charles Augustin Coulomb.[14] To account for this law of force, Coulomb introduced a new physical magnitude, which we now call the *electric charge*. The electric charge is the measure of a fluid that is the cause of the electric phenomena. The force between two charged bodies, Coulomb found, is proportional to the charge of each of them and falls with the square of their distance, just like the gravitational force.

The concept of the electric charge demonstrates the shift of emphasis characterizing modern science. While the notion of a fundamental "fluid" is preserved, its defining quality has also a quantitative component. In other words, the fluid is subject to precise measurement and treatment by mathematical methods. In the process, one forgets the subjective undertones of the notion of quality and talks about *physical quantities*. To quote the most influential late nineteenth-century textbook on electricity:

> Every expression of a Quantity consists of two factors or components. One of them is the name of a certain known quantity of the same kind as the quantity to be expressed, which is taken as a standard of reference. The other component is the number of times the standard is taken in order to make up the required quantity. The standard quantity is technically called the Unit, and the number is called the Numerical value of this quantity.[15]

Clearly, a physical quantity is not a plain number. Mere numbers are meaningless without the choice of a unit. However, units serve more than simply providing a standard of reference. One meter and one kilogram are entirely different objects: they refer to different properties of things. These properties are not distinguished numerically, but qualitatively. A physical quantity, therefore, also denotes a physical quality.

However, contemporary science deliberately avoids the use of the word "quality." It brings too many memories of older philosophies of nature, from which the new science had tried so hard to distance itself. Scientists are therefore trained to employ more sanitized words in their discourse. Still, the only way to permanently exile quality from physics, as Descartes had realized, is in a fully mechanistic theory of the world that uses only geometrical terms that describe the motion of matter in space. Nineteenth-century physics was far from that ideal, and modern physics has moved even further away. Concepts such as the electric charge (and for some people the inertial mass) are not geometric or motional but refer to *intrinsic* attributes of physical systems. Physical quality is, by far, the most reasonable expression to describe these attributes, but modern science, valuing reserve more than simple language, prefers to denote them (rather misleadingly) as *internal properties* of a physical system.[16]

The electric force may be either attractive or repulsive, and for this reason it was thought that there exist two types of electric fluids. These were originally named vitreous and resinous.[17] The two fluids are identical in every respect, but the force between two charged bodies is repulsive if they carry the same type of fluid and attractive otherwise. The crucial property is that when a body of vitreous character is mixed with one of resinous character, the magnitude of the resulting force is always smaller than the one exerted by either of them.

A mathematical trick comes to our help at this point. Rather than assuming two different qualities, we can postulate a single one, whose measure takes both positive and negative values. We denote the measure of the vitreous quality as *positive* electric charge and that of the resinous quality as *negative* electric charge. We can then use the properties of addition and multiplication of numbers that are either negative or positive (called *real numbers* by the mathematicians) to explain the phenomena without the assumption of two fundamental qualities. For this purpose, we use the convention that a negative force is attractive and a positive force repulsive. Hence, the use of negative numbers (relatively new in the eighteenth century) as values for physical quantities reduces the number of basic assumptions one must make to account for the phenomena. This strategy is very common in physics and reached its absolute apex in the modern theory of quantum fields. Rather than assuming many fundamental qualities, each of them characterized by a positive number that has a direct interpretation in terms of measurements, we define a single physical quantity, which is represented by a *generalized mathematical object*.

Coulomb's law is similar to Newton's law of gravity. In gravitation, the role of the charge is played by the mass, but the gravitational force is only attractive. This implies that mass cannot take negative values. This is a good thing because mass is a measure of inertia. A negative mass would violate every principle of Newtonian mechanics, let alone of everyday experience. Imagine a body that moves in the opposite direction from which we would push it: this happens only in cartoons.

Another similarity in the role of mass and electric charge—first realized by Benjamin Franklin—is that they are both conserved. In all closed systems, in which electric phenomena take place, the total amount of charge (considered as taking both positive and negative values) remains constant.

The nineteenth century began with an invention that dramatically transformed studies of electric phenomena. Alexandro Volta, a professor at the University of Pavia, created the battery. The voltaic pile, as the battery was named in his honor, allowed experiments with electric charges in a prolonged and sustained motion, what we nowadays call electric currents. It eventually led to one of the greatest surprises in the history of physics—that electricity is deeply related to magnetism.

Magnetism was named after Magnesia, an area rich in a metallic substance with strange properties. That substance attracted some other metals, and when it took the shape of a rod—or a needle—it tended to align in the direction from north to south. It had been tempting, during the eighteenth century, to interpret magnetic phenomena in analogy to electric ones, as being due to two types of fundamental fluids that possess the magnetic property. Only one difference seemed to exist between these theories: magnetized pieces of matter always have a part of positive magnetic charge and one of negative magnetic charge of equal magnitude (these parts are known as the two poles), so that the total magnetic charge of a material body is zero. Poles always come in pairs: magnets are always *dipoles*.

In spite of all similarities, people did not see that magnetism is in any way related to electricity. No less an authority than Coulomb had categorically rejected this supposition. One can then imagine the astonishment of Hans Christian Oersted, a Danish scientist who, in the course of a public demonstration on magnetism, saw a compass needle responding to the presence of a current-carrying wire. His audience was hardly impressed, but for Oersted this was a revelation. Moving electric charges apparently behaved like magnets! Detailed experiments soon followed. Oersted found that the needle's response depended on the strength and direction of the current, as well as on the distance between the needle and the wire.

Oersted's discovery was destined to make the concept of magnetic charge redundant. In later years it would be verified that magnetic charge is nothing but moving electric charge, and that the magnetism of metals is due to microscopic electric currents, whose average does not vanish, in contrast to ordinary nonmagnetic matter. This explanation of magnetic charge was suggested immediately after Oersted's experiments. However, it became a complete physical theory only much later, in the 1920s. This explanation was possible only after the development of the modern atomic theory and statistical mechanics.

The Oersted experiments provided the impetus for a persistent inquiry about the nature of these mysteriously intertwined electric and magnetic forces. The answer came from an unexpected source, a bookbinder turned into physicist, Michael Faraday.

2.7 Faraday and the Field

Others, again, are not content unless they can project their whole physical energies into the scene, which they conjure up. They learn at what a rate the planets rush through space, and they experience a delightful feeling of exhilaration. They calculate the forces with which the heavenly bodies pull at one another, and they feel their own muscles straining with the effort. To such men momentum, energy, mass are not mere abstract expressions of the results of scientific inquiry. They are words of power, which stir their souls like the memories of childhood.
    —James Clerk Maxwell, Address to the British Association

Michael Faraday is a unique figure in the history of physics; one might say that in his career he broke all stereotypes of how physics is practiced. Whereas most scientists enter research after an extensive period of education and training, Faraday was self-taught. He came from a poor, working-class family, and he had to work as a bookbinder apprentice from the age of thirteen. However, the sciences touched something deep in his psyche. He found in them an anchor of objectivity, almost a mechanism of survival for an imagination that was equally captivated by the encyclopedia and *Thousand and One Nights*. He read everything he could find at any moment he could spare from work, and one day the opportunity presented itself. In 1812 Faraday found the time to follow the lectures of Humphry Davy, a major figure in the study of the relation of electricity to chemistry. He kept notes from these lectures, cleared and bound them, and sent them to Davy, asking for a job in a scientific field. Davy overcame any reservations he might have had and offered Faraday the opportunity to work as a laboratory assistant. This act of trust was something that Faraday never forgot. Even when Davy turned against him later, Faraday did not waver in his feelings of gratitude.

Faraday's lack of formal education implied that he could only become an experimentalist. He had a complete ignorance of higher mathematics, but in a man of such imagination, this proved a benefit rather than a drawback. He never censored his ideas to fit them to the bounds of the mathematical systems of his times, and for this reason he was able to see much farther and much more clearly than any of his contemporaries. However, he did not let his imagination run unchecked. Facts were important to him. What he lacked in mathematical rigor and coherence, he compensated for by an untiring and incessant experimentation, framing, checking, and discarding one hypothesis after the other, until he reached a point of clarity, of true vision. The manual work and his personal involvement in all these experiments gave him a special affinity for the workings of things.

Faraday's experiments dealt with practically every single manifestation of electricity and magnetism in physical phenomena. These would be sufficient to guarantee him fame as one of the greatest experimental physicists ever. However, his achievements in theory—made in spite of his handicap in mathematics—were equally important. Faraday envisioned the concept that underlies all twentieth-century physics. That was the *field of force*.

One particular observation excited Faraday's imagination more than any other. When he dropped iron trimmings around a magnet, these were not distributed randomly. They formed curves, which started from one magnetic pole and ended at another. Initially, Faraday thought of these lines as simple representations of the directions of force. However, in his later years, when he was increasingly drawn to the contemplation of the nature of things, he started viewing those curves as something real that existed independently of their source. Perhaps, he thought, these *lines of force* are the manifestations of a medium that fills space and is responsible for the transmission of the electric and magnetic forces, and perhaps even of gravity. There was a name for this medium, *ether*, the

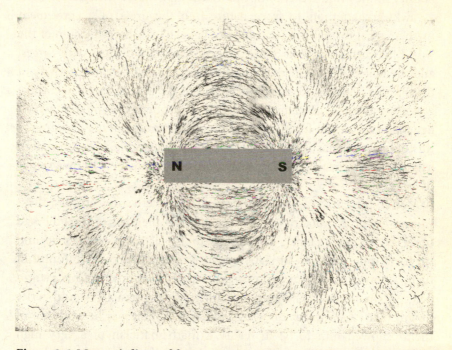

Figure 2.6 **Magnetic lines of force.** Iron trimmings spread around a magnet provide a visualization of the magnetic field's lines of force. This photo shows the lines of force associated with a long rectangular magnet: they are more densely concentrated around the magnet's poles.

ancient name of the fifth element from which the heavenly realms were constructed.

The idea of the ether was nothing new. Newton had toyed with it because he could not bring himself to accept his own idea of gravitational forces acting at a distance. Supporters of the wave theory of light had also introduced a kind of ether as a carrier of their light waves. But Faraday's ether was radically different. Newton and his mechanicist successors thought of the ether as material, consisting also of particles that move in empty space—only much smaller than the atoms of the ordinary matter. If one visualized atoms of ordinary matter as pebbles, the ethereal atoms were more like grains of sand.

Faraday would have nothing of the picture above. He believed neither in the atomic theory nor in mechanicism. He thought of them as awkward attempts to force nature into the dress of a convenient mathematical formalism. He could not believe in a sharp division of the world between space and matter. Space was for him filled with forces, and the material substances are nothing but places of intense activity of these forces. "Matter fills all space," he wrote, "or at least the space, where gravity extends. . . . Because gravity is a property of matter, which depends on a certain force, and this force structures matter. According to this opinion, matter is not simply mutually penetrable, but every atom extends, say, to the whole solar system, preserving however its own centre of force."

Matter, for Faraday, was a continuing and everlasting process; there was no better demonstration of this fact than the lines of force he had observed in the magnet. He saw them in all sorts of electric and magnetic phenomena—in static and in moving charges. He found in them a powerful tool for expressing his ideas, a language of his own in place of mathematics. Where a mathematical physicist would write down equations, Faraday drew lines of force, and the way he made use of them "showed him to be in reality a mathematician of a very high order—one from whom the mathematicians of the future may derive valuable and fertile methods."[18] The lines of force could expand, contract, be cut, and be reconnected, magnetism, electricity, and even light being related to their vibrations and flux.

Faraday's theories were in a state of perpetual flux: he kept developing his ideas and did not refrain from using many different and contradictory images to describe the same thing. Nature is always changing in its appearance, and it should not be trapped in images any more than it should be trapped in words or symbols. Faraday occasionally employed the image of the ether to describe the medium that carries the lines of force, but he would urge the greatest caution against it. He did not desire a sharp separation between matter and space, but neither did he desire a complete abolition of all differences. He could see the lines of force existing in empty space, even though he completely disagreed with the concept of an action at a distance.

Faraday did not employ the word "field." He might have been familiar with its use in the mechanics of fluids, but from the very beginning he avoided the description of electricity and magnetism in terms of hydraulic concepts and images. It is easy to imagine the ether as a kind of fluid. If a ball floats on a pond of water and then another ball is thrown in, the second ball will create ripples, which propagate and cause the first ball to move. Hence, the two balls act upon each other through the mediation of water, and the water's flow lines play the role of Faraday's lines of force. This picture is remarkably similar to the description of electric and magnetic forces by the theory of electromagnetism that was built by Faraday's successors: the geometric description is almost identical. Faraday perhaps would have endorsed this image, but without the water. The mediator should not be so crude a thing as a fluid—even an extraordinary fluid that shared none of the properties of ordinary matter. The lines of force are the real things, and the mediating fluid is just a mental image. Perhaps the lines of force represented a *strain on space*—this was Faraday's last idea—but they definitely do not describe the motion of some underlying substance. This would give too much ground to mechanicism. Faraday was astonishingly prescient on that matter. He saw much farther not only than his contemporaries, but also than the generation that followed him. His idea of the independent existence of the lines of force is in remarkable agreement with the modern understanding of the field concept.

Faraday's intuition and breadth of vision can be compared among his predecessors only with that of Newton (even though there can be no comparison at the level of their concrete achievements). They shared a discerning eye, a skillful arm, and an inexhaustible patience in the execution of experiments, and they could see deeply and afar in their theoretical speculations. But here their similarities end. Unlike Newton, Faraday had a deep reserve toward the bonds in imagination that follow from an unconditional surrender to the language of mathematics. In fact, this was the reason for his success: he constructed new concepts because he was not concerned with the necessity to express them with mathematical rigor. His intuitions cut through into the essence of the electromagnetic phenomena. However, these intuitions were not sufficient by themselves. Modern physics is quantitative, after all: it was necessary that someone express these concepts in a sharp and precise mathematical language. That person was James Clerk Maxwell.

2.8 Maxwell's Synthesis

> And sight
> Kindled afresh, with vigour to sustain
> Excess of light however pure. I look'd;
> And, in the likeness of a river, saw
> Light flowing, from whose amber-seeming waves
> Flash'd up effulgence, as they glided on
> —Dante Alighieri, *Paradise*

In 1856 the scientific journal *Transactions of the Cambridge Philosophical Society* published an article titled "On Faraday's Lines of Force." The author was a young Scottish scientist by the name of James Clerk Maxwell. Maxwell's aim was to provide a mathematical language for the lines of force. He was not the first to have done so. William Thomson, once a critic but eventually a strong advocate of Faraday's ideas, had preceded him, but he had failed to provide a complete picture.

Maxwell was a practical person as well as a man of broad imagination. He realized that mere mathematical demonstrations would not easily convince the skeptics about the reality of the fields of force. He therefore contrived to employ a model that would be easy to visualize. For this purpose, he introduced a helpful mental device. He assumed that the carrier of the lines of force was a fluid. He called it ether—as many others had done before him. However, this fluid was very different from ordinary fluids. It carried no mass, so it was not material in the strict sense of the word. Maxwell did not believe in the actual existence of this fluid. It was a convenient metaphor that lent itself nicely to mathematical abstraction. It allowed him to identify Faraday's lines of force with the fluid's flow lines, thus uniting Faraday's sharp intuitions with the powerful mathematical techniques of fluid mechanics. In contrast to the single field of velocities of ordinary fluids, the ether was characterized by two fields: one responsible for the electric and one for the magnetic phenomena.

The ether, according to Maxwell, was "carrying" lines of force of both magnetic and electric type. Two lines of the same type could never cross. From each point of the ether, only one electric and one magnetic line of force could pass. Their direction denoted the direction of the electric and magnetic field, respectively.[19] Maxwell also found a clever device for representing their magnitude. He compared a line of force to an elastic tube, whose cross-section may change in size and shape from point to point. The speed of the ether's flow in each type tube, he assumed, would represent the strength of the corresponding field. We know from ordinary fluids that the flow is faster in thin tubes and slower in thick ones, so a thin tube would represent a strong field and a thick tube a weaker one. Of course, all talk about tubes was nothing but an analogy, a visualization of the ether as similar to the human body, full of fine veins and arteries. However, this picture provided an excellent translation of the mathematical formulas into concrete objects of everyday life, and consequently it was easier to grasp.

Maxwell then asked a simple question. Assuming we take a tube and follow it as it extends in space, how does it change in form? First, its shape may be deformed—its cross-section may be a square at one point and a circle at another. Second, the tube may expand or contract, while keeping its shape. The rate of expansion or contraction is given by a number (in fact a function) that Maxwell called the field's *divergence*.

Finally, the tube may be twisted without any change in the shape of its cross-section. Maxwell, very appropriately, referred to the mathematical quantity that measures this twist as the field's *curl*. With this language, Maxwell was able to translate all Faraday's qualitative explanations of electric and magnetic phenomena into the language of numbers. Faraday was greatly pleased. "In the beginning," he wrote, "I was almost scared, when I saw so much force expended on the object, but then I saw to my surprise that the object could take it just fine."[20]

If Faraday thought in terms of images, physical analogy was Maxwell's most powerful tool. As in the tube model of the lines of force, Maxwell sought the similarities between all too different types of physical phenomena. Even if some of his analogies may seem rather far-fetched and irrelevant to the modern physicist, Maxwell had a discerning eye that allowed him to distinguish the essential from the superficial, and the skill to transform this essence into concrete mathematical equations.

It took Maxwell ten years to complete his investigations of electric and magnetic phenomena. An important issue he had to deal with was the relation of fields to the electric "fluid." Faraday had tried hard to remove the concept of the electric fluid from physics. He hoped to explain it away as a manifestation of his lines of force. Maxwell shared the same opinion. Nevertheless, their attempts to find such an explanation were in vain. They eventually resigned themselves to the idea that the electric charges should be treated—at least provisionally—as having an independent existence. This assumption proved very convenient in the construction of a mathematical description of the electric and magnetic phenomena.

More important to Maxwell's development of ideas was the concept of energy, which had been recently introduced in the study of thermal phenomena (see sec. 2.3). What role did energy play in the electric and magnetic phenomena? Numerous experiments had demonstrated the transformation of heat and mechanical energy into electric or magnetic energy. If that were the case, Maxwell asked, where exactly is this energy stored? The older theories would probably look for it in the charged bodies, in the circuits, in the magnets, and so on. But in what form would energy be stored there? Maxwell eventually came to the conclusion that there was only one object capable by its nature of storing all this energy: the field. The growing realization that all forms of energy are mechanical energy moved his thoughts even further. If the fields carry mechanical energy, then they are fully dynamical in the sense of Newtonian mechanics and should be subject to specific laws of motion.

Maxwell was successful in finding laws of motion for the fields. In doing so, he realized that the electric and magnetic fields are deeply intertwined in nature. They are two aspects of a single field: the *electromagnetic field*. This field is determined by specific laws of motion, which Maxwell expressed in the form of four simple mathematical equations:

- The first equation was an affirmation of the fact that the electric charges act as the source of the electric field.[21]
- The second equation stated that magnetic charges do not exist, namely, that there is not such a thing as a source of the magnetic field.
- The third equation stated that a magnetic field that changes in time generates an electric field with twisted lines of force, even if no electric charges are present.
- The fourth equation stated that there exist two causes for the twist of the magnetic lines of force: electric charges in motion and the presence of an electric field that changes in time. The important point is that a magnetic field can exist even in the absence of moving charges, as long as it is accompanied by an electric field that changes in time.

These equations, simple, elegant, and economical, describe completely the kinematics and dynamics of the electromagnetic field.[22] If one adds to them the principle of conservation of the electric charge and the laws that determine how the fields act on electrically charged bodies,* then one may describe without any ambiguity any single phenomenon observed in relation to electricity and magnetism.

Maxwell did not stop there. His equations predicted the dynamical behavior of fields even in the absence of electric charges: a changing

---

*The electric field acts on all charged bodies, while the magnetic field acts only on moving ones, with the force depending on the direction and magnitude of the body's velocity.

electric field generates a magnetic field and vice versa. Hence, if one generates a time-varying electric field, this will generate a time-varying magnetic field, which itself will allow the time-varying electric field to survive, and so on. The electromagnetic field would consequently persist even in the absence of matter. In fact, Maxwell's equations predict that the electric and magnetic fields propagate in space in the form of waves. These waves are *transverse*, in the sense that the fields oscillate on a plane vertical to the direction of the wave. Remarkably, the equation that guided the propagation of these electromagnetic waves was identical in form with that of the wave theory of light that had first appeared in the work of Fresnel (see sec. 2.4). Moreover, Maxwell's theory predicted the speed of these electromagnetic waves, which had been measured some ten years earlier in the experiments of two German physicists, Friedrich Wilhelm Georg Kohlrausch and Wilhelm Eduard Weber. Its value agreed so closely with the speed of light[23] "that it seems we have strong reason to conclude that light itself is an electromagnetic disturbance in the form of waves propagating through the electromagnetic field according to electromagnetic laws."

Maxwell's equations unify not only the electric and magnetic phenomena, but also the optical ones, and they emphatically affirm the wave nature of light. A few years after Maxwell's death (1879), Heinrich Hertz experimentally produced and detected electromagnetic waves of very low frequency, and he verified that their properties are similar to those of light. After that, it took only a little time before it was accepted beyond any reasonable doubt that light is indeed an electromagnetic phenomenon.

Maxwell's equations summarize in compact form one of the two great achievements of nineteenth-century physics (the other being the development of statistical mechanics, in which Maxwell also played a major role). They constitute the final chapter in a long period of study of the electric and magnetic phenomena. They are so complete and self-contained that nothing has been added to them ever since. They not only survived the great revolutions of the twentieth century—quantum theory and relativity—they lay at the center of these breakthroughs.

Relativity is fully compatible with Maxwell's equations, while for quantum theory they simply have to be translated into the suitable language. In effect, modern engineering of electrical and electronic devices, from simple lamps to mobile phones, employs one form or another of Maxwell's equations.

2.9 The Triumph of Mechanicism

Pray look better, Sir . . . those things yonder are no giants, but windmills.
—Miguel de Cervantes, *Don Quixote*

The nineteenth century witnessed the high point of the mechanistic worldview. Slowly, but at a steady pace, all physical phenomena succumbed under the efforts of mechanistic explanation. Mechanicism had the support of a powerful mathematical formalism and was able to capitalize on the ever-increasing success of the atomic idea in chemistry and in the explanation of the thermal phenomena. But the greatest success of mechanicism lay in its ability to neutralize its opponents by assimilating them into its body. The energy principle is a typical example. Energism was initially conceived as an alternative to the mechanistic ideal. It was soon realized that all forms of energy can be reduced to mechanical energy (see sec. 2.3). What is more, the incorporation of energy into the Newtonian theory made the troublesome concept of the force redundant. Newton's laws could be written without forces, in a way that emphasized that the *motion of matter is a result of its being endowed with energy.*

However, theoretical success was not the sole reason for the triumph of mechanicism. Scientific research does not take place in a cultural vacuum. The societies of western Europe were changing in a way that made the idea of a rigid necessity in a machinelike universe very compelling. Mechanicism in the nineteenth century was more than a philosopher's pastime or a scientist's method. It had become—explicitly or implicitly—the dominant ideology, and its influence had spread to all strata of society. The reason lay in the complex social phenomenon that we usually label as the *Industrial Revolution.* Nineteenth-century Europe was the place from which the new technology grew, slowly but persistently transforming the everyday life of all social classes and geographical places. A person living in this time found every aspect of life strongly affected by the advent of the new technologies. This explains why, in the course of the nineteenth century, "progress" became the new god and the "machine," or rather the machines, were his numerous incarnations.

The machine was not, however, without opponents. In reaction to the "soulless and inhuman" universe of mechanicism, the perception of a rigid domination of cold reason in the physical sciences and the creation of an impersonal society due to the industrial transformation, people were prone to affirm human emotions, the value of the internal life of humans, and the creativity of life. However, as mechanicism dominated the field, its opponents had to play a defensive game, reacting to its initiatives rather than taking initiatives on their own. It is well known that a purely defensive play is uncreative and often clumsy. Dissenters feared their opponent too much, and their response was, more often than not, unbalanced and extreme. The affirmation of human emotions often degenerated into shallow sentimentalism, the emphasis on a person's internal life into a blind subjectivism and a rejection of all science and rationality, the celebration of nature's living powers into a fascination with the occult, the desire for spirituality into dry spiritism. There were, of course, many brilliant and balanced voices in art, philosophy, and literature whose vision transcended the bounds of their age. However, they were the exception rather than the rule. As the nineteenth century was coming to a close, Western societies were literally being plagued by a wave of conspiracy theories, secret societies, pseudo-mystical traditions, magic, and spiritism, which had enough momentum to survive (and even flourish) until the present day.

The sciences were, of course, the powerhouse of the mechanistic ideal. This ideal was not to the liking of many scientists, but they could do nothing but accept it in the practice of their work. Mechanical explanations were usually more concrete, mathematically sounder,

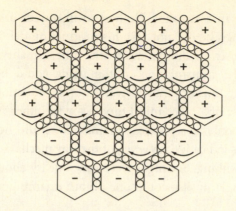

Figure 2.7 **Mechanical model of ether.** Maxwell constructed a model of the ether consisting of hexagonal molecular vortices with idle wheels between them. Idle wheels are small particles that couple adjacent vortices. This model played an important role in Maxwell's construction of the electromagnetic theory of light and helped him achieve the unification of electromagnetism and optics. However, it did not appear in the final papers of his synthesis. This figure is a reconstruction of Maxwell's original drawing in *The Scientific Papers of James Clerk Maxwell*, edited by W. D. Niven (Cambridge: Cambridge University Press, 1890).

and on this ground more convincing. The only alternative was to see the mathematical theories that describe the physical phenomena solely as a convenient and powerful tool, without thinking too much about the actual physical reality that lies beneath. This attitude ran contrary to the gut feeling of almost any scientist, and for this reason it was not widespread. It nonetheless became quite popular among the future generations of physicists, mostly because of the overwhelming abstraction involved in the formulation of the most successful twentieth-century physical theory, namely, quantum mechanics.

The theory of the electromagnetic field provides the most telling demonstration of the dominance of the mechanical ideal. Maxwell's theory was built on Faraday's insights, which constituted the most ambitious refutation of mechanism during the nineteenth century. But Maxwell—like most of his contemporaries—could not believe that the fields propagate in a vacuum. He had to introduce a substratum, the ether. Even though he did not take his original model of the ether as a fluid very seriously, he did believe in the ether's reality. He apparently believed that the transmission of the fields could be explained in mechanical terms. His original derivation of the equations of electrodynamics made use of a mechanical model for the ether (see fig. 2.7). He used it perhaps only as a physical analogy, for it did not appear in his later work. It was indicative, however, of the way mechanicism dominated the scientists' thinking. Scientists were very uneasy about nonmechanical explanations. How else could one obtain unambiguous and precise laws of motion?

Maxwell did not have to assume the existence of the ether to justify his equations. He was, perhaps more than anyone else, familiar with Faraday's intuition that fields exist in the vacuum. Yet he found it too hard to move beyond the idea of the ether. It was simply too difficult to swim against the current of the era. To a person familiar with the developments of twentieth-century science, such an overwhelming emphasis on mechanical explanation may seem rather weird—or even grotesque. Maxwell's model of the ether, with its wheels and rotating vortices, feels closer to a factory's assembly line than to a theory of fundamental physics. Admittedly, it is too easy to be judgmental about the past, knowing as we do the solution to the problems faced by the earlier generations of scientists. However, the story of mechanicism and its domination provides a lesson too valuable to leave uncommented upon: it reminds us that, as times change, all things become relative— sometimes even what is considered as concrete and settled knowledge.

The energy principle, together with the theories of electromagnetism and statistical mechanics, seemed to fulfill the dreams of Galileo, Descartes, and Newton, of a mathematical science that would describe all different aspects of the physical world. Certainly, there were still gaps in the picture. What exactly are the atoms? Do they have an internal structure? Are the molecules of chemistry identical to the ones of the kinetic theory? What does the ether consist of? How does the ether relate to atoms? These and many others were questions that physicists were trying to answer. Nevertheless, it was widely believed that science had uncovered the basic laws of nature. It would be simply a matter of filling in some few details, were the words of William Thomson, Lord Kelvin, one of the most influential and respected physicists of his time. Nothing could have been further from the truth. For the world was ready to witness the advent of the theory of relativity and quantum mechanics, which literally tore that neat picture into pieces.

3

A NEW ARENA IS BUILT

SPECIAL RELATIVITY AND THE NOTION OF SPACETIME

3.1 The Coming of the Twentieth Century

For some years past, the entire world has been in a state of unrest, mental as well as physical. It may well be that the physical aspects of the unrest, the war, the strikes, the Bolshevist uprisings, are in reality the visible objects of some underlying mental disturbance, world-wide in character. . . . This same spirit of unrest has invaded science.
— C. Poor, *New York Times* interview

The beginnings of the twentieth century in Europe left an imprint on historical memory as the "Belle Epoque." Indeed a beautiful epoch it was, as the European Powers were supremely dominant on the planet and there had been no serious war in Europe for the last twenty years or so. For the people who participated in the prosperity, it was an era of self-confidence and optimism for the future.

On the forefront of this gratification lay a sense of superiority, not only material but also cultural. Western civilization reigned supreme because of its technology and the dynamic transformation of social life its use entailed. The overall success was based—it was often claimed—on a spirit of inquiry and adventure, which found one of its greatest expressions in the overwhelmingly successful scientific world-view. At the root of this worldview (both historically and structurally) lay Newtonian mechanics. From its birth it had been hailed, very perceptively, as one of the greatest achievements of human spirit. It had captured the essence, the blueprints of the inner working of physical objects. We may not know yet what the nuts and bolts of matter are, but whatever they may prove to be, Newton's laws will describe them. Moreover, all the development of science during the previous centuries pointed to the real possibility of reducing all physical phenomena to mechanical ones. There were dissenters, of course, but their arguments were stifled in the overwhelming success of the mechanistic paradigm. In science, as in politics, success can mute even the most reasonable of oppositions.

Certainty and optimism were violently torn away during the first quarter of the twentieth century. The First World War and its horrors killed all trust in the rationality of modern society, and the Russian Revolution started the chain of events that would remove Europe from the position of dominance and create the contemporary world. These world-shaping events made the voice of the intellectual dissenters heard more loudly. The critique of the modern tradition intensified and even became mainstream: in the arts, philosophy, psychology, and human sciences there was a burst of creativity unparalleled since the seventeenth century. Perhaps a small thing among these earth-shaking developments, it was found that the great paradigm for the physical world that had either dominated or subtly underlain intellectual discourse during the last two centuries did not describe reality that accurately, after all. It was supplanted by a new physics that was based on two pillars: relativity and the quantum theory.

3.2 Reference Systems and Inertial Frames

Whatever motion comes to be attributed to Earth must necessarily remain imperceptible to us and as if non-existent, so long as we look only at terrestrial objects; for us inhabitants of the Earth, we consequently participate in the same motion. But on the other hand it is indeed just as necessary that it display itself very generally in all other visible bodies, which being separated from the Earth, do not take part in this movement.

—Galileo Galilei, *Dialogue Concerning the Two Chief World Systems*

The theory of relativity was born out of electromagnetism. Its roots are much older, though. They stretch back to the time of the Scientific Revolution. We should recall that the mechanics of Newton was based on the notions of absolute space and absolute time: space contained matter in a similar way to a vessel containing water, and all change in the world took place with respect to an external time, whose flow was not affected by any physical events.

In retrospect, one may say that the postulate of absolute time and absolute space in Newtonian mechanics was forced by the necessity to describe nature in terms of numbers. In mechanics, one needs to specify the locations and motions of bodies in terms of numbers, something that can only be achieved using coordinates. The mathematical notion of coordinates corresponds to the physical notion of a *reference frame*. A reference frame is a set of physical rules that assigns numbers as coordinates for the location of any physical phenomenon. A reference frame may be defined, for instance, by the floor and walls of a laboratory. The coordinates of a particle are determined by measuring its distance from two (nonparallel) walls and the floor. This is a good reference frame for objects in the room. For events on a larger scale, other reference bodies are more appropriate: a complicated calculation involving the geometry of a sphere allows us to assign numbers (latitude and longitude) to any point on the surface of Earth. If we decide to study the motions of the planets, it is more convenient to consider a coordinate system based on the Sun.

To describe a phenomenon we need to specify not only where, but also *when* it occurs. We use clocks for this purpose. Clocks are physical systems, some parts of which follow a periodic motion. This periodic motion serves as a unit for the measurement of time. If, for instance, the second hand of a wristwatch rotated six times during the time I was having breakfast, I would say that my breakfast lasted six minutes. A mechanical wristwatch is good for measuring seconds, minutes, and hours, but for the measurement of timescales of the order of a century or of one millionth of a second we need something different. The natural clock for centuries has been the rotation of Earth around the Sun, while for millionths of a second the consideration of atomic processes is more appropriate.

However, we need something more than the possession of good clocks in order to specify the flow of time. Unless we have a rule to translate between the readings of different clocks, our measurements of time are incomplete and disconnected. This is the reason most societies in history placed a high priority on the exact determination of the number of days in a year. Years and days are units that refer to the periodic motions of two different "clocks." The former refers to the periodic motion of Earth around the Sun, while the latter to the rotation of Earth around itself.

It is in general a complex task to find a set of rules that translates between the units corresponding to two very different clocks— comparing, say, the period of an atomic process with that of Earth's motion around the Sun. Nonetheless, the postulate of an external, immutable time in Newtonian theory guarantees that such a rule will always exist, because all clocks reveal the uniform flow of absolute time.

Concerning the reference systems for space, we also need a rule that translates the numbers that specify a point in one reference frame into other numbers for the same point in another reference frame. This rule is likely to be very complicated because the bodies that define one reference frame may move with respect to the ones that define the other. A (sitting) passenger on a train is at rest with respect to a reference system fixed on the train, but he or she is in a state of motion with respect to a station. Still, even if we find a general rule for the translation between different reference frames, we still have to face an important question. Does the choice of a reference frame make a difference in our physical theories?

According to the laws of Newtonian mechanics, a body acts upon another through forces that cause changes to the bodies' accelerations. If no force acts upon a body, the body moves with a *constant speed along a straight line*. However, speed is something that depends on the reference frame. If a body moves with constant velocity along a straight line with respect to a reference frame 1, and the reference frame 1 moves with constant velocity in a straight line with respect to a reference frame 2, then the body moves with constant velocity in a straight line with respect to frame 2. Hence, according to Newton's law there are no forces acting upon the body on reference frame 2. If, however, the motion of frame 1 relative to frame 2 involves acceleration or rotation, the body does not move with a constant speed along a straight line with respect to frame 2. Hence, from the perspective of frame 2, a force acts upon the body. However, no force acts on the body according to reference frame 1. We then have two choices: either the notion of the force depends on the reference frame or there exist *preferred* reference frames.

Newton thought that it would be disastrous if forces depend on the choice of a reference frame: the relation between cause (force) and effect (motion) in his theory would be then lost.[1] Moreover, how would he write the very simple law of Universal Gravitation if its form changed from one frame to another? For this reason, he accepted the existence of a preferred class of reference frames, which define absolute space and absolute time. If something does not move with respect to these frames, it is motionless in an absolute sense. Conversely, something moves absolutely if it moves with regard to the preferred frames of reference. Most important, the simple rule for the force of Universal Gravitation is valid only in the preferred reference frames.

Newton's laws of motion, however, are still valid in a reference frame that moves with constant velocity in a straight line with respect to these preferred reference frames: no spurious forces arise there. Such frames were named *inertial* because the principle of inertia—Newton's first law of motion—is valid only with respect to them. In all other frames, the laws of motion differ: *pseudoforces* appear, that is, bodies accelerate without any force acting on them. A typical example is the "force" we feel pushing us back when we sit in an accelerating car. Nothing really acts upon us. We feel a force only because we follow the car's accelerating motion, which defines a noninertial reference frame.

It is in fact impossible to distinguish between inertial and absolutely static frames. We obtain information from the world by interacting with it, and we interact only by means of exerting or receiving forces. Such forces satisfy Newton's laws of motion, and they have the same form in all inertial frames. There is no way to identify which of all inertial frames corresponds to absolute rest. Hence, it is impossible to distinguish whether a physical body is in a state of absolute rest or not. That is, this was the situation until the late nineteenth century, after the electromagnetic waves predicted by Maxwell's theory were detected.

People believed that the electromagnetic fields propagate on the ether, and that the ether was supposed to be present everywhere. Its distribution in space should be homogeneous because Maxwell's equations did not refer to any special point or direction of space. It was natural, then, to think that the ether defines the reference frames of absolute rest. Our only way of interacting with the ether is through electromagnetic waves. Could we perhaps use electromagnetic phenomena to identify the absolute motion of physical bodies? Earth definitely moves with respect to the Sun and with respect to the ether: thus, it defines a moving reference frame. Since the ether's distribution is homogeneous in the frame of absolute rest, the electromagnetic waves should travel with the same speed to all directions. However, from the perspective of a reference frame that moves with Earth, the speed of light should be different in different directions. If we could measure these speed differences in a controlled way, we could identify Earth's absolute motion.

Such were the thoughts of the American physicist Albert Abraham Michelson, a brilliant experimentalist who had been studying the implications of Maxwell's theory for the properties of light. He conceived an experiment that would allow him to test his idea. Its essence was the following. One sends two beams of light toward different directions, in such a way that they are reflected upon mirrors that are placed at equal distance from their common source. The reflected beams will return to their origin, but as light travels with different speed in different directions, the profiles of the corresponding waves will not be identical. This

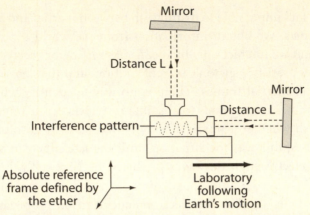

Figure 3.1 **A sketch of the Michelson-Morley experiment.** We send two light signals from our laboratory, one parallel to Earth's motion and one normal to it. At equal distance from the source, we place two mirrors, which reflect the light back to our laboratory. According to Newtonian mechanics, the speed of light is different in the two directions, so the reflected waves will not arrive simultaneously. The wave profiles at a moment of time will be different, and hence we will observe interference. The study of the interference pattern will enable us to determine the laboratory's absolute velocity.

implies that the waves will interfere, showing a pattern similar to that appearing in Young's two-slit experiment (see sec. 2.4). It will then be possible, Michelson thought, to infer from the interference pattern the difference in the speed of light in the two directions, and consequently to determine Earth's velocity as it moves with respect to the absolute reference frame defined by the ether.[2]

The problem with Michelson's experiment is that the speed of light is much larger than the speed of Earth's motion. This implies that the difference in the speed of light in the two directions is very small. Hence, the profiles of the two light signals will be almost identical. Thus, it would be very difficult to discern the interference pattern. Nonetheless, the experiment was designed and performed in 1887, Michelson collaborating with Edward Williams Morley. The experiment involved the highest precision possible for the technology of its day, and there was a certainty that the accuracy they had achieved would be adequate for the purpose.[3] The result was shockingly fruitless. Interference was observed, but it was much less than what would correspond to the orbital motion of Earth around the Sun (which is estimated through astronomical methods). In Michelson and Morley's words, "it appears . . . reasonably certain that if there be any relative motion between the earth and the luminiferous ether, it must be small."[4] In fact, the observed interference was so small that it could be attributed to experimental error. The results of the experiment were fully compatible with Earth being motionless, and this seemed to be the simplest conclusion.

Clearly, an explanation had to be found. The first one that comes to mind is that Earth is at rest with respect to the ether; it does not move. That would be preposterous! Notwithstanding that Copernicus's ghost would stir angrily from his rest, such an explanation would be a mockery to all progress in physics during the past two centuries. What cosmic conspiracy could force Earth to remain at rest, while being subjected to the (empirically well-established) forces of Universal Gravitation?

The second explanation was that the ether is not as "ethereal," as physicists tended to believe. It interacts with ordinary matter, so it also affects phenomena that are not electromagnetic. In effect, the ether behaves like an ordinary fluid, which resists any body that moves within it. The Irish physicist George Francis Fitzerald suggested that the ether's resistance to the motion of physical bodies would exert a force on any physical body that would cause it to shrink along its direction of motion. He estimated a law for this contraction, which would be compatible with the results of the Michelson-Morley experiment. The distances from the mirrors to the source had to contract in accordance with the change of the speed of light. This way the time-interval it took light to travel from the source to the mirror and back would be the same in both directions. Hence, the wave profiles would be identical, and no interference would be observed. Many of his colleagues scorned Fitzerald for his implausible idea. Nonetheless, a few years later Fitzerald's idea was rediscovered by Henrik Antoon Lorentz, a Dutch physicist who had gained great respect because of his extensive study of the action of the electromagnetic field on particles. This length contraction, the Fitzerald-Lorentz contraction as we call it now, turned out to be real, but its justification went much further than anything either of its discoverers had foreseen.

3.3 Einstein's Solution

> An invasion of armies can be resisted, but not an idea whose time has come.
> —Victor Hugo, *History of a Crime*

The third explanation for the results of the Michelson-Morley experiment was simpler, elegant, and, consequently, more radical. It was developed in a paper titled "On the Electrodynamics of Moving Bodies," which appeared in the June 1905 edition of the prestigious journal *Annalen der Physik*. Its author was a rather obscure twenty-six-year-old semi-amateur physicist who was working as a "technical expert third class" in Bern's patent office. He had written three other papers on different topics in the same year. Their combination widened the horizons of physics to a degree unmatched by the work of any other person in such a short period. The author's name was Albert Einstein, and his work marked the birth of the special theory of relativity.

Einstein had already been working for three years in the patent office. He had been unable to find an academic position after his graduation as a teacher of physics and mathematics from the Federal Polytechnic of Zurich. The job in the patent office was a good solution in the face of financial difficulties. It allowed him plenty of free time, which he devoted to his favorite pastime, the study of physics.

Einstein's genius was not simply the result of him possessing an agile mind and a lively imagination. What really set him apart as a scientist was his character. He was singularly persistent in his investigations. He would follow his intuition all the way to the end, even if the route were laden with immense difficulties. This behavior was not due to stubbornness, arrogance, or mere self-confidence. It was simply very difficult for Einstein to take any form of intellectual authority very seriously. He could not accept even an idea that had the support of the vast majority of the scientific community, if that idea could not be ascertained by means of a clear argumentation. He literally *could not* accept them—they simply left no imprint in his mind. This is why his encounter with the highly authoritative teaching of the sciences in his training almost killed in him any desire to work in physics.

If we realize this aspect of Einstein's character, it becomes easier to understand the freshness of his approach to the Michelson-Morley experiment. That experiment was not in fact the main motivation for his work on the "Electrodynamics of Moving Bodies." Einstein's work was primarily motivated by his disappointment in the status of the electromagnetic theory. Already at the age of sixteen—spurred by readings of popular science—Einstein was wondering how electromagnetic waves would appear to an observer that moves with the speed of light. They would appear to the observer motionless, and this would seem to imply that their vibrations would be frozen. However, this behavior was out of the question in Maxwell's theory. Something was missing there, something he could not pinpoint at that stage.

By 1905 Einstein had learned more physics and had grown confident in his understanding. Frozen waves cannot exist in Maxwell's theory, he thought. Hence, even if we move at the speed of light, we should see that electromagnetic waves do not change their character. He then experienced a leap of intuition: his toying with the propagation of light crystalized into a simple principle. The speed of light should be the same in all reference systems, or at least in all inertial ones. This idea cut through the knot of the Michelson-Morley experiment. Light moves with the same speed in all directions irrespective of Earth's motion, hence interference between light rays that have moved in different directions is not possible. One should then forget the ether and all ideas of absolute space and absolute motion; in light of the new principle, they are simply irrelevant.

As it turned out, Einstein needed only one more principle to make his ideas into a concrete theory. It was a simple one. *The laws of physics are the same in all inertial reference frames*, a statement that was true already in Newtonian mechanics.[5]

Where did Einstein's two postulates take him?

In Newtonian physics, one knew how to translate the coordinates of one frame of reference into the coordinates of another that moves with constant velocity in a straight line with respect to it. This translation involves two rules. The first rule is that velocities have to be added or subtracted: we may consider, for example, the case of a person moving within a train, which itself moves with respect to the station. If the person moves at 1 kilometer per hour with respect to the train and the train moves at 40 kilometers per hour with respect to the platform, then the person moves at either 41 or 39 kilometers per hour with respect to the platform according to whether he or she moves in the direction of the train or against it.[6] The second rules states that clocks can be arbitrarily synchronized. If two clocks (of identical make) start at the same time, they are going to run at the same rate, independent of their state of motion. These rules were called Galilean transformations in retrospect, as a tribute to Galilei, who had first discussed the relativity of motion.

Michelson and Morley relied on the Galilean transformations for the conclusion that light moves with different speed in different directions. Einstein's first postulate rejected this conclusion. The Galilean transformation laws were then incompatible with Einstein's perspective. They had to be replaced by better ones, which would not fail to preserve the speed of light in all directions. It turned out there was only one possible answer, and Einstein did not have a hard time finding it. In a nutshell, the content of the new transformation rules was the following.

- The way we transform velocities from one reference frame to another is not obtained by a simple addition but by a more complicated expression, which incorporates the fact that the speed of light is the same in all reference frames. This transformation cannot be defined if the relative speed between two inertial frames is larger than the speed of light. This is a strong indication that *the speed of light is the maximum speed attainable in nature.*
- Clocks cannot in general be synchronized. Two clocks of two different reference frames register different time intervals when they measure

the duration of the same event. Hence, there cannot be such a thing as absolute time, or if it does exist, it cannot be used to effect clock synchronization, so it is a useless concept. Two observers can synchronize their clocks only if they communicate with each. However, no signal they can use to communicate travels faster than the speed of light. For this reason, any two observers moving with different speeds will have a hard time synchronizing their clocks. The notion of simultaneity, therefore, depends on the choice of the frame of reference. This fact leads to a remarkable new phenomenon known as *time dilation*: a clock that is at rest with respect to an event will measure for the event's duration a number that is always smaller than the one measured by a moving clock. Hence, for example, if it takes the passenger of a moving train three minutes to get to the bar, an observer on the station will record a longer time. If the train moves with an extremely high speed (much faster than any conceivable vehicle on the surface of Earth), the observer on the station may measure ten minutes, or even twenty. The faster the train runs, the longer the time the observer will record. In the extreme case that the train moves with the speed of light, the observer will record an indefinitely large time interval.[7]

- The length contraction, first discovered by Fitzerald and Lorentz, is valid in special relativity. The length of an object looks always larger to the person that is at rest with respect to it than to somebody in a state of relative motion.

- When the relative speed of two inertial frames is much smaller than the speed of light, the new rules cannot be distinguished from the old ones. In this sense, the previously accepted theory is recovered at the limit of small velocities. The reason we had not observed phenomena like time dilation or length contraction is that the speed of light is much larger than any speed we can observe on the surface of Earth (even the fastest airplane flies with less than 1/100,000 of the speed of light).

Einstein had not been the first to derive the correct transformation rules. Lorentz had written them a few years before Einstein,[8] and they carry Lorentz's name for that reason. However, Lorentz did not see in them anything but a convenient mathematical trick. He could not bring himself to abandon the idea of absolute time, a step that was necessary for these rules to make physical sense. As Einstein recalled fifteen years later:

> I had wasted almost a year in fruitless considerations, with a hope of some modification of Lorentz's idea, and at the same time I could not but realize that it was a puzzle not easy to solve at all.
>
> Unexpectedly a friend of mine in Bern then helped me. That was a very beautiful day, when I visited him and began to talk with him as follows: "I have recently had a question, which was difficult for me to understand. So I came here today to bring with me a battle on the question." Trying a lot of discussions with him, I could suddenly comprehend the matter. Next day, I visited him again and said to him without greeting: "Thank you. I've completely solved the problem." My solution was really for the very concept of time, that is, that time is not absolutely defined, but there is an inseparable connection between time and the signal velocity. With this conception, the foregoing extraordinary difficulty could be thoroughly solved. Five weeks after my recognition of this, the present theory of special relativity was completed.[9]

Even in the renunciation of absolute time, Einstein had not been the first. The French mathematician Henri Poincaré had written in 1902: "There is no absolute time; to say that two durations are equal is an assertion which has by itself no meaning . . . we have not even the direct intuition of the simultaneity of two events measured in different places."[10] Perhaps the discussions of Einstein with his friend brought this prescient passage back to his memory.[11] Poincaré had literally prophesized many of the developments in twentieth-century physics, but like many prophets, he failed to acknowledge the fulfilment of his prophecies. He provided an elegant proof of the Lorentz transformations only a few weeks after Einstein, but he did not realize their implications about the nonexistence of absolute time. In fact, Poincaré never accepted the basic physical principles of the theory of relativity.

What distinguished Einstein's work from that of all his predecessors was not so much the introduction of new concepts or the derivation of new results, but his achievement of bringing together so many different components toward a synthesis of overwhelming clarity. Indeed, clarity of vision was the greatest of Einstein's gifts.

3.4 The Union of Space and Time

Motion is the process, the transition of Time into Space and of Space into Time: Matter, on the other hand, is the relation of Space and Time as a peaceful identity.
—Georg Wilhelm Friedrich Hegel, *Philosophy of Nature*

Einstein's new theory stated explicitly that measured lengths and time intervals are *relative* with respect to the inertial frame employed in the measurement. For this reason, it was thought that the name "relativity" would be most appropriate.

The name went down well: it resonated with the spirit of the times. Absolute structures in science could be mentally identified with absolutist structures in society (political absolutism, moral, social, or racial intolerance), and their downfall in science would suggest that old social certainties and prejudices would have to be swept aside. For others, relativity in nature suggested or licensed relativism in the social, intellectual, or moral sphere. For yet others, relativity was identified with subjectivity. Many an artist tried to relate the awakening of the subjective element in twentieth-century art with Einstein's theory. These "interpretations" of the theory of relativity were proposed not only in the beginnings of the twentieth century, when the theory was still young; they appear repeatedly from people of very different backgrounds and convictions. Natural science together with its applications play such an important role in modern societies that people look often to scientific theories for a rationalization of their beliefs.

With the only exception of quantum mechanics, no other theory of physics has so stimulated the imagination of the public as the theory of relativity. In that light, it is rather ironic that the name "relativity" is very inaccurate for the physical theory that Einstein developed, and that the images it evokes strongly misrepresent its physical content. The reason for this is the following. Whenever in physics and mathematics we have a description of a structure in terms of relative objects, *we can always introduce an absolute description at another level.* This is a statement of such a universal validity that it is practically a law of thought; it is not a law of logic, though, because the exact procedure by which this "absolutization" is achieved varies from case to case. In general, an absolute description is more elegant, but it may not be very useful for practical purposes. In the case of "relativity," the absolute description was so overwhelmingly successful that it was immediately thought to contain a very accurate representation of reality. Einstein himself preferred the name "*theory of invariants*" for his theory, but by the time this new name was proposed, the term "relativity" had been universally accepted, and it would not change.

If space and time are not absolute by themselves, they become absolute when they are joined together—thus forming *spacetime*. This insight came from Hermann Minkowski, a German mathematician who had been one of Einstein's teachers in Zurich. Minkowski realized that the Lorentz transformations correspond to a kind of "rotation" in a mathematical space, each point of which corresponds to a point of physical space *and* an instant of time. This mathematical space he called *spacetime*: it had four dimensions, three for space and one for time. Spacetime is the arena in which physical processes unfold, and for this reason, its points were called *events*.

The most important benefit of the spacetime description is that it allows us to talk about the relation between events in a way that does not depend on the choice of reference systems. To see this, we consider two events, and we set down an inertial reference frame by which to describe them. We then calculate the distance between the events with respect to that frame: let us denote this distance by the letter L. We may also determine the time interval T between these events. As an example, consider the coordinate system fixed on Earth, the first event being the lift-off of a rocket on Wednesday, and the other the arrival of the rocket on the Moon on Saturday. In this case, the number L is the distance from Earth to the Moon, and T is the time interval of three days, both measured by us on Earth.

Minkowski realized that the number $L^2 - T^2 c^2$, where c is the speed of light, is independent of the choice of inertial frame. This quantity, called the *line element*, is essentially the difference between the square of the distance L minus the square of the distance light would have traveled within time T. Even if we change the reference frame used in the measurement of the distance L and the time interval T between two events, the value of the line element remains the same.

We may assume that the two events under consideration correspond to the emission and the reception of a physical signal. In the rocket example, the signal is the rocket itself. The signal traveled through a distance L within a time interval T. Since its speed could not have been greater than that of light, the distance L must be smaller than the distance $c\,T$ that light would have traveled in the time interval T. The line element would consequently be negative. It follows that if a physical signal connects two events, *the corresponding line element takes negative values*. One of these events must be later in time with respect to the other: the rocket arrives at the Moon *after* its lift-off from the surface of Earth. It is then possible to express the relationship between past and future through the line element, without choosing a specific frame of reference. The converse property also holds. A negative line element between events implies that there exists at least one reference frame, in which one of these events "takes place" after the other. Such events are called *timelike* separated.

When the line element between two events is positive, no physical signal can connect them: any such signal would have to travel faster than light. Hence, none of these events lies in the future of the other. In fact, these events are simultaneous (their time difference T is zero) in some reference frames. Such events are called *spacelike* separated.

It is also possible that the line element is zero. In this case, only signals moving at the speed of light can connect the two events. Such events are said to have a *lightlike (or null)* separation.

From the arguments above, we conclude that the line element on Minkowski spacetime encodes information about both readings of clocks and measurements of length in a way that does not depend on the reference frame, even though the measurements of length and time considered separately depend on the reference frame.[12] This result may sound, perhaps, too abstract. Einstein certainly thought so. He initially dismissed the spacetime picture as "superfluous learnedness," but he soon recognized his mistake. The mathematical abstraction involved in the definition of the line element proved a benefit rather than an obstacle to the development of his physical intuition.

One may visualize the geometry of Minkowski spacetime through the so-called light cone diagram (see fig. 3.2). To draw this diagram, one first selects a coordinate system. Its origin, which we denote by O, is some arbitrary spacetime event. The vertical axis represents distance in time from the event O multiplied by the speed of light, and the horizontal axis represents spatial distance from O in a specific spatial direction x (the other two spatial directions are suppressed). The two lines at 45 degrees correspond to the propagation of light rays and are known as the light cone. This name arises because in two spatial dimensions the area spanned by all light rays passing through the origin is a cone. Particle trajectories correspond to curves on this diagram. Since physical particles travel more slowly than light, all trajectories passing through O lie within the cones in the areas labeled as I and II. In other words, any event in I and II is timelike separated from O. Since we assume that time goes forward in the direction from the bottom of the page upward, region II represents events in the *past* of O; region I, events in the future of O. Regions III and IV represent events that are not causally connected to O, that is, they are spacelike separated from O.

We must be careful when we talk about the geometry of spacetime. Our intuitions of geometry rely on our sense experience: we perceive space as extension, and we understand its properties guided by the shapes of objects in our everyday life. The geometry of spacetime is completely different. It is only with respect to spacelike directions that we should comprehend distance as analogous to our experience of extension. For the timelike directions, our mental imagery should be in terms of duration rather than extension. This difference in the perception of the directions of time and space is reflected mathematically in the positivity or negativity of the line element. Moreover, we really have no sense experience what "distance" is in a null direction. It is only by

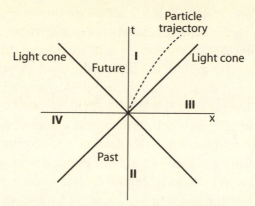

Figure 3.2 **The light cone diagram.** The region labeled I lies at the future of the observer *O*, while the region II lies in its past. The points in regions III and IV can never contact the point *O*. The thicker lines represent the light cone, that is, the region of spacetime that is spanned from all light rays that pass from *O*. All particles move with speed slower than light, hence if their trajectory passed at some time from *O*, it must lie wholly within regions I or II.

mathematical thinking, namely, proofs and constructions, that we can develop a perception of its properties.

While space and time have no absolute existence in relativity, the theory did not divest itself of all absolute concepts. Now *spacetime is the absolute and immutable structure*. When space and time were separately considered as the background of physical phenomena, they defined the notion of absolute rest. The existence of the inertial frames, in which the laws of motion take the same form as in the frame of absolute rest, was a minor embarrassment. When spacetime is taken as the background, inertial frames substitute absolute rest as the fundamentally absolute notion. In this sense, Einstein's theory was not that radical an overthrow of Newtonian physics: it was a reformation rather than a revolution. However, this is true for the particular theory of Einstein, which has come to be known as special relativity. In general relativity, in which gravitational phenomena are also included, spacetime loses its absolute character: it becomes a dynamical quantity, subject to laws of motion and change by its interaction with matter. This *is* a radical overthrow of previous ideas about space and time: the distinction between kinematics and dynamics originating from Newton's theory is lost, to mention the least of the changes. However, the topic of this book is the structure of matter, which in modern physics is most adequately described by quantum field theory. This theory relies on special relativity. Thus, it can be meaningfully defined only with reference to an absolute spacetime. All attempts to reconcile it with the insights of general relativity have been so far unsuccessful. General relativity remains an isolated example setting out how our theories for the world should be, but we cannot yet incorporate its principles into the rest of physics. In particular, we have found so far no way to reconcile the principles of general relativity with the rules of quantum theory. We shall return to this issue many times in this book.

3.5 Mass Is Energy!

> We must assume that nothing exists in itself, but all things of all sorts
> arise out of motion by intercourse with each other; for it is . . .
> impossible to form a firm conception of the active and passive element
> as being anything separately; there is no active element until there is a
> union with the passive element, nor is there a passive element until there
> is a union with the active element.
> —Plato, *Theaetetus*[13]

I mentioned in the previous section that special relativity accepts the notion of a background structure for space and time, and in this sense it does respect the Newtonian edifice. There is also another context in which it inherits the structure of Newtonian mechanics: its concept of dynamics is identical. Newton, or any physicist that came after him, would have little trouble arriving at a version of the theory of relativity if he had any reason to suspect that the speed of light is constant in all reference frames. This is not to underestimate Einstein's genius: but really, his power as a physicist, in combining an otherworldly intuition with strong perseverance and dedication, reaches its full fruition in the general theory of relativity. Special relativity involved essentially one novel idea, namely, that Maxwell's equations make full sense only if the speed of light is the same in all frames of reference. Once Einstein had grasped this idea, he could follow the general lore of Newtonian mechanics. In general relativity, it was a continuous effort of eight years during which he made many breakthroughs, each of them of the same significance as the theory of special relativity itself.

Special relativity modifies only the *kinematical* background upon which the Newtonian theory is defined. Newton's three laws of motion remain practically unchanged; they only have to be reformulated in a way that respects the novel kinematics. This reformulation leads to many remarkable results and insights, but it does not change the most fundamental consequence of Newtonian mechanics: the *deterministic* character of the laws of motion. If at a moment of time (now this moment is defined with respect to a specific reference frame) we have full knowledge of the properties of a physical system, we can predict with certainty any of its properties at any subsequent moment of time. There is also no change to the fact that bodies act upon each other by causing change on their *accelerations* rather on than their velocities or on any other geometric quantity. I think that the most accurate name for the theory of special relativity would be "mechanics in absolute spacetime" in contrast to "mechanics in absolute space and in absolute time," which would best describe the Newtonian theory.

Special relativity may be compatible with the mechanistic view of the world, but it greatly undermines many of its achievements. In particular, relativity makes the notion of the ether obsolete and useless and in this sense renders impossible the mechanistic interpretation of fields that dominated nineteenth-century thinking. This implies an important change of perspective. Rather than the field being a property of a medium, we have to think of the field as the medium itself: *the field is itself material and not a property of matter*. This is a very great conceptual breakthrough, invaluable for the development of twentieth-century physics. The notion of matter cannot any more be restricted in the concept of impenetrable bodies that extend in space, as most interpretations of the atomic theory implied. Matter comes in many forms; the field form is as real and as important as that of atoms. It seems as though relativity resurrected the key physical ideas of dynamism about matter, even though it did not entirely remove mechanicism from its position of dominance. Indeed, one cannot rule out a fully mechanistic description of the material Universe, as long as the "nuts and bolts" of this "machine" are compatible with the geometry of spacetime.

What did Newtonian dynamics look like when it was forced to lie on the bed of special relativistic kinematics? To answer this question, we need to recall that the speed of light is the maximum attainable in nature. No matter how much we accelerate a body, it will never reach the speed of light. As the body's speed grows, so does its inertia, that is, its resistance to external forces. In fact, the inertia must increase in such a way as to tend to infinity when the speed reaches the speed of light. However, the quantitative measure of inertia is the mass. Hence, mass should depend on the body's velocity. This is quite remarkable by itself, but it is not the end of the story. When we accelerate a particle through forces, we increase its energy by a specific amount, which turns out to be proportional to the velocity-dependent *mass*. This implies that the energy is itself a measure of inertia, like mass. This property is expressed by the famous Einstein law of equivalence between energy and mass $E = mc^2$, which is perhaps the only equation of physics to have achieved a cult status.

When a body is at rest, its energy does not necessarily vanish. It takes a constant value, the *rest energy*, or *rest mass*. This *quantity is not any more a measure of inertia*. It is rather an intrinsic property of the body, being the same in any place or time and with respect to any reference frame. Two particles with the same speed but with different rest mass have different values of total energy because the energy of each particle is *proportional* to its rest mass. In contemporary particle physics, the word "mass" is almost exclusively reserved for the rest mass, while the word "energy" refers primarily to the measure of inertia. We will adopt this usage from now on.

The only common feature of energy and mass in nineteenth-century physics was the fact that they were both conserved. Energy was thought

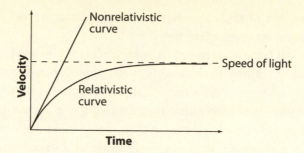

Figure 3.3 The speed of a particle when accelerated by a constant force. In the nonrelativistic case, the speed increases indefinitely, while relativity imposes a limit equal to the speed of light.

to be the active principle of nature and mass the passive one. Matter was brought into motion by the forces through which energy was transferred. We saw that many physicists and philosophers perceived energy as the key concept that would liberate science from the materialistic, mechanistic worldview. Einstein's work was quite a shock to the few remaining adherents of this view because the identification of energy and mass removes any distinction between the cause of motion and the resistance to motion.[14] The ability of a body to change the state of motion of another is identical to its ability to resist change. It therefore becomes impossible to disentangle matter from the causes of its motion. Matter changes its state of motion, not because an external principle or force acts on it, but because it is in its nature to do so.

Relativity brings out a concept of matter more dynamic than the one of the Newtonian theory. It infused matter with all the vigor and power contained in the concept of energy. Energy, as a result, became a shadow of its former self. After Einstein, it does not represent any more the active principle of nature. It becomes a mere number that measures the quantity of matter, namely, the potential of a physical body to effect change or to resist change. Contemporary physicists use the word "energy" as referring to something passive, a kind of clay that can be plastered to take any shape or form; "it is like a mould for all things, which changes and takes the shape of all that enter it, and this is the reason it seems different every time."[15]

If a physical body accelerates, its energy becomes increasingly larger, and it tends to infinity as the body's speed approaches that of light. A particle can therefore never attain the speed of light if it starts at a lower speed. This does not imply that no particles move with the speed of light. Such particles may well exist, if their mass (meaning now their rest mass) is zero. However, they can never move at a lower speed.[16] It is therefore impossible to see them at rest. We conclude that the particles of the relativistic world fall into two categories: ones moving at the speed of light and ones moving at a lower speed.[17]

We shall return to the different categories of particles in the theory of relativity in the next chapter. We will first address a very important issue. Einstein's second postulate was that the laws of motion must be the same in all inertial reference frames. Supposing we could find such laws,[18] how should we describe them in the spacetime language? To answer this question, we should first recall the notion of velocity in Newtonian mechanics. We determine velocity by specifying its magnitude and direction. For this reason, we represent velocity with an arrow. The arrow's length represents the magnitude of the velocity, and the arrow's direction the velocity's direction. Overall, one needs to provide three numbers, each giving the length of the projection of the arrow in one of the axes of a reference frame. These three numbers form the *vector* of velocity. The magnitude of this vector is unchanged under rotations in space.

Minkowski spacetime has four dimensions, so a vector is constructed out of four numbers. Three of these numbers refer to spatial directions and the fourth one to the time direction, all making reference to a specific reference frame. Nowadays the time component is referred to as the "zero-th" direction. If the magnitude of this four-vector is unchanged under Lorentz transformations (the analogue of rotations in spacetime) then the four-vector is a genuine geometrical object in Minkowski spacetime the same way a vector is a geometric object in our familiar three-dimensional space. Hence, any law of physics expressed in terms of four-vectors (or objects that can be derived from four-vectors) is, indeed, invariant under the change of inertial frame.

Remarkably, many physical quantities of nonrelativistic mechanics combine to form four-vectors. The laws of motion for these quantities may be written in a geometric form without taking the radical step of introducing new physical quantities for the explanation of systems that are already well understood. The most important example is the so-called *energy-momentum vector*: its zero-th component is the energy of a physical body, and its spatial components are the components of the body's momentum.

To summarize, Einstein's idea about the constancy of the speed of light had dramatic consequences at the level of fundamental physics: all concepts should refer now not to the absolute time and space of the Newtonian theory, but to a new absolute structure that integrates them—spacetime. Spacetime is not only an arena, a background of physical events. It also acts as a censor: it permits only processes that respect its structure. We shall see this in detail in the next chapter, in which we elaborate on the concept of symmetry.

THE SYMMETRY BENEATH

SYMMETRY IN PHYSICS—SPACETIME SYMMETRIES

4.1 Symmetries in Physics Are Hidden

The hidden harmony is superior to the apparent.
—Heraclitus

While the special theory of relativity brought a revolution in our understanding of time and space, it left most conceptual distinctions of Newtonian mechanics intact (see sec. 3.4). In particular, it did not affect the realization that we have two types of knowledge for a physical system: the *laws of motion* (also called dynamical laws) and the *initial conditions*. The laws of motion determine which of all conceivable motions of a physical body is possible. However, in a specific physical situation we need to identify only one of the possible motions. For this purpose, it is necessary to have some knowledge about the body's state at a specific moment of time—we need a "photograph" of its profile in terms of position and velocity. This knowledge constitutes the *initial conditions* of the physical system.*

---

*We use the expression "initial conditions" in a rather generic sense because the prediction of a system's motion in the future is the most common situation of interest. In general, the information

Since Newtonian mechanics is a deterministic theory, the knowledge of the laws of motion and the initial conditions allows one to determine completely the behavior of the physical system at all moments of time.

In general, the initial conditions of a physical system are arbitrary and chaotic. A photograph of the planets in the solar system at an arbitrary moment of time would reveal nothing of the simplicity of the law of Universal Gravitation. We would only see large lumps of matter dispersed around the Sun seemingly at random. The same is true in any physical system we observe in everyday life, from the waters of a river to the grains of sand on a beach. The initial conditions contain all irregularity and randomness of the world as it appears to our senses. However, a fundamental physical theory must be built on very simple rules. The human mind would otherwise be unable to cope with it. For this reason, modern science does not possess a *theory* explaining initial conditions; it can only determine initial conditions through observation and experiments.[1] In other words, we have theories that allow us to determine a class of possible motions for a given system, but to select one motion among them for a specific physical situation we need some observational data.

In stark contrast to the initial conditions, the laws of motion manifest regularities, and modern physics identifies these regularities as the *symmetries* of a physical system. I emphasize that the notion of physical symmetry does not ordinarily refer to the behavior of a physical system as it appears to our senses. The regularity or the simplicity of a physical system's motion *does not imply the existence of a physical symmetry*: such regularities may be due to very special initial conditions. A satellite may move in a highly symmetric orbit (e.g., a circular one), but this is simply the effect of the fine-tuned initial conditions that were imposed by the engineers that launched it. Conversely, a behavior that appears chaotic

---

needed for the determination of a system's motion may also take the form of *final conditions*, for example, if we know its state at some time *t* and we want to identify its past motion. Alternatively, it may take the form of *boundary conditions* if we know the state of the system on a closed surface in spacetime and want to identify the motions in the spacetime region enclosed by the surface.

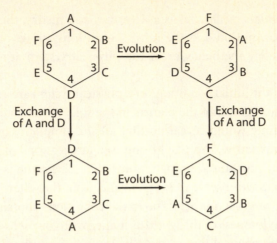

Figure 4.1 **Symmetry of dynamical laws.**

and random may be compatible with a fundamental law of motion that is characterized by a high degree of symmetry. Modern physics realized that it is this second type of behavior that is more permanent and fundamental in nature. Symmetry that exists beneath the observed phenomena is more important than the apparent.

Below is an example that demonstrates more precisely the concept of symmetry to a law of motion (see fig. 4.1). The reader is advised to follow it in some detail because it clarifies the way the concept of symmetry appears in modern physics. We consider a circle and choose six points on its circumference at equal distances from each other—thus we construct a hexagon. These points are the hexagon's vertices. We label them 1, 2, 3, 4, 5, and 6. We also have six people who can occupy these positions, one and only one at each position at a moment of time. Let us call the six people A(nn), B(asil), C(arol), D(iana), E(rol), and F(ay). We may then assign one person to each of the six points. For instance, A goes to 1, B goes to 2, C goes to 3, D goes to 4, E goes to 5, and F goes to 6. Every such assignment defines a *state* or *profile* for this system. We assume that the profile changes in time according to a rule that plays the role of the law of motion for this system.

The following rule is very convenient: once every minute (one minute being our *time-step*), each person moves to the position immediately next to him or her in a clockwise sense. Hence the person sitting at 1 goes to 2, the person sitting at 2 goes to 3 and so on, while the person sitting at 6 goes to 1. This law of motion is simple and has a high degree of symmetry. It is invariant under the *arbitrary exchange of the position of any two persons*. We now proceed to demonstrate this in detail.

For example, we consider the exchange of A with D. If the initial profile is A1, B2, C3, D4, E5, F6, the exchange transforms it into D1, B2, C3, A4, E5, F6. If we evolve this profile by one time-step, according to our law of motion we obtain the profile F1, D2, B3, C4, A5, E6.

We next consider an alternative procedure. We first evolve the initial profile with the law of motion and apply the exchange afterwards. The evolution of the initial profile yields the profile F1, A2, B3, C4, D5, E6. We next exchange A with D, and we obtain the profile F1, D2, B3, C4, A5, E6. This profile is identical to the one we obtained when we first performed the exchange and then took the evolution law into account. This property persists for any initial configuration: one needs either patience to try them all or some mathematical skill to prove it by argument. It makes no difference to the final profile whether we first perform the exchange or we first evolve according to the dynamical law, as long as we start from the same initial profile. This property is the defining feature of a symmetry to a law of motion. The law of motion is then said to be *invariant* under the symmetry operation. In the example above, the symmetry operation was the exchange of A and D, but we can easily see that the law of motion described in figure 4.1 is invariant under any exchange of people.

We next consider an example of a law of motion that does not have this symmetry of exchange. The corresponding rule is the following. Every minute each person moves to the position immediately next to him or her in a clockwise sense. But D cannot stay in the position 2. If she gets there, she has to exchange position with the person immediately next to her in a clockwise sense. This law is not invariant under the exchange of positions. To see this, we go back to our previous example. Starting from the same initial condition, the exchange of A and D gives D1, B2, C3, A4, E5, F6. When we evolve with the new law of motion, we obtain F1, B2, D3, C4, D5, E6 (here we use the special rule that D is incompatible with position 2). The opposite procedure gives, after evolution, F1, A2, B3, C4, D5, E6 and, after the exchange, F1, D2, B3, C4, A5, E6, which differs from the result we obtained from the previous procedure. So this dynamical law is *not* invariant under an exchange of people. Clearly, this is an effect of the law being more complicated, as it contains more complex instructions than the symmetric laws. This is again an example of a more general fact: *a complicated law has less symmetry than a simple law.*

To summarize, in modern physics the word "symmetry" refers primarily to the invariance of the laws of motion under specific transformations. A state of a physical system at a moment of time is usually chaotic and irregular, and it does not reveal the underlying symmetry. The identification of physical symmetries proceeds through the study of dynamics.

4.2 Noether's Remarkable Theorem

> We still share the belief of a mathematical harmony of the universe. It has
> withstood the test of ever-widening experience. But we no longer seek
> this harmony in static forms like the regular solids, but in dynamic laws.
> —Hermann Weyl, *Symmetry*

We return to the simple example of figure 4.1. Suppose we observe the six people exchanging their positions and we record faithfully each exchange. If we go on doing this for some time, we will be able to make a guess about the laws that govern this exchange with a fair degree of certainty. The same is true for all macroscopic physical systems, in which we can have a precise and accurate observation of all their motions. Such was the case of the planets in the solar system: the study of their motion was detailed enough to enable Newton to discover the law of Universal Gravitation. Once we know the laws of motion, the study of their symmetries is a simple mathematical exercise.

In modern physics, however, we study physical systems on the tiniest of scales, for which we cannot have a detailed observation of their motions. Our knowledge of their precise state or profile and its evolution in time has typically many gaps, and it does not allow a determination of the corresponding law of motion. In this case, the identification of the symmetries comes first, and the laws of motion are deduced by the condition that they satisfy these symmetries. But how can we identify the symmetries in the first place, given that we have only a very coarse description of the evolution of microscopic systems?

The solution comes from a deep mathematical theorem, which was established by the mathematician Amalie Emmy Noether in 1918. She was working at that time as an unsalaried member of staff at the University of Göttingen. The content of her theorem is remarkably simple to state. If the laws of motion of a system are invariant under a particular transformation, then there exists a specific physical quantity, whose value remains constant in time. In other words, *the presence of symmetries implies the existence of constants of motion*.[2]

Noether's theorem suggests an alternative procedure for the discovery of the symmetries to the laws of motion. We should study our experimental data with the purpose of identifying physical quantities that do not change in time. The determination of as many as possible of these quantities enables us to determine the system's symmetries. We may then infer the laws of motion themselves. The important point is that we *need not observe the full details* about the behavior of the physical system at all moments of time, something that is practically impossible. We need much less information than that—only what relates to the conserved quantities.

To appreciate the significance of Noether's theorem, we need to make a distinction between two different categories of physical principles. On one hand, we have the results that originate from the collection of a large number of observational data, which are processed by the scientist's intuition and lead to the statement of empirical laws. Empirical laws are then collected, in the formation of statements of more general validity. At the end of this procedure, we have the work of genius, which manages to explain the empirical results in terms of a small set of basic principles. This way a great theoretical synthesis is achieved, which allows the description of a large number of physical phenomena. This category includes Newton's law of Universal Gravitation, Maxwell's equations, special relativity, quantum mechanics—all these creations we refer to as fundamental physical theories.

On the other hand, we have results that are not tied to particular physical phenomena or even scientific fields. These are often solely mathematical in nature, and they provide an insight into the fundamental structure of physical theories, usually unifying descriptions from different branches of science. These results are deemed so basic and universal that they are expected to be valid even when they are removed from their original context. Moreover, they persist even when this context is proved inadequate, inaccurate, or even downright false.

When physical theories of the first type are modified, the changes can usually be isolated in specific components of the theories and do not really cripple our trust in the present understanding of nature. A principle of the second kind is, however, so deeply immersed in our understanding of nature and in the way we perform and interpret experiments that it is almost inconceivable (to the practitioners of the art) that it might fail. In any scientific revolution, it would be the last bastion to fall. Such principles are the conservation of energy and momentum; the principle of inertia (any body resists a change in its state of motion by an external agent); the basic principles of statistical mechanics (the thermal properties of matter are determined *solely* by the dynamical properties of its basic constituents). The last of these basic principles is the theorem of Noether: any symmetry in the dynamical law corresponds to a conserved quantity. This statement is a supporting pillar to all recent advances in fundamental physics, not only at the theoretical, but also at the experimental level. As we will see in chapter 8, the interpretation of most experimental results in high-energy physics relies on the relation between symmetries and conserved quantities.[3]

4.3 Space and Time Translations

The difference between one event and another does not depend on the mere difference of the times or the places at which they occur, but only on differences in the nature, configuration, or motion of the bodies concerned.

—James Clerk Maxwell, *Matter and Motion*

One of the most fundamental assumptions of physics is that the laws of motion are the same at all places and at all moments of time. In other words, relationships of cause and effect have to be the same always and everywhere. If a given force accelerates a particle by a given amount here and today, the same force should accelerate the same particle by the same amount there and tomorrow. In the absence of such an assumption, it would be impossible to understand the relations between different physical events.

When we try to implement this principle into concrete physical theories, we run into difficulties. How do we determine that the force we act upon a body is the same in two different points of space and time? To compare these forces, we must measure them. For this purpose, we need two frames of reference—one for each measurement. These reference frames have to be inertial. Otherwise, pseudoforces will appear (see sec. 3.1), and these will complicate and perhaps even invalidate any comparison. We must therefore specify two reference frames at rest with respect to each other that differ only in their origin in space and in time. If the effect of the forces were identical in the two reference systems, it would be possible to ascertain that the laws of motion remain unchanged under changes in the origin of an inertial coordinate system. In other words, we would be able to verify that Newton's laws of motion respect the symmetry of space and time translation.

A symmetry according to Noether's theorem corresponds to a conserved quantity. The conserved quantity associated with space translation is the *momentum*. To see this, we may consider a straight and homogeneous road that goes on endlessly in an unchanging landscape—like a highway in the American desert. In our idealization, two points along this road cannot be distinguished by virtue of the surrounding landscape. Since the road was assumed to be infinite in length, the distance of the point from the "beginning" is also irrelevant.

We next assume that a car travels along this road, and that the total force acting on this car vanishes. Newton's first law then implies that the car's velocity is constant. This car defines an inertial frame, whose origin follows the car's motion. In effect, the velocity of the car causes (or we can say it *generates*) the translation in space for the associated reference frame. It then seems natural to say that the generator of the translation in space is the car's velocity. In fact, the generator is the momentum, namely, the car's velocity multiplied by its inertial mass. This is true in both relativistic and nonrelativistic physics, only in the first case the inertial mass also depends on the car's velocity. In absence of external forces, the velocity does not change and the momentum is constant.

If a bump exists on the road, the car's speed does not remain constant. It changes when the car hits the bump. The existence of a bump in a specific place along the highway implies that the road is not the same everywhere and that the symmetry of space translation is lost. Therefore, a breakdown of the symmetry implies a failure of conservation for momentum.

The conservation of momentum holds not only for a single body, but also for systems of many bodies, and it implies the vanishing of external forces. Any external forces arise from physical bodies that lie outside the system under study. If we enlarge our system to include all such bodies, we will eventually consider a system upon which no external force acts (perhaps this will be the whole Universe). The symmetry of translation in space implies that the only origin of forces is the interactions between bodies (or perhaps fields). If, for example, empty space exerted a force on material bodies, we would never find a sufficiently large system at which no force is acting.[4]

The symmetry of translation in time is more difficult to visualize; space gives the feeling of extension, which is perceived through the sense that is responsible for the orientation of our bodies. It is therefore more concrete than the feeling of duration associated with time. Still, the symmetry of time translation may be summarized in the intuitive statement that the laws of motion for a physical system do not depend upon the instant of time one applies them in order to determine the future evolution of the system. For example, the law of Universal Gravitation today is the same as it was yesterday, and it will remain the same tomorrow.

Remarkably, the conserved quantity associated with the symmetry of time translation is the *energy*. Since, in relativity, time refers to a specific reference frame, so does the conserved quantity that corresponds to time translation. Hence, the energy of a physical system depends on the choice of the reference frame. The same also holds for momentum. However, as I mentioned at the end of section 3.5, energy and momentum are intertwined by Lorentz transformations, and they combine to form the energy-momentum four-vector. Their corresponding symmetries may also combine into the symmetry of *spacetime translation*, which states that the laws of motion are independent of the spacetime origin of the inertial reference frame that is employed in their formulation.

However, energy is more than a conserved quantity of mechanical theories. It was first introduced in the study of thermal phenomena (see sec. 2.3), and in that context it possessed some features distinctly different from those it possesses in mechanics: *Energy determines whether a physical system is stable or not.* Any physical system tends to reach a state of thermal equilibrium with its environment. It emits energy if the environment is colder, and it absorbs energy if the environment is hotter; and these processes continue until the system's temperature becomes equal to that of its environment.

In particular, if the system is surrounded by vacuum, it will only be able to emit energy, because the vacuum has no energy to give it back. The system's internal energy will therefore decrease. If no state of minimum energy exists, the system continues to emit energy indefinitely and never settles down. If we could isolate a single system with that property, we would be able to extract indefinitely large quantities of energy. We do not encounter this behavior in nature. In fact, it would be very worrying if we did because it would allow for the existence of sudden and uncontrollable bursts of energy with no apparent external cause. It is unlikely that anything stable would ever form in the Universe if this could happen. Even one single isolated system not possessing a state of minimum energy could radically affect the rest of the Universe because it could be used to channel indefinitely large amounts of energy to its environment. (Clearly, this would be very helpful in solving the energy problem on our planet!)

To guarantee the stability of matter, we must assume that all physical systems possess a state of minimum energy. From relativity, we know that energy is a measure of inertia, and on physical grounds it can only be positive. This suggests the postulate of *energy positivity*: the total energy of a physical system is always positive, except perhaps for the state of minimum energy, in which the energy may equal zero.[5] This principle is an essential part of modern theories of physics: we shall see in chapter 7 that it constitutes one of the basic assumptions of the quantum theory of fields.

4.4 The Poincaré Symmetry and the Origin of Particles

In the tanner's spindle, straight and spinning motion are the same.
—Heraclitus

The symmetry of spacetime translation refers to the possibility of changing the origin of the coordinate system in spacetime without affecting the laws that determine the motion of a physical system. What happens if we keep the origin fixed and rotate the axes? In that case, there exist two possibilities. The first is that we keep the time axis fixed and rotate the axes of space. This is simply a spatial rotation; we do not need to use relativity to account for this.

The other possibility is that we rotate without keeping the time axis fixed. As Minkowski realized, this amounts to going from one inertial frame to another one that is in a state of motion in relation to it. This transformation is what we earlier called a Lorentz transformation (see sec. 3.3). It is customary nowadays to designate both types of rotations in spacetime as Lorentz transformations: one refers to the original transformations of Lorentz as *pure Lorentz transformations* or simply as *boosts*. The conserved quantity associated with spatial rotations is called *angular momentum*, while the one corresponding to boosts does not have a name of its own: unlike angular momentum, it rarely appears in practical applications, and we cannot measure it in a straightforward way.

Spacetime translations and spacetime rotations jointly constitute the *Poincaré symmetry* of spacetime. The name is a tribute to the great French mathematician Henri Poincaré, who derived the corresponding transformations just one week after Einstein presented his work on relativity. The conclusion from our previous discussions (see secs. 3.3 and 4.3) is that the law of motion for any physical system that evolves in Minkowski spacetime has to be invariant under the Poincaré symmetry. This conclusion has far-reaching consequences, which I describe below.

Even if the symmetries of a physical system are the only thing we know about it, we can still understand many aspects of its structure. There is a precise mathematical procedure that allows us to do so. We can decompose any system with a given symmetry into smaller systems that still possess this symmetry. We then decompose these smaller systems into yet smaller ones possessing the symmetry. We repeat this procedure until we arrive at some fundamental systems, which cannot be further decomposed. These are the *irreducible systems* for this symmetry. Irreducible systems are in a sense similar to atoms; any other system characterized by the *same symmetry* is built up from their combinations.

The study of the irreducible components of the Poincaré symmetry yields a simple and striking result: *These irreducible components describe free particles*, that is, pointlike bodies that move in the absence of forces. Detailed mathematical analysis demonstrates that two numbers distinguish between different types of particle. The first such

TABLE 4.1
Poincaré transformations and conserved quantities

The Poincaré Symmetries	
Transformation	Conserved quantity
Time translation	Energy
Space translation	Momentum
Rotation	Angular momentum
Boost	No name

number is the particle's rest mass: this can be either positive, in which case the particle always moves with speed slower than that of light, or zero, in which case the particle moves with the speed of light.

The second number always takes nonnegative values, and it represents a new concept: it refers to a physical property that is not related to the particle's position or velocity, and it has a resemblance to an intrinsic spinning. It is therefore called *spin*. To get a preliminary visualization of spin, consider the particle as carrying an arrow that can spin around pointing toward any direction of space. This arrow corresponds essentially to spin, and the number that distinguishes between different particles corresponds to the arrow's "length." However, one should not take this picture literally. Particles do not really carry any arrows. Moreover, they are supposed to be pointlike, and it is meaningless to talk about the self-rotation of a single point: only extended objects can self-rotate. Hence, while spin is mathematically similar to a rotation, it is unlike any rotation that we can ever perceive with our senses. Nonetheless, as we will see in section 6.1, it does exist as a physical property of particles.[6]

The fact that the irreducible components of any system with Poincaré symmetry correspond to particles is a remarkable demonstration of the power of the concept of symmetry. However, one should not get carried too far from this and reach hasty conclusions. After all, this is only a mathematical statement. It implies that any possible motion or process of matter that is compatible with the symmetries of Minkowski spacetime can be analyzed into simpler motions that ultimately refer to particles with spin. That such an analysis is always possible does not necessarily imply that these particles exist. The analogy between words and letters is useful for the clarification of this point: any word is analyzed into letters, but a letter by itself does not form a word, and it does not typically appear isolated in speech. Similarly, the irreducible components may appear in nature only as parts of large globules, clusters, or even fluids, having no independent existence as physical particles.

Before elaborating on the features of irreducible systems for the Poincaré symmetry, we first note that any physical system is characterized

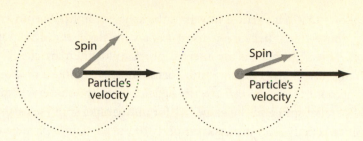

Figure 4.2 **Representation of the particle with spin.** We visualize the spin as an arrow (gray) attached to the particle. The arrow's length is always fixed. The tip of the spin arrow moves on an imaginary sphere. As the particle's velocity (the black arrow) increases, the spin arrow tends to align itself with the direction of motion.

by its *degrees of freedom*. The degrees of freedom are a set of physical quantities that provide a complete description of the system's profile at every moment of time. For example, for a particle with no spin we need to specify its position (three coordinates) and its velocities or momenta (three components), so the total number of degrees of freedom is six.

I described earlier a particle with spin as a point that moves in space carrying a rotating arrow. In the absence of external forces, this direction remains unchanged during the particle's motion. If for some reason the spin changes direction, the tip of the arrow moves on the surface of a *sphere*. Two numbers (we may think of them as the latitude and longitude of the sphere) suffice to specify any point of a sphere. If we add them to the six degrees of freedom that are needed to account for the three coordinates of position and the three of momentum, we see that a particle with spin has in total eight degrees of freedom.[7]

The description above is valid only for particles that possess nonzero rest mass (henceforward called *massive*). As the particle's speed increases, the arrow gradually tends to align itself with the axis of motion, pointing either at the same or at the opposite direction as the particle's motion. At the limit that the particle achieves light speed, the axis of the spin arrow and the axis of velocity coincide.

It follows that if a particle travels at the speed of light, its spin arrow cannot move freely because it must always lie along the direction of motion. This alignment is an internal property of the particle: it does not change under any Poincaré transformations. We define a new quantity to account for it, the *helicity*: the helicity takes the value $+1$ for the case of parallel spin and the value -1 for antiparallel spin (see also sec. 6.3). Since helicity is an intrinsic property of the particle, a particle with positive helicity cannot develop negative helicity any more than it can change the value of its rest mass or its spin.

As explained in section 3.4, particles with zero rest mass (*massless*) travel on the light cone. The light cone's geometry is rather unusual, and we cannot apply our intuitions from ordinary three-dimensional space to it. A careful mathematical analysis is required. The first outcome of this analysis is that a massless particle with spin is characterized by *six* degrees of freedom rather than the eight of the massive ones. The reason is that (as explained earlier) the spin arrow always lies along the direction of motion. Hence, it cannot move freely. This constraint implies that massless particles have two degrees of freedom less than massive ones.

However, the spin degrees of freedom do not disappear without a trace in the massless particles. They are intertwined with the degrees of freedom that describe the particle's position. It follows that it is not straightforward to talk about the position of a massless particle. From one point of view, a massless particle does not look like a particle at all—it cannot be localized at a specific point of space. One may describe it equally well as a sheet that moves along the light cone with a spinning motion. However, the geometry of the light cone is strange: from the perspective of an observer that does not move with the speed of light, this rotating sheet looks like a particle moving along the light cone.[8] This sounds weird, a particle being equivalent to a sheet. It is very weird indeed, but nonetheless true! It is difficult to get used to such a picture. It is a blessing that in most appearances of massless particles in physics, we need not worry too much about such details because they can all be swept under the carpet of a successful mathematical formalism—that of quantum field theory.

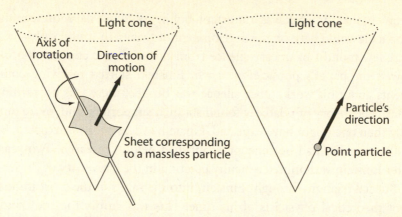

Figure 4.3 **The massless particle with spin.** The position of a massless particle is not well defined because the position degrees of freedom are intertwined with the degrees of freedom related to spin. For this reason, there exists an alternative description, besides the conventional description of massless particles as point-like objects that move with the speed of light. A massless particle can be represented by a sheet, which moves along the light cone and rotates around an axis that is also contained in the light cone. Both descriptions are equivalent; in fact, the sheet picture behaves better under changes of the reference frame. This ambiguity is a direct consequence of the unusual geometry of the light cone.

4.5 General Relativity

There is the name and the thing; the name is a sound, which sets a mark on and denotes the thing. The name is no part of the thing nor of its substance; it is an extraneous piece added to the thing, and outside of it.
—Michel de Montaigne, *The Essays*

The classification of the irreducible physical systems corresponding to the Poincaré symmetry, described in the previous section, is perhaps the last step in the elaboration of the theory of special relativity. It took us a long way from that morning in the spring of 1905 when Einstein placed his manuscript on the electrodynamics of moving bodies in an envelope, sealed it, and mailed to the *Annalen der Physik* journal. He knew at that moment that his work was important, but he could not have foreseen the impact it would have on the development of twentieth-century physics.

A few months of uneasy silence from the academic community followed the paper's publication. Then, Einstein started receiving comments about his work. Once silence was broken, there was no turning back. The theory of relativity found staunch supporters, and, more rapidly than one might have expected, Einstein's ideas became accepted. By 1907, relativity had become an established field of research. Two years later Einstein was offered a faculty position in the University of Zurich.

Special relativity brought Einstein into the ranks of the most important theoretical physicists of his time. This was hardly the end of the road. The gravitational phenomena did not fit in special relativity: Newton's law of Universal Gravitation was not the same in all inertial reference frames. It was 1915, after many years of continuous effort amid crisis in his personal life and a war that had broken Europe apart, that Einstein achieved the marriage of gravity and relativity within the arguably most elegant theory ever to appear in the physical sciences: general relativity. This theory removes the need for a background, immutable spacetime of his special theory. It places spacetime on the same footing with all other physical objects: it changes and interacts with matter.

I shall not elaborate on the general theory of relativity. The present book is about the modern scientific theories of matter, while general relativity is a theory for space and time. Matter enters the picture there, but very hesitantly; it prefers to stay in the sidelights. General relativity is not its proper arena, yet. The reason is that twentieth-century physics came to a description of matter based on a different perspective, the collection of ideas and methods known as the quantum theory. So far, quantum theory has not been reconciled with general relativity. Our theory for space and time and our theory of matter are simply very different—one might even say incompatible.

Quantum theory has been reconciled with special relativity, grudgingly perhaps, but very successfully. Special relativity still carries the baggage of an absolute spacetime, and the resulting theories still see matter as something that changes with respect to a preexisting and unchanging vessel, even if we know that general relativity has made this picture obsolete. Still, this is what we now have, and we have to make do with it.

However, an acquaintance with the basic concepts of general relativity will prove necessary in later discussions about the limitations of the quantum theory of fields (see chapter 10). For this reason, we conclude this chapter by listing these principles in a concise form.

- Physical phenomena take place in a spacetime of four dimensions, one dimension for time and three for space. However, the geometry of this spacetime is not predetermined and absolute like in special relativity. Spacetime changes and is affected in its evolution by the presence and motion of matter. Conversely, the motion of matter is affected by the geometry of spacetime.

- The spacetime of general relativity looks like Minkowski spacetime at short scales, but it is different at large scales. Its geometry is curved: free particles follow curved lines instead of straight ones, and this deviation from straightness is what we perceive as gravitational forces.

- The key symmetry of the theory arises from the principle of *general covariance*, which states that the laws of physics are the same in all reference frames and not only in the inertial ones. General covariance uniquely specifies the laws of motion for both the geometry and the matter.

- General relativity is not characterized by the symmetry of time translation because the notion of time depends on the geometry, and the geometry changes dynamically. For this reason, the concept of energy is in general not defined in general relativity.

5

THE MACHINE BREAKS DOWN

THE DEVELOPMENT OF
QUANTUM MECHANICS

5.1 The Birth of Quantum Theory

It was then that I began to look into the seams of your doctrine. I
wanted only to pick at a single knot; but when I had got that undone,
the whole thing ravelled out. And then I understood that it was all
machine-sewn.

—Henrik Ibsen, *Ghosts*

One of the greatest triumphs of nineteenth-century physics was
the development of a theory that explained many properties of
macroscopic physical systems in terms of the laws of motion of their
microscopic constituents, namely, molecules (see sec. 2.3). This theory
was named *statistical mechanics* because it used statistics to describe the
chaotic motions of molecules, which cannot be directly observed.

We may recall from our earlier discussion that statistical mechanics
arose out of Maxwell's kinetic theory of gases (see sec. 2.3). This theory
provided a description of gases that was in good agreement with experiment. However, it failed to explain the behavior of matter in its other
states, most notably the properties of metals and other solids. This failure

was initially attributed to inherent faults in the theory. But as time went by, the principles of statistical mechanics were set on a firmer ground—most notably in the work of Ludwig Boltzmann and Josiah Willard Gibbs—and its failure to account for metals seemed more and more baffling. It turned out that the problem was due not to an inadequacy of statistical mechanics, but to an inadequacy of the sacrosanct Newtonian laws of motion. Eventually, the properties of solids were explained by a new theory that replaced Newton's mechanics, namely, quantum mechanics.

However, it was not through the study of solids that Newtonian mechanics were overthrown. The blow came from another direction, a seemingly minor phenomenon that lay at the borderline of electromagnetism and statistical mechanics: the radiation of *black bodies*.

A body is black if it absorbs all light and reflects nothing back. Usual black objects (like a black dress) are not perfectly black because they do reflect some light (a black dress might have a shade of brown or red). A hole provides a much better example of a black body. Consider, for instance, a wooden box with thick walls and a cavity inside. We assume that the walls of the cavity are quite rugged. Suppose then that we open a small hole in one side of the box. Any light that enters the cavity will be absorbed by the rugged surface. If a light ray is not absorbed after its first contact with the cavity's wall, it will be reflected, but then it will be absorbed the second time, the third time, or perhaps later, but definitely before it finds its way back to the entrance. Hence, a "hole" constructed in this way really absorbs all light that falls on it. It looks pitch black.

However, black bodies turned out to emit electromagnetic waves.[1] These waves are not visible because their frequency is smaller than the ones that our eyes can detect. One could explain the radiation of black bodies using elementary arguments from thermodynamics. The box is never isolated from the rest of the world. It exchanges heat with its environment, until they both reach the same temperature. We saw in section 2.3 that temperature is related the internal energy of a physical system. In a black body, the carrier of energy (and hence of temperature) is the electromagnetic field. However, if the electromagnetic field has nonzero energy, Maxwell's equations imply that some kind of electromagnetic waves must be present in the cavity. These waves will eventually find a

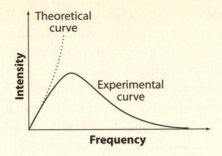

Figure 5.1 **Theoretical predictions vs. experimental results for black body radiation.** One measures the amount of energy emitted at each frequency of the black body radiation. Maxwell's theory together with statistical physics predicts that the energy always increases with frequency. However, the experimental data show that the energy grows up to a specific frequency and then drops slowly to zero.

way out of the hole and we will observe them. Detailed mathematical analysis reveals that the behavior of these waves does not depend on any other physical parameter but the box's temperature.

The black body was an ideal system for the study of the relation between electromagnetic and thermal phenomena because its theoretical treatment is simple: it involves only elementary concepts of electromagnetism and statistical mechanics. Many experiments were therefore performed to measure the black body radiation. The results of these experiments brought out a huge surprise. The distribution of energy among the various frequencies of the emitted waves—the *black body's spectrum*—diverged dramatically from the theoretical predictions.

If a theoretical model fails to explain the experimental results, it is usually an event of little consequence. Models usually involve many assumptions, and these may easily conceal theoretical errors. If, however, the model is very simple, it employs only basic assumptions of well-established physical theories, and its predictions involve only one single parameter, then its disagreement with the experiment starts looking a bit more serious. Moreover, if the prediction deviates from the experimental results not just 10 or 20 percent (or even 90 percent), but amounts to a completely different qualitative behavior of the physical quantities involved, then we have a serious problem. There is an important mistake in our understanding of the system. This may be good news because the identification of such mistakes provides deeper insight into the structure of the physical laws. The lack of interference in the Michelson-Morley experiment and the unexpected behavior of the black body radiation were both problems of this type. Their resolution led eventually to the overthrow of the picture of physical reality that Newtonian physics had drawn.

Not suspecting the forces that were to be unleashed by a solution, a significant number of theoretical physicists attempted to explain the riddle of black body radiation in the late 1890s. The solution came from Max Karl Ernst Ludwig Planck, a forty-two-year-old professor at the University of Berlin. Planck's field of expertise was thermodynamics, the study of thermal phenomena. When he chose that field at the beginning of his career, he had been warned to do something else because theoretical physics was coming to a dead end: all major discoveries had already been made. His reply was that that he did not look for new discoveries, but only to understand the existing foundations of physics. This feels most appropriate for the person who made the first step in the direction that revolutionized the physical sciences.

Planck realized soon enough that "the only way to recognize how this could be done [i.e., solve the problem of the black body radiation] was to start from a definite point of view. This approach was opened to me by maintaining the laws of thermodynamics. . . . It seems to me that [they] must be upheld under all circumstances. For the rest, I was ready to sacrifice every one of my previous convictions about physical laws."[2] His expertise on thermodynamics served him well. It allowed him to make an educated guess about the correct behavior of the black body spectrum. This guess turned out to be in remarkable agreement with the experimental results. However, a mere guess was not good enough. He had to justify it in terms of known physics. But no matter how hard he tried, he could find no such justification: his result seemed to be completely unwarranted.

Eight weeks followed, with Planck working more intensely than ever before. He could see a way to solve the problem, but the solution scared him. It involved an assumption so alien to the contemporary under-standing of physics that it could not possibly be true. Nonetheless, there seemed to be no other way. "It was an act of desperation," he later recalled, "a theoretical explanation had to be supplied at all cost, what-ever the price."[3]

Planck's solution was the hypothesis that a black body does not emit energy continuously, but in discrete pieces. There exists a minimum value of energy, a *quantum* as Planck named it from the Latin word for "discrete," and the values of the emitted energy can only be multiples of a single quantum. Not all quanta are the same, however. They differ according to the frequency of the corresponding wave. To be precise, a quantum's energy is proportional to the wave's frequency. The constant of proportionality between energy and frequency was later named after Planck; it is called *Planck's constant*.

One may imagine how many of Planck's colleagues would react on hearing his ideas. "Energy emitted in lumps? Energy proportional to the wave's frequency? This is incredible, if not ridiculous. We should try to explain the phenomena, but not at the expense of simple common sense. It is impossible for any system that satisfies Newton's laws or Maxwell's laws to possess discrete values of energy. Not only does this explanation violate all understanding that physicists had accumulated about electro-magnetism, it also suggests a violation of the most cherished principles of physics."

Planck surely expected such remarks when he put forward his expla-nation. He probably shared some of these misgivings himself. He did not abandon his ideas, though. His instinct told him that he had come across something very important. His son later reported that his father, on long walks through a forest, would intimate his feelings that he had made a discovery comparable only to that of Newton.

Planck developed a strong argument in support of his results. His the-ory explained the phenomena, and it relied on well-established thermo-dynamic principles. He had a good name and experiment on his side. Eventually his voice was heard. Still, not many people were convinced. The majority marveled how such a weird assumption could lead to the correct results, and they tried to explain it away by means of more mainstream physics. Nonetheless, Planck's idea survived within the scientific community and, like the seed thrown randomly in the air, only had to find fertile ground before it grew and bore fruit.

It did find fertile ground, in fact the most fertile possible. Four years after Planck had proposed his mysterious quanta and less than one year before the first paper on special relativity, Albert Einstein realized that the quantum hypothesis had further consequences. I mentioned in section 3.3 that he had been deeply fascinated with the strange properties of the electromagnetic waves. One of the issues that had been troubling him was the apparent contradiction between the continuous nature of the electromagnetic fields and the discrete character of the atoms. How can an extended and continuous object, like a light wave, interact with a discrete object that is localized in space, as an atom is supposed to be? This problem did not trouble his contemporaries very much, but for Einstein the dichotomy between continuity and discreteness was baffling. When he first heard of Planck's theory, he realized that the energy quanta touched exactly upon this sore spot of electromagnetic theory. This was a shock for him. As he later wrote (on learning of Planck's theory): "it was as if the ground had been pulled out from one, with no firm foundation to be seen anywhere, upon which one could have built."[4]

Einstein's thoughts on the meaning of quanta gradually matured. He concluded that energy is emitted in quanta because the quanta are themselves physical things: particles. It would then be reasonable to assume that electromagnetic waves consist of such particles: "When a light ray is spreading from a point, the energy is not distributed continuously over ever-increasing spaces, but consists of a finite number of energy quanta that are localized in points in space, move without dividing, and can be absorbed or generated only as a whole."[5] Einstein was content to designate his postulated particles as light quanta, but the name that eventually stuck on them was *photons*, carriers of light.[6] Einstein emphasized that photons were different from the light corpuscles of Newton's theory. They were somehow intertwined with the electromagnetic waves: they carried energy proportional to the wave's frequency—something completely unjustifiable for ordinary particles.

People were rather skeptical of photons. One might have accepted Planck's quanta of energy as a hypothesis, but quanta of light! Wouldn't this be a return to an outdated way of thinking? But Einstein's reason for introducing them was very concrete. Quanta of light can do things that electromagnetic waves cannot—and those things we can observe.

To explain the point above, we first consider a seemingly unrelated argument. Let us assume that we dig a deep hole in the ground and place a football in it. We then start throwing pebbles at the ball. If we throw a small one, the ball will not move. If we throw simultaneously two of them, again the ball will not move. Even if we throw one hundred small pebbles together, the ball may move a bit, but it will not get out of the hole: the pebbles spread as they move toward their target, and they are not very effective. We then start increasing the size of the pebbles. Eventually we will find a pebble (in fact a stone) that is big enough to take the ball out of the hole. What is the lesson from this? A small pebble contains a small amount of energy. Using a large number of pebbles one may transfer a large amount of energy, but it is much more efficient to transfer the same amount of energy in the form of a single stone.

We next compare the photons to the pebbles we threw at the ball in the hole. According to Planck, the energy of a quantum is proportional to the frequency of the corresponding wave. This implies that there will be jobs that electromagnetic waves with high frequency can do that ones with low frequency cannot, even in cases that the latter carry collectively more energy. A rock can break a window, while a shower of pebbles cannot.

The simple analogy above contains the essence of Einstein's explanation for the *photoelectric effect,* a phenomenon that had been puzzling physicists ever since its discovery by H. Hertz in 1888. The photoelectric effect involved the interaction of electromagnetic waves with matter. If one directs light toward a metallic substance, nothing happens no matter how strong the beam is (i.e., how much energy it carries). One then increases the frequency of the radiation gradually. Still nothing happens. Suddenly, after a certain frequency is reached, an electric current goes through the solid (charged particles are also emitted). The value of this critical frequency varies from solid to solid, but it does not depend on the total energy of the electromagnetic field.

Einstein's explanation of the photoelectric effect was the following. There exist microscopic systems—perhaps the atoms themselves—in which the particles that carry the electric charge (the electrons) are trapped (see sec. 5.2). These particles are liberated by light quanta of sufficiently large frequency that fall upon the atoms. Quanta of small frequency achieve nothing, the same way that pebbles cannot take the ball out of the hole. Atoms are so small that no matter how many photons we bundle together, it is practically impossible that two photons will simultaneously collide with the same atom. In different metals, the depth of the hole is different, and for this reason, one needs different "size" of quanta in order to get the electrons moving in different solids.

A problem being solved by an idea may be a coincidence, but when two problems of completely different character are solved, then this idea has to be taken seriously. Still, the photon was a difficult concept to accept. The idea that light consists of particles had been thoroughly rejected by nineteenth-century physics on the basis of sound experimental work. Why should anyone resurrect such ghosts in the twentieth century? In 1916 the American Robert Andrews Millikan, one of the foremost experimental physicists of that time, obtained results for the photoelectric effect that fully corresponded to those predicted by Einstein's theory. He still refrained from adopting the hypothesis of light quanta, which he considered as "bold, not to say reckless," and flying "in the face of thoroughly established facts of interference."[7] It was only in 1923 that the idea of the photon became unreservedly accepted. The person responsible for that was Arthur H. Compton, who demonstrated experimentally that electromagnetic radiation of very high frequency scatters off electrons in a way that can only be accounted for by a particle description.

The main misgiving Einstein's contemporaries had in accepting the notion of light particles was that they did not understand how these particles relate to the electromagnetic waves predicted by Maxwell's theory. After all, the wave theory of light had passed through rigorous tests of confirmation. It seemed as though both behaviors—wave and particle—were equally valid. The introduction of the photon had given birth to the fundamental principle of quantum theory—that of the duality between particles and waves. However, this principle could only be accommodated by an all-new physical theory.

5.2 New Windows to the World

> The crumbling of the atom was to my soul like the crumbling of the whole world. Suddenly the heaviest walls toppled. Everything became uncertain, tottering and weak. I would not have been surprised if a stone had dissolved in the air in front of me and became invisible. Science seemed to be destroyed.
>
> —Wassily Kandinsky, *Reminiscences*

Planck's quanta were hardly the most important discovery of the time— or so the majority of his contemporaries thought. Unlike relativity, which came to life as a full-fledged and logically structured theory, quantum theory went through the development of a human child: it had a long infancy and a troubled adolescence before reaching adulthood and maturity. In the 1900s it was still an infant. It consisted only of an idea: this seemed to work, and it had disturbing and exciting consequences. However, the hot topics of research were different: relativity, the consolidation of statistical mechanics, and—most importantly—the growing power of technology to offer us a glimpse into the world of the atom.

In the previous section, I referred to the electron as an electrically charged particle that is responsible for the electric currents that appear in the photoelectric effect. Joseph John Thomson had discovered it in 1897. Like any experimental discovery of great significance, the discovery of the electron started from an accident. In the 1860s the German experimentalist Julius Plücker was carrying out a series of experiments studying electric discharges in gases at very low pressure. To remove the air from the glass tubes he used for this purpose, Plücker employed a new, very efficient gas pump. He observed that when almost all the air was removed, a greenish glow appeared near the source of negative electric charge (known as the cathode). Apparently something was coming out of the cathode—some kind of rays that carried negative electric charge. They were naturally named *cathode rays*.

The cathode rays remained a puzzle for about three decades. The study of their properties proved to be a difficult task for experimentalists even of the caliber of Heinrich Hertz. J. J. Thomson was the first to succeed in that effort. He studied the motion of the cathode rays in electric and magnetic fields and realized that the cathode rays consisted of a new type of particle. They were neither molecules nor perturbations of the ether, as current theories supposed them to be. Thomson was able to determine the charge and mass of these particles.[8] They were negatively charged, and the value of the charge was extremely small, much smaller than any other observed in nature. Could charge be of a fundamentally discrete nature, a property of particles rather than one of fluids?

Even more interesting was the fact that the new particles seemed to be much smaller than atoms. Did we have here a new state of matter, a much finer subdivision than had ever been thought possible? A remarkable conclusion it was, and not one that people would easily accept. Thomson recalled how his colleagues reacted in his first presentation of these results: "At first there were very few who believed in the existence of these bodies smaller than atoms. I was told long afterwards by a distinguished physicist that had been present in my lecture that he thought I had been pulling their leg."[9] Nevertheless, the existence of the new charged particle—the electron—was widely accepted within a few years.

The discovery of the electron signals the birth of modern *elementary particle physics*. We should remark here that the word "elementary" in the expression "elementary particle" is very appropriate: we have seen that the word "element" traditionally referred to irreducible *qualities* that are inherent material objects. In spite of the novel context, this notion is not lost in modern physics. In the electron's case, this quality is the electric charge—understood as an intrinsic defining property of the particle.

Other "rays" were also discovered during this period. William Konrad Roentgen discovered the X-rays; Henri Bequerel, Marie Sklodowska Curie, and Pierre Curie discovered three different types of nuclear radiation, which they called alpha, beta, and gamma rays (see sec. 8.1). These rays stirred the imagination of scientists and public alike because they traversed most ordinary forms of matter, including thick metallic plates. Physicists perceived them as a tool that would unlock the behavior of matter on the tiniest of scales, thus opening the gates of the hitherto inaccessible world of the atom. Indeed, the hopes were soon to be fulfilled.

However, the most intriguing object was the electron. Its method of generation—as well as the photoelectric effect—suggested that it was an ingredient of the atom. If that were the case, atoms consist of smaller objects and have an internal structure. They were, therefore, not worthy of their name (atom meaning something that cannot be split). One had to redefine the atoms of chemistry as the *stable* lumps of matter, whose structure accounts for the chemical phenomena.

If the electron is part of the atom's internal constitution, other particles with opposite charge have to be present in the atom too, so that the total charge of an atom is zero. It was conceivable that the positive-charged particles were much smaller and finer than the electrons, so that they would practically correspond to a continuous medium. An atom consisting of electrons that swim in this positively charged medium may very well be stable. This was the basic idea of a model for the atom, which was proposed by J. J. Thomson and Sir William Thomson in 1903. An atom of this type would definitely not be a dense and impenetrable body. It would be similar to a drop of water, a bubble of air, or a lump of jelly. The last image proved more suggestive, and this description of the atom was referred to as the "jelly" model.

What happens if one shoots a bullet toward a big lump of jelly? If the lump is very large, it may trap the bullet. Otherwise, the bullet will escape with some loss of energy. Perhaps the bullet's direction of motion will also change by a small amount. Still, no matter how many times one performs this experiment, the bullet will never ricochet off the jelly and come straight back to the shooter. It was then quite a surprise when an experiment that involved the scattering of the newly discovered alpha rays on thin films of gold revealed exactly this behavior.

The alpha rays consist of rather heavy particles that carry positive electric charge. These particles should be able to cruise through individual atoms with little trouble; at least that was what most people expected.

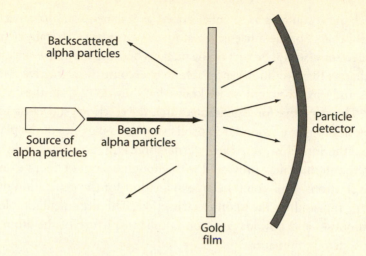

Figure 5.2 **Rutherford's experiment.** Alpha particles are directed toward a thin film of gold. If the jelly model of the atom were correct, the alpha rays should all cross through the film with very small deflection from their initial direction. The deflection of the alpha particles was much larger than expected, and moreover backscattering was observed. To explain this result, one would have to assume the existence of a hard core in each atom, the nucleus. This experiment was crucial for the formation of the modern picture of the atom.

What happened was exactly the opposite: some of the alpha particles were scattered backwards after their collision on the gold film. This experiment had been carried out in Ernest Rutherford's laboratory in Manchester.[10] Rutherford, perhaps the greatest experimental physicist of his day, realized that these results completely invalidated the jelly model of the atom. He could think of only one explanation for this backscattering: there must be a region in the interior of the atom in which a very large percentage of its mass is concentrated. This region, Rutherford thought, would be the atom's *nucleus*.[11] The nucleus has to carry positive charge so that it repulses the alpha particles strongly. When the alpha particle collides with the nucleus, the same thing happens as when we shoot a bullet on a piece of iron; the bullet scatters wildly. If, on the other hand, the particle never approaches the nucleus, its motion is little affected.

Each atom contains as many electrons as it is necessary to cancel the positive charge of the nucleus: the atom can then be electrically neutral. The electrons cannot be part of the nucleus because they have the opposite charge: they would never be able to overcome the attractive electric forces and leave the atom, as we know that they do (e.g., in the cathode rays). The attractive force guarantees that the electrons stay close to the nucleus: they move around it in orbits similar to the ones of the planets around the Sun. However, if an external force (e.g., an electric field) acts upon the atom, it may provide it with enough energy to liberate one or more electrons. This could be an explanation for the generation of the cathode rays and for the photoelectric effect. This argumentation led to the famous—and nowadays all too familiar—picture of the atom as a solar system in miniature.

The nucleus is much smaller than the full atom (by almost one part in ten thousand). On the other hand, the mass of the nucleus is much larger than that of all the electrons combined (much more than a thousand times larger than an electron's mass). It follows that an atom consists mostly of empty space, and that its mass is practically contained in the nucleus. Electrons contribute little to the total mass of the atom, but their role is crucial. It is only through electrons that one atom may interact with another in the processes that lead to the formation of molecules. The chemical properties of the elements depend, therefore, on the motion of the atomic electrons. The number of electrons in an atom was eventually identified with the atomic number, which had appeared in Mendeleev's classification of elements. People hoped that the detailed study of the orbits of the electrons would provide the key to explain all chemical phenomena. This way the whole body of chemistry would be reduced to a description of particles interacting through electromagnetic forces.

This high hope was to come true eventually, but not just yet. No sooner had the "solar system" theory for the atom been proposed than it ran aground. Maxwell's theory predicts that any electrically charged body emits electromagnetic waves when it is accelerated. It therefore loses energy. The motion of the electrons around the nucleus is due to the attractive electric force that the nucleus exerts upon them. Force implies acceleration, hence the atomic electrons must emit electromagnetic waves. Hence, they lose energy. This implies that the electrons will gradually lose their freedom to stay away from the nucleus: their orbits will become shorter and shorter, and eventually they will all crash onto the nucleus. Hence, the theory of electromagnetism predicts that an atom cannot be stable. Elementary calculations showed that the collapse of electrons would take only a very short time. But if atoms are not stable, chemical phenomena can never arise. This means that there is something fundamentally wrong in the "solar system" description of the atom.

Physicists understood that the problem lay at the level of applying electromagnetic theory to the physics of atoms. As these discussions took place in 1913, a well-informed physicist might remember that two important problems of physics had been solved about a decade ago (or at least they seemed to have been solved) by postulating the existence of quanta, and that in one case (the photoelectric effect), electrons were involved. A perceptive physicist might notice that the quantum hypothesis implies a violation of the laws of electromagnetism as written down by Maxwell, exactly what is necessary for the new model of atom to make sense. However, it took an uncommonly brilliant physicist to combine the quantum ideas with the atomic model, to resolve the issue of stability of atoms, and, in addition, to solve an apparently unrelated problem: why hydrogen emits electromagnetic radiation the way it does. This brilliant mind was the Danish physicist Niels Hendrik David Bohr.

5.3 The Adolescence of Quantum Theory

Thus to describe the properties of matter as well as those of light, waves and corpuscles have to be referred to at one and the same time. The electron can no longer be conceived as a single, small granule of electricity; it must be associated with a wave and this wave is no myth. . . . And it is on this concept of duality of waves and corpuscles in Nature, expressed in a more or less abstract form, that the whole recent

development of theoretical physics has been founded and that all future
development of this science will apparently have to be founded.
—L. de Broglie, Nobel lecture

Bohr was a young theoretical physicist, who at the time of Rutherford's
discovery had just completed his doctorate. He was working at Rutherford's laboratory, and he knew from first hand the new model for the
atom and its problems. The new ideas about quanta had made a deep
impression on Bohr. Unlike Einstein and Planck before him, he believed
that the barely nascent quantum theory provided a window to a new
view of reality, radically different from that of established physics. He
was always keen on demonstrating the fundamental inability of the
present theories to describe the microcosm.

Besides the quantum ideas and Rutherford's model for the atom,
Bohr had further information to take into account. This information
came from *spectroscopy*, the study of the electromagnetic radiation
emitted by physical bodies. Many pure substances emit electromagnetic waves, which are characterized by discrete values of the frequency. These values do not depend on any external circumstances, but
they are specific and always the same for each substance. In particular,
this was the case with the simplest (lightest) atoms like hydrogen and
helium.

According to the quantum hypothesis, the frequency of an electromagnetic wave corresponds to the energy carried by each individual
photon. The spectroscopy data imply that the photons emitted by the
atom carry only specific, discrete values of energy. It is then natural to
assume that the electrons orbiting around the nucleus are the source
of these photons: the emission of a photon corresponds to the electron
moving from one orbit into another with less energy (closer to the
nucleus). Energy conservation implies that the photon's energy equals
the difference in energy between these two orbits. But if the emitted
photon's energy takes only discrete values, the possible transitions
between different orbits must be constrained—perhaps not all conceivable orbits can be realized in nature. Bohr followed this line of reasoning
and concluded that the electrons move only in specific orbits around the

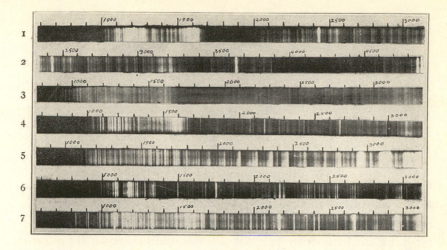

Figure 5.3 Discrete atomic spectra. This plate includes seven photographs of the lines corresponding to different wavelengths of electromagnetic waves emitted from hydrogen (photos 1–4) and nitrogen (5–7). The handwritten numbers refer to the wavelength in units of 1/10,000,000 of a millimeter. One can see clearly that only waves with specific discrete values of wavelength are emitted from these substances and that there exist values of the wavelength for which no radiation is emitted. Reprinted with permission from J. J. Hopfield, *Phys. Rev.* **20**, 573 (1922). Copyright 1922 by the American Physical Society.

nucleus. All other conceivable orbits are forbidden. Hence, the electron may avoid falling to the nucleus, if the orbit corresponding to this fall is one of the *forbidden* ones. The task was then to find a simple model that would enable us to identify the allowed orbits.

Bohr considered the simplest of atoms, the hydrogen atom, which contains only one electron. He found out that he needed only one assumption in order to account for the observed phenomena. One should postulate that one of the physical quantities that characterize the motion of the electron around the nucleus takes only discrete values. Through trial and error, he realized that this physical quantity was the angular momentum, the conserved quantity associated with the symmetry of spatial rotations (see sec. 4.4). One needs only assume that the

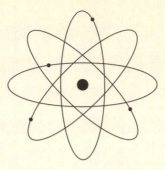

Figure 5.4 **The Rutherford-Bohr picture of the atom.** The atom contains an area with high concentration of mass, the nucleus, which is populated by heavy, positively charged particles. Electrons move around the nucleus in orbits that are characterized by specific discrete values of energy. The allowed orbits possess values for angular momentum, which are an integer multiple of Planck's constant.

electron's angular momentum takes values that are integer multiples of Planck's constant; then everything comes out like magic. One demonstrates that an electron only moves in specific orbits, that it cannot fall to the nucleus, and that its energy takes discrete values. Moreover, the energies of the photons that are emitted when the electron jumps from one orbit to another correspond to the observed frequencies with a high degree of accuracy.

That was it! In one stroke the Rutherford model of the atom was vindicated, the observations of spectroscopy were accounted for, and, more important, the new ideas about the quantum nature of the world became mainstream. After this third spectacular success of the quantum principle, nobody would doubt again that a new physics radically revising the Newtonian theory would have to be constructed.[12] The latter came slowly to be referred to as *classical physics*, in a (perhaps nostalgic) reminiscence of its era of dominance.

However, there was a price to pay for the new theory. "There appears to me one grave difficulty in your hypothesis," Rutherford wrote to Bohr shortly after he had read a first draft of Bohr's paper, "how does an electron decide what frequency it is going to vibrate at when it passes from one orbit to the other? It seems to me that you would have to assume that the electron knows beforehand where it is going to stop."[13] Rutherford had discerned the most important problem of the young quantum theory—a problem that was going to persist even in its final formulation. The quantum laws seemed to violate all concepts of causality and determinism of classical physics. However, what Rutherford saw as a problem, for Bohr was a necessary sacrifice on the way toward an all-new perspective on the physical world.

In spite of its success, Bohr's model worked only for the hydrogen atom. Its basic postulate did not provide a general physical principle that could describe other physical systems, for example, more complicated atoms. Furthermore, the origin of the discreteness of physical quantities remained unknown. Bohr himself regarded his theory as only a preliminary account. "It indicated a way in which it appears possible to bring the spectral laws in close connection with other properties of the elements, which appear to be equally inexplicable on the basis of the present state of science."[14]

Bohr always kept the greatest picture in his mind: his model was but a small piece toward the creation of a radically new physical theory. He did not view the transition of the electron from one orbit to another as a continuous process similar to the ones of classical physics. He thought that it was a *jump*, an unpredictable and noncausal process, for which the ordinary imagery of motion in space and in time is inadequate. He "did not try to bridge the abyss between classical and quantum physics, but from the very beginning of his work, searched for a scheme of quantum conceptions, which would form a system just as coherent, on the one side of the abyss, as that of the classical notions on the other side of the abyss."[15]

The following years—the 1910s, the years of the Great War— witnessed several efforts to understand the origin of Bohr's postulate. Most of the research in quantum theory attempted to apply Bohr's, Planck's, and Einstein's insights to other physical systems. These systems were either microscopic ones, like other atoms, or ones that consisted of a large number of atoms, like solids. In these efforts there were many successes, but even more failures. It took some time before a new physical principle was discerned. This came from the work of a doctoral student at the Sorbonne, Louis de Broglie.

By the early 1920s Einstein's idea of photons had been broadly accepted. Nonetheless, the duality of light as wave and as particle remained a great mystery of quantum theory. There existed simply no rule about the use of one or the other aspect. It was said jokingly that the practice of quantum theorists consisted in describing light as a wave on Mondays, Tuesdays, and Wednesdays and as a particle on Thursdays, Fridays, and Saturdays.

De Broglie thought deeply on that problem. The novelty of his approach arose from his fascination with the concept of time and the role this would have to play in the new physics. He was wondering how it would be possible to reconcile the notion of time in relativity with the concept of the light quanta. "The difference between the relativistic variations of the frequency of a clock and the frequency of a wave is fundamental," he wrote. "It had greatly attracted my attention, and thinking over this difference determined the whole trend of my research." Photons move by definition with the speed of light—hence they are massless particles. De Broglie realized that the issue of rest mass was of secondary importance as far as the relation of photons to waves is concerned. Perhaps this relation is also applicable to massive particles, such as the electron. In his words, the "determination of the stable motion of electrons in the atom introduces integers [in Bohr's model of the hydrogen atom], and up to this point the only phenomena involving integers in physics were those of interference and of . . . vibration. This fact suggested to me the idea that electrons too could not be considered simply as particles, but that frequency (wave properties) must be assigned to them also."[16]

Hence, de Broglie came to the most general statement of the duality between particles and waves: "Any moving body may be accompanied by a wave and it is impossible to disjoin motion of body and propagation of a wave."[17] It is not only light that exhibits particle behavior. Objects that we usually think of as particles (like electrons) exhibit distinctive wave behavior. Waves and particles coexist; they are inseparable, the wave playing the role of a guide to the particle's motion.

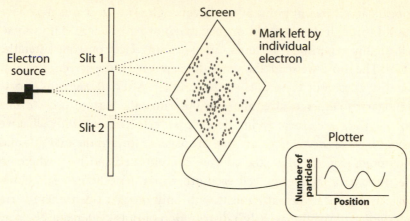

Figure 5.5 **The two-slit experiment.** A source emits electrons toward a wall with two slits. Each individual electron passes through one of the two slits and leaves a mark on the screen placed behind the wall. We have no way of predicting from which slit the each electron passed, or of predicting the place where it will land on the screen. As the number of electrons grows, we observe a periodic pattern of black and white regions. Indeed, if we plot the number of electrons vs. the position on the screen, we see a wavelike behavior, similar to one that would have been obtained if a wave had been directed toward the two slits.

De Broglie's hypothesis of "matter waves" was soon verified by experiment.[18] Matter waves should exhibit all features of ordinary waves— interference the most important among them. Indeed, the archetypal demonstration of the wave behavior in particles is the *two-slit* experiment, which is similar to Young's demonstration of the wave character of light (see sec. 2.4).

The two-slit experiment refers to the following situation (see fig. 5.5). A source emits electrons, which pass through a curtain with two holes in it. We place a screen behind the curtain, at which every individual electron leaves a burning mark like a dot. We assume we have such control in the emission of electrons that we can actually send them one after one. We let the first electron go: it will pass through one of the slits and will leave a dot on the screen. We send a second electron; another dot appears at a different point of the screen. (The two electrons may have passed through the same slit or they may not have. There is no way to tell in this experiment.) We continue sending electrons one after the other. The more we send, the more dots appear on the screen. Initially, the distribution of these dots seems random. After 10,000 runs, a pattern starts to emerge; after about 50,000 runs, the pattern is clear. The dots on the screen form an interference pattern. This is exactly what we would have observed if a wave had propagated toward the two slits and interfered. Regions where the dots are concentrated and areas where no electrons have landed succeed each other in a periodic manner. However, if we repeat the experiment with only one slit open, there is no interference. We observe only a distribution of dots centered at the part of the screen that lies immediately behind the open slit.

What can we infer from these phenomena? Electrons individually behave as particles even though their motion is not predictable—we do not know through which slit they cross. However, they seem to behave *collectively* like a wave because they exhibit an interference pattern. This is weird because we think that we know how particle beams behave. As beams spread during their motion, we would simply expect two concentrations of dots, one behind each slit. Instead, we observe this utterly inexplicable wave behavior.

In the early days of quantum theory, the technology did not allow the construction of an electron source that would emit individual electrons: large numbers of them were lumped together in a single beam. One could not then see the individual dots building up toward the wave pattern. It was therefore natural to make the strong statement that an *individual* electron behaves like a wave, or that each particle is accompanied by a wave that drives its motion. In fact, de Broglie proposed a law that relates the properties of the particle to the properties of the wave. The wavelength of the wave that corresponds to a particle is inversely proportional to the particle's momentum. As the momentum of a particle is proportional to its mass, heavy particles correspond to waves with small wavelength. The relation between momentum and wavelength turns out to be correct also for photons, which have zero rest mass. De Broglie's proposal seems to be universal as far as the correspondence between particles and waves is concerned.

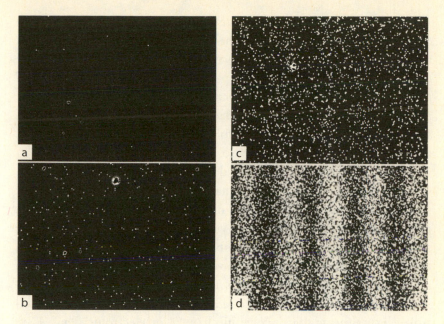

Figure 5.6 **Buildup of the interference pattern in a two-slit experiment.**
These photos show the recordings of electrons in the screen in the two slit
experiment: (a) 8 electrons, (b) 270 electrons, (c) 2000 electrons, (d) 60,000
electrons. Reprinted by courtesy of Dr. Akira Tonomura, Hitachi, Ltd., Japan.

Could one interpret de Broglie's proposal by saying that there are no
particles in nature, but only waves? Most definitely not: we call an object
a "particle," when it behaves like a particle, that is, like a discrete, pointlike object that moves in a specific trajectory. However, beams of particles do exhibit wave behavior. This not only is inexplicable by Newton's
laws, but it remains inexplicable even if we employ statistical arguments
for the behavior of the particles in a beam. The way out of this dilemma
is simply to accept it as a new principle of nature. This is the *particle-
wave duality:* under different circumstances, the same object can behave
either as a particle or as a wave. This does not imply that the same object
is *both* a particle and a wave. Its behavior depends on the context,
namely, on the specific experimental situation at a given moment of
time. The particle and the wave aspect never appear together.[19]

It is necessary to exercise some caution toward a catch in the principle of particle-wave duality that (for the reasons explained earlier) was not apparent in de Broglie's time: the wave behavior appears only when we take large collections of particles (particle beams), while *an individual particle will always behave like a particle*. Individual particles have the *potentiality* to behave like waves when large numbers of them are considered, but it does not logically follow that they individually exhibit such behavior. In fact, they never do.

The relation between particles and waves has one very important consequence. From the most rudimentary observation (when we just see something with our eyes) to the most complicated experiment that probes the structure of the micro world, one basic principle stays the same. We may extract information about a physical system only through the study of material objects that have interacted with it. The extraction and manipulation of information is automatic with our senses: our nervous system is built around it, and so is our mind. However, in controlled experiments that involve mechanical or electronic devices, we can extract information only if we have some prior knowledge about the laws that govern the bodies we use to probe the physical system.

From quantum theory, we learn that particles correspond to waves, and that they are characterized by a wavelength. If the wavelength is larger than the dimensions of the system we probe, we do not expect to get any detailed information about this body's inner structure; the particle hardly notices it. Only if the wavelength is of the same order of magnitude as the scale characterizing the body's inner structure should we expect this interaction to provide us with fine details. To give an analogy, we cannot use a car mechanic's screwdriver to repair a finely crafted wristwatch; we need a screwdriver that can "see" the fine structure of the watch's mechanism. Hence, if we want to study physical systems at small scales, we need to use particles with appropriately short wavelength. Short wavelength corresponds to high momentum, and consequently to high energy. These considerations lead to the fundamental principle of modern high-energy physics: *the deeper one wants to probe a physical system, the higher the energy of the probing particles should be*. This was hardly evident before the discovery of the particle-wave duality.

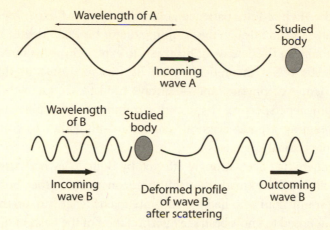

Figure 5.7 **Wavelengths and extraction of information.** We use waves to probe the inner structure of a particular physical object. When the wave falls on the object, the wave's profile is deformed in accordance with the object's inner structure. The wavelength of A is larger than the relevant distances. For this reason, its interaction with the constituents of the probed system affects it little. Therefore, wave A will not give us any information. On the contrary, wave B has much smaller wavelength. Its profile will be substantially deformed after its encounter with the probed body. It will therefore allow us to extract a large amount of information about the object's internal structure.

5.4 Heisenberg's Revolution

Imagination is an almost divine faculty which, without recourse to any philosophical method, immediately perceives everything: the secret and intimate connections between things, correspondences and analogies.

Charles Baudelaire, *New Notes on E. Poe*

De Broglie's discovery took place in 1923. Let us move forward to 1925, in Göttingen. A young theoretical physicist named Werner Heisenberg had recently finished his doctoral degree and was working on quantum theory under the supervision of the mathematical physicist Max Born. Heisenberg was an unusual type of physicist by twentieth-century standards. He had little patience about the details of experiments. He had barely scraped through the examination for his Ph.D. degree because he was unable in the defense of his thesis to explain how the microscope worked. His was a theoretical mind: he was interested in the most abstract issues of physics, and he always tried to see them through a philosophical perspective. His scientific training had not prevented him from obtaining a good classical education—his father was in fact a teacher of classical languages.

Heisenberg had been strongly affected by Bohr. They had met briefly in 1922, and both had made a great impression on each other. Bohr saw a brilliant young man who had the audacity to challenge him on the thing he was supposed to know best, the interpretation of the nascent quantum theory. Heisenberg saw in Bohr a great physicist with a deep philosophical thought and a worldview that appealed strongly to his character. Bohr thought that the quanta were to bring a revolution in the very concepts through which we understand the world, and Heisenberg, only twenty years of age at the time, was all too eager for a revolution against the "*mechanistic materialism*" of contemporary science and the "*devaluation of spirit*" in modern mass societies. Heisenberg and Bohr had kept up a lively correspondence ever since. Heisenberg slowly realized the need to seek the deeper origin of the semi-empirical laws that Planck, Einstein, and Bohr had uncovered. He felt that this was more important than looking for other applications of these principles. Heisenberg shared Bohr's opinion that the deepest and most urgent problem was to identify the level at which the laws of classical physics should change, in order to account for the quantum phenomena. It was thought at that time "that the way out of this difficulty lies in the fact that there is one basic assumption of the classical theory which is false, and that if this assumption were removed and replaced by something more general, the whole of atomic theory would follow quite naturally. . . . However, one has had no idea what this assumption could be."[20]

For many people, the most promising direction lay in a change of the *dynamics*. One should look for a new law of motion that would allow the derivation of the quantum phenomena. Heisenberg thought that this change should take place at the level of *kinematics,* a much more radical approach. At the kinematical level lie the definitions of the most basic physical quantities like position and momentum. These quantities refer directly not only to the way we carry experiments and interpret their results, but also to the most basic level of our perception of the physical world. The corresponding definitions are essentially geometrical, and as such they rely on our fundamental intuitions about the nature of physical events taking place in space and in time. Moreover, at the root of all these definitions lay the basic premise of modern science: physical quantities can be described by numbers. Experiments determine these numbers, and by doing so they provide us with information about the structure of the world.

Heisenberg conceived that a theory for the quantum phenomena needed to question these fundamental premises. What nerve, one may say, and rightfully so. However, Heisenberg was not a lone figure; his frame of thinking was representative of his times. It was the 1920s. The whole of Europe was reeling from the painful memories of the Great War.

Intellectual activity mirrors and interprets but only occasionally directs the emotion of a person and of a people. No surprise, then, that the aroused emotion of the 1920s led to a burst of creativity, unparalleled in the recent past. Earlier questioning of old certainties reached the highest level of intensity. The fruits were seen not only in the physical sciences—even though general relativity and quantum theory would be sufficient by themselves to prove the point. They appeared in psychology, with the spectacular diffusion of the new ideas of Siegmund Freud, Carl Gustav Jung, and their disciples; they appear in painting, with the renunciation of the photographic representation of reality and the intensification of the search for the form underlying external appearances; they appear, finally, in philosophy, which saw the culmination of a long process of criticism of the foundations of the Western intellectual tradition.

The developments in philosophy are particularly relevant to our discussion. The stark criticism of the roots of the Western philosophical tradition—a term that could mean different things to different people—had started already in the nineteenth century. The movement of Romanticism in the arts had tried to restore the emphasis on a person's inner life beyond the rational conscious thought that characterized the previous generations of the Scientific Revolution and Enlightenment. Romanticism emphasized the subjective, the irrational, the imaginative, the personal, the spontaneous, the emotional, and the transcendental, all the aspects of life that were threatened by the rationalizations and compromises characterizing contemporary society. The Romantics challenged the view of the world as a machine. They proposed a different metaphor: the world as an organism, all its parts being interconnected by links that reflect its unity. Nature was perceived as a dynamic reality, ever changing and unfettered by any rules. Romanticism shaped the European cultures during the first half of the nineteenth century and provided a vessel for the national awakening for many European peoples, but it left practically no permanent imprint on scientific thought. It is true that individual scientists were affected strongly. The energy principle, for example, arose partly from the Romantic emphasis on the living interconnections of all aspects of the physical world. Still, at the end of the day the success of Newtonian mechanics was too overwhelming to allow scientists too much free ground. The human type idealized by Romanticism was the artist. No wonder it failed to change the ideal of the era it had succeeded, the scientist.

Romanticism was a reaction against the forces that were giving birth to a new society. Being a reactionary movement, it was difficult to keep a proper balance—it was easy to move toward extremes, even toward absurdity. Still, it did provide a cradle for dissent against the political, moral, and intellectual ideas that slowly gained dominance during its time. Romanticism nurtured a philosophical tradition, which undertook a sharp criticism of the dominant worldview. A common theme in this tradition was the criticism of the worldview that had arisen and dominated in Western Europe ever since Scholasticism: the dominance of rationality and logic in the human psyche and consequently in humans' attitude toward the world. The critics emphasized that human rationality is based on a more basic and primitive level of being and living in the world. They gave different names to this fundamental level, according to their experiences, their characters, and their micro cultures. This philosophical criticism did not leave the scientific worldview untouched. The perception of a world as a nicely ordered, predictable place was viewed as an externalization of the unbalanced domination of the rational and conscious part of the human psyche upon the totality of a human's inner life. Perhaps the perception of a perfectly ordered and predictable world was nothing more than a projection of a desire to have things as stable and unsurprising as possible. However, the all too successful Newtonian mechanics described the world as such a place: a new science—and the demand was ripe for such—should break away from all basic concepts of classical physics.

Bohr was the first person to realize that quantum theory was the branch of science that could lead to a radically new worldview. He had been strongly influenced by the philosophy of his compatriot, Søren Kierkegaard. Kierkegaard had stressed the conflict between thought and reality, emphasized a person's choice in all acts including the rational ones, and spoken about the participation of the observer in the process of observation. These ideas, we shall see, resurfaced in Bohr's later conception about the meaning of quantum theory. Heisenberg, on the other hand, was mostly captivated by Greek philosophy. The view of many levels of reality in the world—particularly strong in Plato—had always been a special point of attraction for him.

In times of upheaval, some people turn to radical thoughts and sharp criticism of the established values. Others feel the need to embrace these values more firmly, reinterpreting them in a way that can address contemporary challenges. This happened also in philosophy, in a process that had started in the middle of the nineteenth century. Science should be placed at the center of all intellectual activity, and its method should be the final criterion for all thought about the world. Rational discourse should stick closer to the well-controlled empirical facts, which should be interpreted with sharp and precise rules of logic. Philosophy and science had to be reconsidered through the light of criticism. All notions that did not have their root in direct experience or in clear logic had to be excised and thrown away as irrelevant to a scientific worldview. All "redundant" concepts and ideas were dubbed collectively as "metaphysical," and they were deemed to have no place in the rational worldview of modern human beings. In some variations of this stance, metaphysical concepts were not to be thrown away completely. They have nothing to do with rational, logical discourse, but perhaps they could be approached by means that did not involve language. Hence, they could not be communicated and could only be experienced inwardly.

This approach is named *positivism*.[21] It has occasionally led to extremities, and for this reason it has gone out of fashion in contemporary philosophy, but at the beginning of the twentieth century it was a force of great influence. Positivism widened the gap between rational knowledge and other forms of human experience. It attempted to draw very clear, sharp rules by which rationality should proceed in order to attain knowledge. Like all attempts at harsh imposition of rules, positivism could not stay in a position of dominance for long. This does not imply, however, that it had no positive influence. Going back to the basics and demanding strict obedience to rules occasionally acts as a gush of fresh water that may revive a movement or a discipline from stagnation.

Einstein himself had flirted very strongly with the positivist worldview in both his theories of relativity.[22] Absolute space and time were a metaphysical concept that had nothing to do with the concrete way we measure extension and temporality through rods and clocks, respectively. However, positivism ultimately postulates a sharp distinction between humans' rational activity and the rest of the world, and Einstein had too strong a sense of the unity of physical reality to be engulfed by positivistic ideals. In any case, the adherents of positivism demanded that all sciences should follow strict rules, which should rely only upon concepts that describe the immediate concrete experience in the form of well-controlled experiments.

Heisenberg, a philosophical mind as much as a scientific one, had been strongly influenced by the radical ideas of contemporary German philosophy, but in his times science had been deeply immersed in the spirit of positivism. His thoughts involved a juggling—in danger of absurdity—and eventually a synthesis of the two fundamentally opposed schools. This uneasy alliance persisted in the final formulation of quantum theory, and it still exists today. The coexistence of these worldviews is very precarious, and it makes the discussion about the meaning of quantum theory verge occasionally on schizophrenia. One may read about the importance of measurement and operational concepts for the understanding of quantum theory (the positivist part). Then, a few lines or pages later in the same textbook or popular science book, one may see the statement that quantum theory leads to a radical overthrow of our cherished concepts of physical quantities or of causality or of physical events, namely, of the very same concepts that are needed in the design and interpretation of any experiment! The interesting thing is that this philosophical schizophrenia has a well-defined mathematical representation in the formalism of quantum theory, and it enters the mind of any physicist who uses it. Most of the time it is accepted as a matter of fact, and the related discussion is swept under the carpet because it seems irrelevant to the theory's physical predictions—or rather, to the predictions that can be verified experimentally with present-day techniques.

Heisenberg postulated that we should formulate physics solely in terms of quantities that refer to physical magnitudes that we actually observe. What do we observe in atomic physics? Our sole direct information comes from electromagnetic radiation, namely, the photons that are emitted by atoms. In Bohr's model, we assume that photons arise in the transitions of electrons between two allowed orbits. All information we obtain from such systems refers to these transitions, these quantum jumps. If our physical theory is to contain only directly observable quantities, its basic objects must refer explicitly to these transitions. Since each orbit can be characterized by an integer that corresponds to the discrete values of angular momentum,[23] a transition is characterized by a pair of such integers. It follows that any physical quantity associated with such a transition is characterized by a pair of integers.[24]

If we take into account the transitions corresponding to all possible pairs of integers, we form an object, which mathematicians call *a matrix*. This is nothing but a table whose entries are labeled by pairs of integers. An important property of matrices is that they behave like individual mathematical objects. One may add, subtract, and multiply any two matrices. In this sense, they are very similar to ordinary numbers. However, matrix multiplication has one important difference from multiplication of ordinary numbers. While the ordering according to which we multiply two numbers is irrelevant to the multiplication's outcome (ten boxes of twenty oranges each contain the same number of oranges as twenty boxes of ten oranges each), in matrices the ordering *does make a*

	1	2	3	4
1	5	4.2	4	−5
2	6	3.9	−4	−4.2
3	2	2+i	2	−9
4	3	3	−2.2	0
5	3	7	9.1	−0.1

Figure 5.8 **Example of a mathematical matrix.** A matrix is a table that has numbers as entries. It is indexed horizontally and vertically by integer numbers. The particular matrix here is 5 × 4; i.e., it has 5 rows and 4 columns. Its entries can be numbers of any type, e.g., integers, negative numbers, and even complex numbers. One may treat a matrix as a single mathematical object: a matrix can take part in addition and subtraction like ordinary numbers. One can also define matrix multiplication via a somewhat complicated rule. Matrix multiplication shares some of the features of ordinary multiplication, except for commutativity: the product AB of two matrices A and B is in general different from the product BA.

difference. The mathematical jargon for this fact is that the multiplication of matrices is *not commutative.*[25]

Elaborating on the remark above, Heisenberg concluded that all physical quantities should be represented by mathematical objects that are characterized by noncommutative multiplication. These were matrices, or, as they were later named, *q-numbers.*[26] Such was to be the fate of position, momentum, energy, and all quantities of classical physics from now on. Heisenberg also identified the principal rule of the correspondence between physical quantities and q-numbers. The product of the position q-number and the momentum q-number differs from the product of the momentum q-number and the position q-number by an amount proportional to the fundamental constant of quantum theory: Planck's constant. Heisenberg's arguments followed a tortuous path and involved an unusual combination of mathematical tools with sound physical intuition. His approach was far from watertight or secure; he was sensing rather than reasoning toward the correct direction.

Of all the great discoveries in the history of physics, Heisenberg's was, in my opinion, the subtlest and least obvious. In comparison, Einstein in his moments of greatest creativity combined his insights into a solidly build structure: he gave concrete reasons for the physical principles of his theory and then proceeded to a derivation of their consequences. His work was that of an architect, a master builder. Heisenberg's creation was closer in spirit to that of a painter, one brushstroke here, another there, with no apparent plan or consistency until the very last moment, when one sees the totality of his conception. It is only the perception of this totality that justifies the idiosyncratic way that the artist used to get there. For people believing that science proceeds linearly in a chain of clear reasoning from established facts, Heisenberg's achievement may look like an accident. The main idea was pure speculation, and it relied upon a "mere" mathematical analogy. However, it cut through the confusion that reigned in the foundations of quantum theory like a diamond through glass. It was so powerful and yet so simple that it had an almost otherworldly quality. Indeed, Heisenberg came to this conception during an attack of hay fever, being in an almost hypnotic state of mind, and he was vacillating strongly about whether his work was a huge achievement or a big delusion.

If Heisenberg's ideas were taken seriously, the whole edifice of classical physics would be shaken from the ground up. Its fundamental physical magnitudes were not to be represented by numbers anymore, but by mysterious mathematical objects with no clear correspondence to immediate empirical understanding or geometric intuition. A new level of physical reality appeared that had been inaccessible to the mechanistic worldview of old. Moreover, this level, by its very nature, may never be fully amenable to rational predictability and control.

Heisenberg's ideas encountered strong suspicion: only people who were sympathetic to a novel revolutionary worldview or people who enjoyed the challenge that the new description would entail to their mathematical skills were immediately stimulated by this work. Niels Bohr was—expectedly—enthusiastic. Max Born—Heisenberg's supervisor—collaborated with

Pascuale Jordan, a mathematician who had overheard Born complaining about the difficulties of matrix calculus in a train and been quick to offer his services as an expert in the field. Born and Jordan placed Heisenberg's ideas on firm mathematical ground. The new theory was appropriately named *matrix mechanics*. The only other person who was immediately captivated by Heisenberg's ideas was a young graduate student at Cambridge University named Adrien Maurice Paul Dirac, who was also destined to leave his stamp on the quantum revolution. Heisenberg soon moved to Copenhagen with Bohr, and their professional relationship became a close friendship. Copenhagen and Göttingen arose as the two centers from which the new physics would spread to the rest of the world.

5.5 The Riposte: Schrödinger's Wave Mechanics

It is good to express a thing twice right at the outset and so to give it a right foot and also a left one. Truth can surely stand on one leg, but with two, it will be able to walk and get around.
—Friedrich Nietzsche, *The Wanderer and His Shadow*

Hardly had six months passed after the publication of Heisenberg's paper than a new and completely independent breakthrough took place. It came from Erwin Schrödinger, a thirty-nine-year old lecturer at the University of Vienna. Schrödinger was, like Heisenberg, a person of broad interests, well versed in philosophy and the classics. But he was more conservative in his research: unlike Heisenberg, his research was not guided by his philosophical beliefs—their effect was more silent and subtle. He did not look for a revolution in physics—revolution was thrust upon him. And as it turned out, it was a revolution that ran counter to that of Heisenberg and Bohr—at least in the beginning.

Schrödinger had taken the ideas of de Broglie about the relation between particles and waves very seriously. If particles behave like waves, he thought, one should look for the laws of motion of these waves. In the back of Schrödinger's mind was a picture from wave propagation in fluid mechanics. In certain flows, a kind of resonance arises, and this results in a discrete behavior of some properties of the waves. This behavior is similar to that of a guitar string that can vibrate only at specific values of frequency. Could something similar be true in atomic physics, that is, were the discrete values of energy in the hydrogen atom due to a resonant behavior of quantum waves?

A brilliant idea, one "springing from true genius," as Einstein exclaimed later. It brought about a spectacular success, but its consequences were baffling. Schrödinger did write an equation. This equation described the propagation of a wave, which was represented by a function denoted by the Greek letter ψ. The equation is nowadays named after him. It was constructed by the requirement that the wave propagates in a way that corresponds to the motion of particles according to Newtonian mechanics. To achieve that, Schrödinger found a specific rule, which relates the energy of particles to the law of motion for the ψ-function.

A specific application of Schrödinger's equation involved the determination of ψ-waves that correspond to the motion of an electron around a nucleus. When he looked for solutions to this equation, which (1) are *stable* in time and (2) describe the electron as *bound* to the nucleus, a remarkable thing happened, much as he had intuitively expected. Such solutions were possible only for certain *discrete values* of the energy. These values—amazingly—coincided with the ones predicted by Bohr's model, which had been rigorously confirmed by the experiment. Schrödinger then interpreted Bohr's orbits as bound solutions of his equation for the ψ-function. Moreover, he demonstrated that the law of motion for the ψ-wave was such that a narrow wave profile would not spread as time goes by. Perhaps, he thought, particles were nothing but waves with such narrow profiles. "It is scarcely necessary to emphasize," Schrödinger wrote, "how much more appealing than the conception of jumping electrons would be the conception that in quantum transitions the energy passes from one vibration to another. The change in the vibration pattern can take place continuously in space and time, and can readily persist as long as the [photon] emission process does."[27]

Physicists of renown such as Einstein, Lorentz, and Planck—and initially also Born—were strongly attracted to Schrödinger's wave mechanics.

It promised to be a tool of immense significance for the study of the micro world, and it eventually fulfilled this promise. One could write down the laws of motion for the ψ-function simply from the knowledge of energy in classical systems. It was, in principle, possible to describe any quantum system that had an analogue in Newtonian physics in a mathematical language that lent itself to a facility in calculations. But at that point the good news came to an end. Schrödinger's favorite interpretation for the ψ-function as a physical wave, in accordance with the ideas of de Broglie, did not survive close scrutiny. If Schrödinger's equation were to be of universal value, it should also describe systems of many particles. There still exists a wave function in this case, but it cannot be interpreted as a real wave that propagates on physical space; it is a wave that "lives" in the abstract mathematical space that contains all possible configurations of the many-particle system.[28]

Another problem of Schrödinger's theory was that the wave function did not take numbers as values—real numbers, that is. Schrödinger's equation necessarily involves the introduction of *complex* numbers, namely, mathematical objects that share all properties of real numbers but also contain the square root of the number -1 and its multiples. While a real number has a direct interpretation in terms of measurable quantities, a complex number does not. This made the meaning of the ψ-function even move elusive.

Heisenberg felt uneasy about Schrödinger's wave mechanics. Waves are continuous, while for Heisenberg the essence of quantum theory was discontinuity. Quanta are discrete objects, and their existence implies that nature does make jumps. Heisenberg followed Bohr in the belief that the world of the quanta ran much deeper than any concepts we may have about space and time. They believed that the discontinuities of the quanta highlighted the necessity for physics to move beyond the Democritean conception of particles moving in the Void. The quantum discontinuity, which had been so elegantly incorporated into matrix mechanics, was threatened by the success of Schrödinger's method.

Heisenberg had a chance to challenge Schrödinger in a talk the latter gave in Munich. There, Heisenberg argued that wave mechanics could not account for any of the discrete quantum objects, like photons. His attitude was perhaps a bit too aggressive, but, in any case, the audience was hostile to his ideas, and Willy Wien (a senior scientist who was Schrödinger's host and who, as an examiner, had almost cost Heisenberg his doctorate) practically told him to sit down and shut up. In Wien's opinion, "all this nonsense about quantum jumps and the atomic mysticism must come to an end."[29]

Nevertheless, Schrödinger's mechanics soon found its limits. Bohr invited Schrödinger to Copenhagen, not only to give lectures on his approach, but also to spend some time discussing the interpretation of quantum theory. Schrödinger accepted. Heisenberg wrote later a lively description of this encounter.

These discussions took place in Copenhagen around September 1926 and in particular they left me with a very strong impression of Bohr's personality. For though Bohr was an unusually considerate and obliging person, he was able in such a discussion, which concerned epistemological problems, which he considered to be of vital importance, to insist fanatically and with almost terrifying relentlessness on complete clarity in all arguments. He would not give up, even after hours of struggling, before Schrödinger had admitted that this interpretation was insufficient, and could not even explain Planck's law. Every attempt by Schrödinger to get around this bitter result was slowly refuted point by point in infinitely laborious discussions.

It was perhaps from over-exertion that after a few days Schrödinger became ill and had to lie abed as a guest in Bohr's house. Even here it was hard to get Bohr away from Schrödinger's bed and the phrase "But Schrödinger, you must at least admit that . . ." could be heard again and again. Once Schrödinger burst out almost desperately: "If one has to go on with these damned quantum jumps, then I am sorry I ever started to work on quantum theory." To which Bohr answered, "But the rest of us are so grateful that you did, for you thus brought atomic physics a decisive step forward." Schrödinger finally left Copenhagen rather discouraged, while we at Bohr's Institute felt that at least Schrödinger's interpretation of quantum theory, an interpretation rather too hastily arrived at using the classical wave-theories as models, was now disposed of, but that we still lacked some important ideas before we could really reach a full understanding of quantum mechanics.

Schrödinger left Copenhagen shaken by Bohr's arguments, but he had not been converted to the rival school. There was one line he would never cross. He could not believe, as Bohr did, that space and time were not fundamental entities. Spacetime was for Bohr a remnant of the materialistic worldview inherent in classical physics that ran as far back as Democritus. He simply found no reason for its existence in the new world order that was being created by quantum theory. Schrödinger would have nothing of this—and in that, he would summon Einstein's support. He believed that one should not abandon the image of physical processes taking place in space and in time because these concepts are essential to our understanding of the world. One cannot simply throw them away. As he wrote in a letter to Willy Wien, "Physics does not consist only of atomic research, science does not consist only of physics and life does not consist only of science. The aim of atomic research is to fit our empirical knowledge into our other thinking. All of this other thinking, so far as it concerns the outer world is active in space and time. If it cannot be fitted in space and time then it fails the whole aim and one does not know what purpose it really serves."[30]

5.6 Conflict and Reconciliation

Exactitude is not truth.
—Henri Matisse

As the year 1926 was ending, two opposing camps were being crystallized concerning the interpretation of quantum theory. On one side were the "rebels," the ones who wanted a complete overthrow of the established concepts of physical theory. They had clear ideas and a program for their implementation. The problem was that the formulation of their theory was cumbersome and difficult to apply in concrete physical systems. The "conservatives," the ones who wanted to preserve as much from the established worldview of physics as possible, gathered behind Schrödinger's results. They were initially the majority. They possessed a method that could be generalized with success to a large class of physical systems, and they shared a vague feeling that they were implementing the very important wave–particle duality of de Broglie. But they lacked clear direction: they had no physical interpretation of their theory that could stand up to close scrutiny.

This situation could not last indefinitely. Open confrontations do not last long in modern physics, especially when there is an abundance of experimental results. One or the other side has to yield, or they have to merge. The latter alternative came true in the case of quantum theory. Schrödinger demonstrated that the matrix mechanics of Heisenberg and the wave mechanics he himself had pioneered were two sides of the same coin. If one knew all solutions of Schrödinger's equation, one would essentially know all allowed orbits of electrons around the nucleus. It would then be possible to construct the matrices that correspond to physical quantities, much as Heisenberg had done in his original work. This Schrödinger demonstrated explicitly, thus proving that, starting from his equation, one may derive Heisenberg's description.[31]

Of equal importance was a proof of the converse, namely, that there exists a well-defined prescription by which the q-numbers of Heisenberg can generate transformations of the ψ-function. Among these transformations, time-translation was in a prominent position; it is generated by the q-number that *corresponds to energy*, and it produces Schrödinger's original equation. It was then concluded that the two formalisms were absolutely identical. Schrödinger was taken aback by his results: they seemed to imply that two altogether different perceptions of the world gave rise to the same physical predictions: "There are today not a few physicists, who . . . see the task of a physical theory to be merely the most economical description of empirical connections between observable quantities. . . . In this view mathematical equivalence means almost the same as physical equivalence."[32]

Still, a mathematical formalism without physical interpretation does not constitute a physical theory: it may occasionally solve problems, but it says nothing about the deepest nature and structure of things. Schrödinger's results were not sufficient for a resolution.

Heisenberg took the next step forward. This was the discovery of the celebrated *uncertainty principle*.[33] Heisenberg scrutinized the way one performs measurements to determine the properties of microscopic particles. "How do we measure a particle's position?" he wondered. We have to use a probe, say, electromagnetic waves. However, an electromagnetic wave cannot give precise information about anything smaller than its wavelength. Hence, we can only determine the particle's position with a degree of uncertainty, which is of the order of the field's wavelength. We could use the same experiment (i.e., an electromagnetic probe) to measure the velocity, or rather the momentum, of the particle we are interested in. If the electromagnetic field were continuous, one would obtain as sharp a specification of the particle's momentum as one wanted. However, the electromagnetic field comes in quanta, each of which carries momentum inversely proportional to the wavelength. If we probe the particle with identical light quanta, it is impossible to specify the particle's momentum with accuracy greater than the momentum carried by a single photon.

It follows from the above considerations that the particle's position can be specified with an uncertainty proportional to the field's wavelength, and its momentum with an uncertainty inverse to the wavelength. Hence, the product of the uncertainty in a particle's position and the uncertainty of its momentum is larger than a constant, which turned out to be a multiple of Planck's constant. It follows that we cannot measure directly the position and momentum of a single particle with *arbitrary accuracy*. Planck's constant determines the limit in the accurate specification of the fundamental physical quantities that characterize a particle.

In the same context, a second expression of the uncertainty principle was identified. This expression referred to the relation between energy and time: the product of the uncertainty in a particle's energy and the time it takes a measurement to be completed is larger than Planck's constant. This implies that there can be neither an instantaneous measurement nor a measurement that determines the value of energy with infinite accuracy.

i. Momentum Measurement

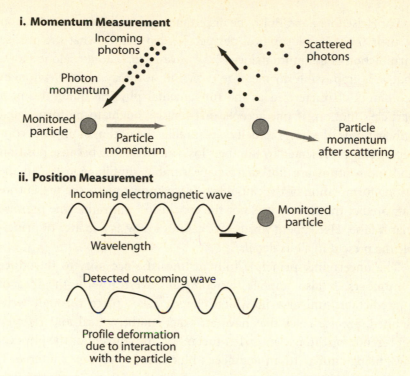

ii. Position Measurement

Figure 5.9 **Measurement by photons and the uncertainty principle.** The first drawing shows a schematic measurement of a particle's momentum. In this case, we exploit the particle behavior of the electromagnetic field. We direct a beam of photons toward the particle we want to monitor. The photons collide with the particle, and we may determine the particle's momentum by measuring the energy carried by the scattered photons. We cannot have a better accuracy than the momentum carried by a single photon. On the other hand, we can exploit the wave properties of the field in order to measure a particle's position. The wave profile is deformed because of the particle's presence, and the particle's position may be surmised from the study of this deformation. In this case, the accuracy cannot be greater than the wavelength of the electromagnetic radiation. The relation between wavelength and momentum then leads to the celebrated uncertainty principle.

For Heisenberg and Bohr, the uncertainty principle was a vindication of their overall philosophical perspective. No matter what the precise formulation of quantum theory is, *the wave-particle duality* alone shows that it is impossible to measure with full accuracy what Newtonian mechanics characterized as the fundamental physical quantities of a particle's motion. If one accepts that physics ought to be formulated only in terms of what is actually measurable, the notion of physical concepts being represented by numbers loses any meaning because position and momentum are not sharply measurable simultaneously. "In the strong formulation of the causal law 'If we know exactly the present we can predict the future,' it is not the conclusion but rather the premise that is false,' Heisenberg wrote. "We cannot know as a matter of principle the present in all its details."

The uncertainty principle fully justifies the necessity to introduce q-numbers. It also implies the complete overthrow of the idea of a predictable universe in which change takes place through well-defined, precise rules that have an exact mathematical and rational representation. To use an earlier metaphor, the concept of the physical laws being similar to an engineer's blueprints is all but shattered. If Leonardo da Vinci and the painters of the Renaissance were the precursors of classical physics with their emphasis on precision of detail and realism of representation, the uncertainty principle brings science close to contemporary art; the sharp distinction between precision in representation and truth is perhaps the defining maxim of twentieth-century art.

"Hold on!" was the immediate answer. "All the uncertainty relation shows is that there is a limit to the accuracy with which we can measure a particle's properties. True, this is a very significant result: it does remove all fantasies about obtaining information with unlimited accuracy from nature. However, this does not mean that physical magnitudes, like position and momentum, are not well defined. The q-numbers are nothing but convenient mathematical tools. There is no reason to think of them as representations of some mysterious—and mystical—structure of reality at the deepest level. The uncertainty principle implies simply that by our nature as material beings we cannot really obtain too sharp information about the state of things. Why should we rush to hasty conclusions about the nature of reality?"

"But how can we write theories," Heisenberg's supporters will retort, "that invoke sharply defined physical quantities, while in reality we never determine them? You are simply imposing false images of your mind to physics, arbitrary mathematical concepts, baseless and meaningless speculations that have no counterpart in reality. You are thinking in terms of your favorite metaphysics, not physics." At a time when positivism reigned supreme, the accusation of a metaphysical attitude was quite effective. "Well, if we are accused of metaphysical attitude, it is because we cherish fundamental intuitions of the mind, which have guided science up to here, intuitions about space and time and number and causality. And you are using these intuitions about physical quantities in all your arguments, but at the end you choose to ignore them and blurt incoherent statements about levels of reality and novel worldviews."

At this point, the discussion invariably degenerates into name-calling. It did then and it does now. There is no agreement about the meaning of the uncertainty relation, and this reflects the lack of agreement about the meaning of quantum theory. One reason is that there exists no unambiguous proof of Heisenberg's uncertainty relation for an *isolated* quantum system, that is, without reference to an act of measurement. All proofs that have been proposed in this direction involve controversial assumptions about the interpretation of quantum theory, and for this reason they become themselves subject to the controversy.[34]

Whatever its implications about the nature of physical reality may be, Heisenberg's uncertainty relation emphasizes the difference between q-numbers and ordinary numbers in the description of physical quantities. This difference lies at the level of kinematics, of the very definition, of the direct way that we perceive physical events. Moreover, if we are to think of physical quantities as being defined only to the degree that we determine them empirically, then perhaps the q-numbers carry inside them the notion of measurement. They would then refer to the actual interaction of an observer with a microscopic physical system, and not to the properties of the system "in itself." Perhaps we should designate the q-numbers as *observables* because they refer only to properties that can be observed.

If the q-numbers refer solely to measurements and not to the properties of the things in themselves, then we need not face the tribulation of explaining how, starting from the q-number structure of reality, we arrive at the description of physical phenomena in terms of ordinary numbers and standard geometry that we employ when we discuss and design our experiments. This is a very important problem in any formulation of quantum theory, which has come to be known as the *quantum measurement problem* (see also sec. 6.2).

The association of q-numbers with measurements highlights the schizophrenia at the root of quantum theory we mentioned earlier (sec. 5.4). Q-numbers were heralded as new objects, which revolutionized the concept of physical quantity and guaranteed that quantum theory describes a level of reality deeper and more fundamental than that of the classical concepts. But if the q-numbers are interpreted in terms of experiments, it is the concept of experiment that arises as fundamental. Quantum theory is then in danger of becoming nothing but a set of rules about the manipulation of experimental designs and outcomes with no direct relevance to the deeper structure of reality. The latter interpretation of q-numbers renders quantum theory into a positivist's dream because the theory makes a direct and necessary appeal to the concept of experiment. These two conflicting attitudes do coexist in quantum theory, and their incompatibility makes impossible a final settlement of what exactly quantum theory tells us about the world.

If the q-numbers refer directly or indirectly to measurements, what is the meaning of the ψ-function? This was the last question that needed to be answered, before a synthesis of all the disparate advances could be attempted. The correct answer had been given by Born shortly after the appearance of Schrödinger's equation. In its initial form, Born's argument was rather technical and perhaps not very convincing, but Heisenberg's discovery of the uncertainty principle allowed him to phrase it in a more intuitive way: the uncertainty relation implies that we cannot obtain sharp descriptions of physical systems. When we run the same experiment many times, we expect to obtain different results.

Since we have lost the ability to predict the behavior of individual quantum systems, the most we can expect is a description in terms of probabilities for the possible alternatives. The ψ-function provides exactly this probabilistic description,[35] and as it turned out this description is fully compatible with the time evolution law provided by Schrödinger.

Born's idea eventually won over its rival interpretations because it provided an immediate relation of the ψ-function to measurable quantities. Needless to say, Schrödinger and his supporters did not like this interpretation at all. It implied that the wave–function was not a real, physical wave, but simply a mathematical object that carries information about probabilities.

Probabilities refer directly to experiments; they can be measured only if we have a large collection of identically prepared systems and perform the experiment in every single one of them. There is a great deal of subjectivity in the determination of probabilities because they change according to the state of our knowledge about the physical system. This implies that the object that describes the probabilities—the ψ-function—has partly a subjective character. It does contain information about the physical system, but not in an objective way. It lies at the interface of physical reality and of our perception of physical reality. In later years, Heisenberg suggested that the wave function may not be objective itself, but it represents an inherent feature of physical systems: a *potentiality* to exhibit one or the other behavior that is not ruled by a strict necessity.

The ψ-function "combines objective and subjective elements," Heisenberg explained.

It contains statements about probabilities or better tendencies ("potentia" in Aristotelian philosophy) and these statements are completely objective as they do not refer to any observer; and it contains statements about our knowledge of the system, which of course are subjective in so far as they may be different for different observers. In ideal cases the subjective element might be negligible compared with the objective one. [Similarly the uncertainties] may be called objective in so far as they are simply a consequence of the description in terms of classical physics and do not depend on any observer. They may be called subjective in so far as they refer to our incomplete knowledge about the world.[36]

Born's interpretation was eventually vindicated by a large number of successful predictions. But this did not mean that everybody agreed on the meaning of the wave function. This question, even more than the one about the uncertainty principle, lies at the center of the discussion on quantum theory. Heisenberg, Bohr, and Born had no problem in accepting the ψ-function as a semisubjective object; they cared more about what it *designated* about the structure of reality rather than any physical reality it might have by itself. But for many physicists (of this and later generations), the idea that the basic object of a physical theory has no objective independent existence was very worrying and signified that something was very wrong in the foundations of quantum theory.

5.7 The Mature Quantum Theory

Come and see how evident each of them is
And neither sight trust more than hearing
Nor the noisy hearing beyond language's precision
And do not deny your faith in any of your senses,
For gates they are to knowledge,
But grasp everything as it reveals itself to you.
—Empedocles

The final confrontation between the "rebels" and the "conservatives" on the issue of quantum theory took place in Brussels in October 1927. All major protagonists in the development of quantum theory met for the fifth Physical Conference of the Solvay Institute with the title "Electrons and Photons" under the chairmanship of Lorentz—in his last public appearance before his death. This was the place to reach a conclusion.

The result was an unmitigated triumph for the Copenhagen school. Born's probabilistic interpretation and Heisenberg's uncertainty principle proved the main weapons for this victory. At the end of their joint presentation of matrix mechanics, Heisenberg and Born set out a challenge: "We maintain that quantum mechanics is a complete theory," they stated. "Its basic physical and mathematical hypotheses are not further susceptible to modifications." Their arguments were so strong, and the success of their theory so overwhelming, that all opposition was intimidated. The only person to pick up the glove was Einstein. He was deeply skeptical of the discontinuities and the lack of causality in the Copenhagen interpretation. Still, he dared not attack it directly. The explanatory power of the new quantum theory placed it in an unassailable position, and Einstein, as he half-ironically admitted, had not fully grasped the details of its mathematical formalism. He tried instead to undermine the confidence in Heisenberg's uncertainty principle, which he perceived to be the theory's strongest weapon. If that was found wanting, the inner coherence of the theory would be destroyed and the whole edifice would sooner or later fall apart.

During the breaks of the conference, Einstein proposed, one after another, elaborate thought-experiments, which would, in principle, register a violation of Heisenberg's principle. Most of his proposals tried to demonstrate that the interaction between the microscopic object and the measuring device were not as forbidding as Heisenberg and Bohr maintained. Bohr immediately realized Einstein's gambit: the uncertainty relation had to be defended at all costs, "otherwise everything would be lost." He took that task upon himself, and he responded admirably to the challenge. One after another, all the arguments that Einstein had proposed were shown by Bohr to ignore in their chain of reasoning certain physical factors. When these factors were taken into account, Einstein's conclusions were invalidated. As a result, the uncertainty principle could not be shaken. The debate between Einstein and Bohr on this issue was incessant: it resumed during the sixth Solvay Conference, three years later.[38] Nonetheless, at the end of the 1927 conference, the majority of physicists had fully adopted the interpretation of the Copenhagen school and the philosophical worldview that went along with it. Einstein, de Broglie, and Schrödinger were not among them. They left the conference, "that witches' Sabbath in Brussels," as Einstein later referred to it, dissatisfied and openly critical.

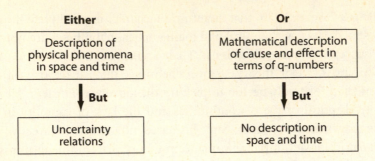

Figure 5.10 **A diagrammatic description of Bohr's idea of complementarity.**
Adapted from W. Heisenberg, *The Physical Principles of the Quantum Theory*
(New York: Dover, 1949).

The Solvay Conference led to a complete vindication for Bohr and for the stance he had maintained toward quantum theory ever since he first started thinking about the topic. His insistence that the rules of classical physics would have to be abandoned proved the most crucial factor in the development of quantum mechanics: we simply cannot imagine where quantum theory would have been led in its absence. The focal point of Bohr's thought was the relation between language—the description of the world—and reality. No description can exhaust the full content of the world; all descriptions are at most partially true. The same holds for the duality between particles and waves. Particles and waves correspond to different ways of interpreting the experimental evidence we collect in our studies of physical phenomena: waves highlight the continuous spacetime aspects of matter; particles the discrete, dynamical ones. These descriptions were for Bohr *complementary*: they could not be simultaneously valid. Indeed, Heisenberg's uncertainty principle provided the perfect demonstration that no situation can arise that simultaneously exhibits the two complementary aspects of a physical phenomenon.

The principle of complementarity is related to the fact that we observe physical systems only by interacting with them. The laws of quantum theory make impossible the sharp separation of the atomic system from the device that measures it. The same system measured in a different experiment will simply manifest different properties. For this reason, we cannot really make any statements about a microscopic physical system without referring to the concrete experimental setup that is employed in the measurement of its properties.

The complementarity of physical descriptions was the key concept of Bohr's philosophy of science. It became a central thesis in the dominant interpretation of quantum theory, in spite of the strong ambiguity inherent in its definition.

Whether people accepted the deepest implications of the complementarity principle or not, there was no doubt that after the fifth Solvay Conference the Copenhagen school had scored a triumph in the mind of the physics community. Its concepts dominated the understanding of quantum phenomena in the generations that followed. However, the Copenhagen interpretation remained alien to the thinking of many physicists. It was difficult to accept that one should talk about the physical system only in reference to a concrete measurement, or that the basic object of such a successful formalism, the ψ-function, is partially subjective. The need was felt that quantum theory should refer directly to the structure of reality; its formalism should describe individual systems in themselves and without any reference to measurements. Hence, there were many attempts to find an objective interpretation, without altering the mathematical formalism of quantum theory. So far, none of these attempts has been entirely successful. Each one of them has its own problems, and for this reason none has been accepted by the majority of physicists. The concepts and ideas of the Copenhagen interpretation— in spite of their perceived inadequacies—still provide the main avenue for the comprehension of the picture of reality provided by quantum mechanics. We must note, however, that the word "mechanics" in the name of the new theory is entirely inappropriate: quantum theory completely dismantles the idea of the world as a machine, even though it fails to provide a clear picture about what we should substitute it with.[39]

After the thrill of the new discoveries faded away, most physicists started forgetting to be concerned about the deeper implications of quantum theory. They had been provided with a very powerful tool, and they wanted only a token interpretation that saved the appearances that the wave function referred to something real, without facing the ultimate consequences of such a stance. Hence, one would be able to use quantum theory as a powerful tool for the study of interesting physical systems, without subscribing to the revolutionary philosophical implications of the quantum principles. This desire dominated the textbooks and the teaching of quantum theory, especially in the postwar years.

Other attitudes existed, though, and they still exist. But only very few people took the stance that the description in terms of the wave function—and of q-numbers—was not the end of the story, and that there may exist a deeper level of quantum phenomena. We saw Einstein, Schrödinger, and de Broglie among them. The majority of physicists were not willing to follow this road; quantum theory was too successful to allow the fomenting of any serious dissent. Such opinions were marginalized, but they never became extinct, in spite of several mathematical theorems that were supposed to prove them impossible or incompatible with the successful predictions of quantum theory.

The basic tenet of the Copenhagen interpretation is that quantum mechanics should be used for the description of physical systems only in reference to a concrete and specific experimental situation. A single experiment never reveals everything there is to know about a physical system, but we will not necessarily learn more if we perform an additional one. Two different experiments—even if they are performed on identical physical systems—refer to two entirely different physical situations: the setup of a measurement that determines a particle's position is very different from that of a momentum measurement. For this reason, one cannot directly compare the information obtained from two experiments of a different type. In Bohr's words: "Evidence obtained under different experimental conditions cannot be comprehended within a single picture, but must be regarded as complementary in the sense that only the totality of the phenomena exhausts the possible information about the objects." This may be the best we can ever expect of a physical theory because physical reality cannot be defined without reference to our participation in it (as Bohr thought). Alternatively, it may mean that quantum mechanics is incomplete because it provides only a statistical description of physical systems, and that a better theory that will describe individual systems will be found someday (as Einstein believed in his later years). On the other hand, one may completely ignore this discussion, "shut up and calculate," for quantum theory provides an explanation and a description of a wealth of phenomena, and this can be achieved without engaging in any "metaphysical" discussions.

In any case, in experiments at the quantum level, we never obtain our data from the study of individual systems. Instead, we perform the experiment in a collection of such systems, which we have prepared in an identical manner. Our conclusions are, therefore, *statistical* by nature. The information about the preparation of the physical system prior to measurement is incorporated into a mathematical object that is known as the *state of the system*.[40] The state is a generalization of the ψ-function. Dirac realized that one could substitute the wave function with a more general mathematical object, which lives in an abstract mathematical space and inherits only one property from the waves: the ability to create interference as in the two-slit experiment. That object is the quantum state. It contains all statistical information that can be physically extracted from measurements on a physical system.

The second basic rule of quantum theory refers to the *observables*, namely, the physical quantities that can be observed. An observable corresponds to a q-number. As Schrödinger realized after he had proved the equivalence between his wave mechanics and Heisenberg's matrix mechanics, a q-number is equivalent to a transformation that takes one quantum mechanical state into another.

The most important predictions of quantum theory involve the probabilities for the possible outcomes in an experiment. We saw that this was the content of Born's interpretation of the ψ-function. This interpretation can be generalized to yield a rule that uniquely determines the probabilities of measurement outcomes, if the state of the physical system—hence the preparation of the experiment—is known. This rule suggests that certain physical variables take only discrete values. A special case of such variables is the energy of the electrons in an atom[41]—as one would expect on the basis of the spectroscopic data.

The fourth basic postulate of quantum theory describes the incorporation of the information we obtain from a measurement into the quantum state. This is necessary in order to describe the physical properties

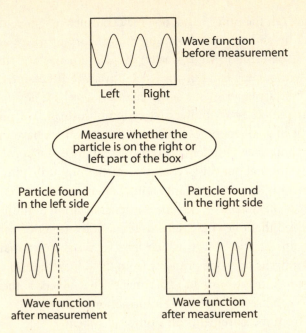

Left   Right

Figure 5.11 Change of the wave function as a result of measurement. We describe a simple measurement performed on one particle that is found in the box. The wavy pattern in the box represents the particle's state (ψ-function). The measurement under consideration determines whether the particle is found in the left or the right half of the box. In this measurement there are two possible outcomes, and there is no way to predict which of them will be obtained. The particle's wave function changes in a discontinuous manner after the measurement takes place, and the form of this change depends on the outcome of the measurement.

of the system after the measurement. Heisenberg explains: "The observation itself changes the ψ-function discontinuously; it selects from all possible events the actual one that took place. Since through our observation our knowledge of the system has changed discontinuously, its mathematical representation has also undergone the discontinuous change and we speak of a quantum jump."

Quantum theory provides a very specific rule about the discontinuous change in the ψ-function, according to the result that has been recorded in the measuring device (see fig. 5.11).

TABLE 5.1

The basic rules of quantum theory

The Structure of Quantum Mechanics	
1 The quantum state (generalization of the ψ-function)	• Contains information about the preparation of the system before the measurement • Incorporates in its definition the concept of quantum interference
2 Q-numbers (observables)	• Describe all possible physical magnitudes that can be measured in a physical system • Are non-commutative objects
3 Schrödinger's equation	The law of evolution for the quantum state, as long as no measurement is performed upon the system
4 Reduction postulate (quantum jump)	The nondeterministic change in the state that arises after a measurement takes place on the system
5 Born's rule	Provides the probabilities for the different outcomes in the measurement of any observable quantity
6 Combination law	Provides a way of describing a composite system in terms of the properties of its constituents

Finally, there exists a rule that governs the procedure of obtaining the description of a composite system if we know the description of its constituents. Thus one may reduce the study of complex systems to that of simple ones.

The above was a brief sketch of the basic rules of quantum theory. The application of these rules in various physical situations allows the description of a huge number of physical phenomena with astonishing success. In particular, they provide the reduction of the whole of chemistry to physics because the totality of the chemical phenomena can be described in terms of electrons moving around the nucleus and interacting through electromagnetic "forces."[42] Quantum theory also explained the physics of solids, of nuclei, and even of the stars. And this was achieved, in spite of the uncertainty, the misunderstandings, and the stark disagreements about its meaning and its implications for the structure of physical reality.

SO FAMILIAR AND YET
SO DIFFERENT

SPIN, QUANTUM PHASES,
AND QUANTUM STATISTICS

6.1 The Discovery of Spin

The days have outnumbered
my fingers and toes.
What can I count with now?
Saying this, the naive girl cries.
—Hla Stavhana, *The Gthsaptaat*

In the previous chapter I described the historical development of quantum theory and the building of the fundamental quantum concepts. This description was not complete. I omitted certain important discoveries, mainly because they did not relate directly to the arguments that led to the eventual formulation of the theory. However, two of these discoveries, namely, the identification of particle spin and the existence of quantum statistics, reveal crucial properties of the quantum world. In particular, they are invaluable for the description of fields in the quantum language. For this reason, I shall devote a whole chapter to the elaboration of their physical significance.

We encountered spin in section 4.4 as a degree of freedom of particles that correspond to irreducible components of the Poincaré symmetry. In quantum mechanics, spin appeared as a *discrete parameter,* which appeared in atomic theory in order to "save" the phenomena.

The huge success of the Rutherford-Bohr quantum model of the atom had motivated research toward an explanation of the chemical properties of the atoms in terms of the electrons' quantum orbits. The number of electrons in an atom was identified with the atomic number, which determined the chemical properties of the elements according to Mendeleev's periodic table. The combination of ideas from chemistry with experimental results from spectroscopy provided a picture about the constitution of atoms, even before the development of mature quantum theory in Heisenberg's and Schrödinger's work.

One should recall that in Bohr's model of the hydrogen atom, the atomic electron could move only in specific orbits, and any other motion was explicitly forbidden. This idea should be applicable to any other atom, the only difference being that one would have to deal with a larger number of electrons. The analysis of the spectroscopic data suggested that the allowed orbits are grouped in shells; each shell contains orbits with neighboring values of energy. Moreover, there should be an upper limit to the number of physically distinct orbits that belong to a given shell: the shell with lowest energy has only two possible orbits. The one with the immediately higher value of energy contains eight, the next eighteen, and the picture becomes rather more complex in shells of yet higher energy. The question then arises: how are the atomic electrons distributed in the allowed orbits? To find an answer, one had to make some guesses that were motivated by the properties of the periodic table of the elements; after all, the chemical properties of matter should be explainable in terms of the atoms' structure.

One may visualize the different allowed orbits in a given atom as boxes that may contain any number of electrons, or perhaps no electron at all. One then needs to find the correct rule that governs the distribution of electrons in these boxes. In principle, this is a matter of trial and error: one tries different rules and then selects the one that provides a structure for the atoms that is compatible with the periodic system of the elements. As it turned out, there was no need to devise any elaborate rules. The simplest one was also the most adequate: starting from the box with lowest energy, place one electron in every box with increasing value of energy until we run out of electrons. Hence, in an atom with sixteen electrons (sulfur), there is one electron in each of the sixteen orbits of lowest energy.

There is, however, a slight complication to the rule above. At most eight electrons can fill the outermost shell of orbits. This is important because the electrons of the outermost shell are the ones responsible for the chemical properties of the atoms. One simple way to see this is the following. Orbits with low energy are typically closer to the nucleus. Hence, the electrons in the outermost shell are farther away from the nucleus; in a sense, they are the atom's ambassadors. When two atoms come close to each other, they "negotiate" primarily through their outermost electrons, the ones with lower energy playing an insignificant role in this encounter. For this reason, atoms with the same number of electrons in the outermost shell follow similar patterns in their interactions with other atoms; they therefore manifest similar chemical behavior. The restriction to a maximum of eight electrons in the outermost shell essentially corresponds to the fundamental period of eight that appears in the periodic table of the elements.

The description above involved a mixture of guesswork with painstaking experimental and theoretical work. Many details were still missing: the ultimate justification had to await the formation of the mature quantum theory, which replaced the image of orbits with the more abstract notion of an *energy level,* that is, a bound solution to Schrödinger's equation with fixed value of energy (see sec. 5.5). Nonetheless, the broad outline of the description turned out to be largely correct, and it forms the backbone of the modern theory of atomic chemistry. However, at this early stage the newborn atomic theory had to contend with a serious problem: the description of the atom seemed to violate a fundamental principle of physics. I explained in sec. 4.3 that any physical system emits energy to its environment and thus decays into states of lowest energy. The lowest energy configuration for any atom is the one corresponding to all electrons occupying the orbits of the lowest energy shell. One would

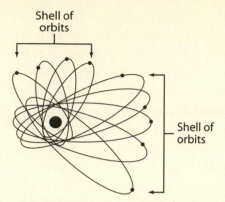

Shell of
orbits

Shell of
orbits

Figure 6.1 **Shells of orbits in an atom with many electrons.** The above is a schematic representation of the fact that the energy levels in atoms of many electrons tend to concentrate in shells, each containing "orbits" with values of energy close to each other. One should note that the language of orbits is not retained in the mature quantum theory, and one speaks about solutions of Schrödinger's equation with definite energy. The essential content of this description remains nonetheless unchanged.

then expect that the vast majority of the atoms would lie on this state. If this happened, it would be impossible to explain the structure of the periodic table of the elements. The latter could only be accounted for by the set of rules explained earlier, and these seemed to be incompatible with any known physical law.

If a scientist cannot account for a strange rule governing the behavior of a physical system, he or she bows to the inevitable and takes it for granted. If the strange rule appears again and again, it should probably be raised to the status of the postulate. If the scientist is perceptive, such a postulate may lead to uncovering a fundamental law of nature. This was the course followed by the Austrian physicist Wolfgang Pauli. He took the rule that accounted for the chemical behavior of the atoms and raised it to the status of a fundamental law. This was the celebrated "exclusion principle": *at most one electron may occupy each orbit.*[1]

The study of Bohr's model and of its generalizations for larger atoms had demonstrated that each orbit could be characterized by three integer numbers: the *quantum numbers* of the orbit. All physical quantities of the orbit—energy, angular momentum, etc—depend on these quantum numbers.* Each quantum number corresponds to the electron's motion in one of the three directions of space. Pauli's exclusion principle is then identical to the statement that there exists at most one electron for each triplet of quantum numbers. The idea was brilliant, but it did not fit the facts exactly right; there was a mismatch between the number of electrons of an atom and the theoretically possible orbits. This looked disastrous for the validity of the exclusion principle, at least initially.

Pauli realized that the mismatch was not arbitrary: the number of electrons in an atom is exactly *twice* the number of orbits that should be occupied if the model for the atom were to explain the periodic table adequately. He then suggested that there is one more quantum number that has *only two possible values*. However, there can be only three possible quantum numbers related to the motions in space. Therefore, the fourth quantum number does not refer to such motions, but to an *intrinsic* property of the electron. This is the number that was later called spin.

Pauli was uncertain whether his idea really made much sense. He admitted to Bohr that "the conception from which I start is certainly nonsense. However, I believe that what I am doing is no greater nonsense than the hitherto existing interpretation of the complex structure. My nonsense is conjugate to the hitherto customary one." A few days later, Pauli received Heisenberg's comments on a postcard (they had been friends ever since their university days). "Today I have read your new work and it is certain that I am the one, who rejoices most about it, because you push the swindle to an unimagined height, by introducing individual electrons with four degrees of freedom, thus going much beyond any idea of mine, you have insulted me for. . . ."[2]

Pauli's idea might have looked absurd at first sight, perhaps even to him, but it proved remarkably successful. Success came later, though: at

---

*One should not confuse these quantum numbers with the q-numbers that appear in quantum mechanics. The quantum numbers are ordinary numbers, the word "quantum" referring only to the fact that the physical quantities that depend on them take discrete values, unlike the continuous values they take in the classical theory.

that time, the notion of this fourth quantum number was puzzling. It had only been introduced to make the orbit bookkeeping work. What was its physical meaning? The most conservative explanation was that spin has something to do with the electron's self-rotation. If the atom was a solar system in miniature, why shouldn't the electrons rotate around themselves as planets do? If this is the case, the electron cannot be pointlike, for it is impossible for a point to rotate. The electron must therefore have a finite size. This implies that its electric charge will not be concentrated at a point but will occupy an extended region of space. This charge distribution will move because of the self-rotation. Hence, it is natural to assume that this motion will depend on the value of the spin quantum number. We know from electromagnetism that magnetic fields act on moving charges in a way that depends on the details of the charges' motion. One should therefore expect that electrons with different values of spin behave differently when placed in a magnetic field. In fact, an experiment performed a few years ago by Otto Stern and Walther Gerlach had captured this behavior.[3]

Stern and Gerlach had studied the motion of atoms of silver in an inhomogeneous magnetic field. A silver atom has a single electron in its outermost shell; the rest of its electrons are distributed symmetrically in the inner shells. Stern and Gerlach believed that this experiment would allow them to demonstrate an effect they called *quantization of space*, which they considered as an important consequence of Bohr's theory. They expected to see the beam of silver atoms splitting into two distinguishable components, and indeed this is what they saw in their experiment. In retrospect, we can say that their interpretation of the result (in fact the motivation for the experiment itself) was completely wrong. But this does not really matter. The important thing is that in that experiment the beam split in two pieces, and this effect was later interpreted as a direct experimental demonstration of spin's existence.

The correct interpretation of the Stern-Gerlach experiment was provided after Samuel Goudsmit and George E. Uhlenbeck, two young Dutch students at the University of Leiden, postulated the existence of spin.[4] Goudsmit and Uhlenbeck proposed that the electron has an additional degree of freedom that can only take two values, and that this

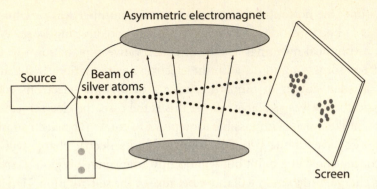

Figure 6.2 **A sketch of the experiment by Stern and Gerlach.** A beam of atoms of silver passing through an asymmetric electromagnet is split into two distinct beams.

degree of freedom essentially renders the electron into a dipole, namely, a small magnet similar to a compass needle. When the electron is found in a magnetic field, the "needle" can point either in a direction parallel or antiparallel to the field.

Goudsmit and Uhlenbeck were motivated by the necessity to explain some seeming inconsistencies in the spectroscopic data of the hydrogen atom—they did not address the Stern-Gerlach experiment directly. However, their results suggested a simple explanation for the latter. The electron in the outermost shell of the silver atom is responsible for the beam's splitting.[5] The force exerted by the magnetic field to the electron (and hence to the atom that contains the electron) depends on the needle's direction. Since there are only two possible such directions, there are two possible ways the magnetic field acts on the atom and hence two possible trajectories—exactly what is observed in the Stern-Gerlach experiment. (Note that if the electron's "needle" were free to point toward any direction, we would have a continuous distribution of trajectories and not two distinct ones.)

The new quantum number had many things in common with the one that Pauli had proposed. However, it was difficult to accept that the magnetic behavior of the electron was due to its self-rotation. Pauli, for

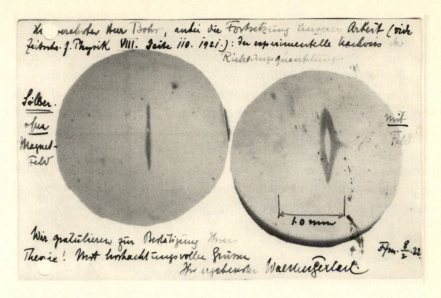

Figure 6.3 Beam splitting on a postcard. A postcard sent by Stern to Bohr with a photograph of the beam splitting. The left photo shows the trace left by the beam on the screen in absence of a magnetic field, and the right photo shows the same trace after the magnetic field had been switched on. Courtesy of the Niels Bohr Archive, Copenhagen.

one, rejected this supposition emphatically. If one assumed that the electron was a small rotating ball, the points on its surface should rotate with a speed greater than that of light. This was unacceptable. Still, the introduction of spin provided a very satisfactory solution to many problems related to the atomic structure and to spectroscopy.

Hence, the introduction of the spin brought about a paradox. Spin is in every aspect similar to a self-rotation. However, treating the electron like an extended object led to a conflict with relativity. The electron behaved as a pointlike particle, at least as far as the up-to-date experiments were concerned. But pointlike particles cannot exhibit any kind of self-rotation.

The contradiction above was resolved, not by any new physical idea but by a development in mathematics. Simply put, spin is a physical magnitude that has all properties of a self-rotation but also makes sense for pointlike objects. We saw this in section 4.4 in our discussion of the Poincaré symmetry of spacetime. Spin is a degree of freedom for particles that had been hitherto unsuspected because it is not amenable to mechanical representation. We can only visualize the self-rotation of extended objects. Nonetheless, our lack of precise visual representation does not mean that spin does not exist. Its existence is a consequence of the Poincaré symmetry of spacetime. As such, spin is fundamentally a geometric object: it is bequeathed to particles by the structure of spacetime. It is described clearly and unambiguously in the language of mathematics, and it is through this language that we can understand and analyze its properties.

6.2 Quantum Phases

The shadows of things are greater than themselves; and the more exaggerated the shadow, the more unlike the substance.
—H. Melville, *Mardi*

We saw in section 4.4 that the existence of spin is a consequence of the symmetries of spacetime. These symmetries are continuous because spacetime is continuous; one may move a body in space or in time continuously, and the same is true for rotations. Then why does spin appear as a *discrete* quantity in quantum phenomena? Position and momentum are also degrees of freedom that characterize particles, but they are continuous even in quantum theory. What is special about spin?

To provide an answer to this question (we do this in sec. 6.3), we must first understand the relation between the physical quantities of classical physics and their counterparts in quantum mechanics.

The first step of any theorist attempting the quantum description of a physical system is to identify the q-numbers that represent the system's basic physical quantities. To this end, the theorist should know what these physical quantities are beforehand. The only way to do so is by using *concepts from classical physics*. For example, to write the q-number that corresponds to position or momentum, one first needs to understand the notion (and the properties) of position and momentum in classical mechanics. Again, to describe spacetime symmetries in quantum mechanics, it is necessary to first define them at the level of the classical theory. In other words, unless we employ the concepts and symmetries of classical physics, we have no idea what the quantum mechanical physical quantities will be. Q-numbers acquire physical meaning only when they are associated with a classical quantity; otherwise they are mere mathematical symbols with no physical content whatsoever.

Niels Bohr postulated a duality between the classical and quantum world as another form of his principle of complementarity. The world is described by classical physics at the macroscopic level. This description is necessary for the design and execution of any experiment, since a measuring device is a macroscopic object whose behavior is largely described by the laws of classical mechanics. Quantum mechanics is not immediately relevant at this level; after all, an experimentalist measures real numbers, not q-numbers. On the other hand, the classical theory is inapplicable in the micro world, which is solely described by quantum mechanics. Any experiment probing the micro world involves the association of quantum objects with classical ones; this is a fundamental fact of nature, in the absence of which no physics is possible. In Heisenberg's words, "Any statement about what actually has happened [in an experiment] is a statement in terms of the classical concepts. . . . The Copenhagen interpretation regards things and processes which are describable in terms of classical concepts, i.e. the actual, as the foundation of any physical interpretation."[6] Another way to state this is by saying that when we talk about past outcomes of experiments (the actual), we must use classical physics, but when we try to predict what may be realized in the future (the potential), we have to use quantum physics.

The coexistence of classical and quantum concepts makes perfect sense, if one accepts that quantum mechanics can be applied only at the level of concrete measurement outcomes. If, in contrast, one thinks of quantum mechanics as describing physical phenomena by themselves, without any reference to measurement, the splitting between the classical (macroscopic) and the quantum (microscopic) worlds is unacceptable, in fact schizophrenic. How can two different (and even conflicting) fundamental theories for the world coexist? All physical objects consist of atoms, which are fully quantum objects. Shouldn't quantum theory also be valid in the macroscopic world?

There have been many attempts to explain the origin of the classical macroscopic world out of the fundamental quantum laws, and hence to get rid of the necessity to use two different languages for the description of the world. However, it seems that in every such attempt there is a gap that either makes the explanation dependent upon questionable assumptions or renders it logically inconsistent.[7] This is indeed one of the most fundamental unresolved problems in our understanding of quantum mechanics. We shall not go into this issue in any detail, because it would take us in directions very different from the ones discussed in this book. In the absence of a better alternative, we will restrict our considerations to the Copenhagen interpretation, accepting the coexistence of the classical and the quantum descriptions. This attitude is fully justified by the general practice of theoretical physicists. Whenever they study a quantum mechanical system, they first write down the corresponding classical physical quantities, and then they attempt to construct q-numbers out of them. This mathematical procedure is referred to as the *quantization* of the classical system.

Physical quantities in classical mechanics (position, momentum, spin, etc.) evolve under deterministic laws of motion. The initial conditions of a physical system uniquely specify the evolution of all relevant physical quantities. If we repeat an experiment many times, each time following the same preparation procedure, we will always obtain the same result. There will be at most an uncertainty due to the possible inaccuracy of reproducing exactly the same initial condition.

Experiments in quantum theory are different on two accounts:

- First, quantum mechanics is not a deterministic theory, so no matter how much we try to reproduce the same initial conditions, we usually obtain different results in different runs of the experiment.
- Second, the behavior of particle beams is similar to that of waves, and interference effects appear often.

If the only difference between classical and quantum theory were the first one, things would be very easy: one could think of an explanation without much knowledge of modern physics. We could simply derive quantum theory from classical mechanics by adding a force component *with random behavior* to the deterministic laws of motion. This was essentially what Epicurus had done 2,300 years ago, when he desired to get rid of the strict necessity of Democritus' atomic physics. However, the second condition implies that things are much less straightforward. The introduction of a random component will *not* produce a wave behavior in the beams of particles. We need something else—as Heisenberg had understood—something that works at the level of the definition of the physical quantities.

The additional quantum structure is a new object that is introduced in the description of every physical system. It looks like a new degree of freedom that supplements the ones known from classical physics. But it is not really a degree of freedom because it cannot be measured in individual systems. It appears only in beams—in collections of individual quantum systems. This object is the *quantum phase*.

The quantum phase is essentially a register, which keeps track of a physical system's history. Its mathematical description suggests that it is analogous to a rod that rotates around an axis passing through its midpoint. The periodicity involved in the rod's cyclical motion is eventually related to the wave behavior of quantum systems. The rod analogy invokes a classical, even mechanical, description. However, the quantum phase is a genuinely quantum object. Unlike any classical physical quantity, it cannot be observed—directly or indirectly—in an *individual* quantum system, but it makes its presence felt when we consider the statistical behavior of a large collection of individual quantum systems.

The quantum phase appears when we compare quantum systems that have had different histories in their past. One may recall, for instance, the two-slit experiment of figure 5.5. The interference pattern appears only when both slits are open. In that case, the electrons that arrive on the screen have *different past histories* because they may have passed through either one of the slits. This difference between their past histories implies that their registers (i.e., their quantum phases)

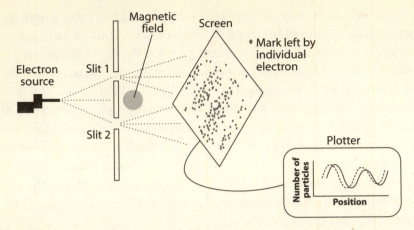

Figure 6.4 **The Aharonov-Bohm experiment.** The experimental setup is similar to the two-slit experiment (fig. 5.5). In this case, however, we place a solenoid wire that generates a strong magnetic field just behind the slits. The introduction of the magnetic field changes the interference pattern on the screen. The plot of number of electrons vs. position on the screen shifts by a small amount (the dotted line corresponds to the zero magnetic field case). The shift in the pattern is known as the Bohm-Aharonov phase. This experiment demonstrates, among other things, that we may change the interference pattern by a controlled change of parameters characterizing the experimental setup. Indeed, changing the strength of the magnetic field, we can see a periodic change in the shift of the interference pattern.

point at different directions. Interference is then a manifestation of the difference in quantum phase between the two alternative histories for the particle's motion. This fact was made more prominent by a famous experiment that was proposed by David Bohm and Yakir Aharonov in the 1950s.[8] If in the two-slit experiment we somehow intervene with the two electron beams after they cross the slits, we will observe a shift in the interference pattern. If we repeat this experiment many times, each time with a different intervention, we can study the physical properties of this shift. It turns out that it shares all mathematical properties of the quantum phase, constituting in effect one of its experimental manifestations.

The realization that the quantum phase is a register of the past history of the system was brought forward by the work of Michael Berry from the University of Bristol in the 1980s, in the latest of the great discoveries about the fundamental structure of quantum theory.[9] Berry's work led to the understanding that the quantum phase is intrinsically a *kinematical* object, which exists irrespective of the laws of motion that describe a physical system. Further work showed that this phase is an integral part of quantum theory, deeply rooted in its mathematical formalism. (It is related to the fact that the wave function takes complex numbers as values—see sec. 5.5.)

It is important to stress (once more) that the quantum phase does *not* appear in individual quantum systems or in collections of quantum systems (like beams of electrons) that had the same history. It appears only when we cause two beams of identical particles with different histories to interfere. It can then be measured by a study of the interference pattern (as in fig. 6.4). However, the actual value of the phase is not itself measured. Only the *difference* in phase between the two separate evolutions of the quantum systems is a physically measurable quantity, and only upon this does the interference pattern depend.

The introduction of the quantum phase allows the direct comparison between the classical and the quantum descriptions of a physical system. This, however, does not imply that quantum theory can be reduced to classical mechanics plus the quantum phase. Quantum mechanics involves many more physical principles: the introduction of q-numbers, the uncertainty relations, Born's probability interpretation, the complementarity principle, and so on. The quantum phase is primarily a *mathematical* structure: it is necessary for the *construction* of the quantum description of a given physical system. It allows the implementation of the principles of quantum theory, but it is not a substitute for them. It does, however, contain essential information about the basic features of specific quantum mechanical systems, and it provides a powerful tool for the description of important physical concepts. I shall elaborate on this point in the remainder of this chapter.

6.3 Spin is Discrete!

> Mysterious, even in broad daylight,
> Nature won't let her veil be raised:
> What your spirit can't bring to sight,
> Won't by screws and levers be displayed.
> —Johann Wolfgang von Goethe, *Faust*

I explained in the previous section that the quantum phase is the mathematical structure responsible for the quintessential quantum behavior of interference. If we find a way to encode its function into the structure of a *classical system,* we go a long way toward constructing the corresponding quantum theory. This is essentially a mathematical problem: can we fit the cyclical structure of the quantum phase into the classical description? The answer is sometimes negative. There are classical systems in which the quantum phase cannot be introduced. This fact provides an answer to the question asked earlier, namely, why is spin discrete in quantum theory? We shall see that the introduction of the quantum phase in the description of physical particles is possible only for specific discrete values of the spin degree of freedom. The actual demonstration of this result involves rather complex mathematics; however, we shall demonstrate it in a visually intuitive way using a mechanical analogue.

We should recall from section 4.4 that a particle with spin can be visualized as an arrow, which moves in space while rotating around its origin. The arrow's length is proportional to the value of the particle's spin. For simplicity we will ignore the particle's motion in space. Hence we will assume that the arrow's origin is fixed, so that the only allowed motion is the arrow's rotation. In that case, the tip of the arrow moves on the surface of a (fictitious) sphere.

The description above is phrased in terms of classical physics. To obtain the quantum description, we have to add the register that corresponds to

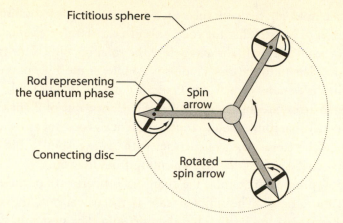

Figure 6.5 **Spin, rods, and connection.** The spin arrow moves on the surface of an imaginary sphere. The quantum phase corresponds to a rod, which rotates together with the arrow, as it is embedded on a disk connected to the arrow. The connection is such that it allows any point of the disk to traverse a distance equal to that of the arrow's tip in its rotation. The physical correspondence is obtained by assuming that the length of the arrow equals the spin's value and the radius of the connecting disk equals Planck's constant.

the quantum phase. This is represented by a rod, which is glued on top of one of the disk's diameters. The disk's radius is assumed to equal Planck's constant, the fundamental unit of the quantum world. The disk is mechanically connected at the arrow's tip, so that the arrow's motion is fully transmitted to the disk's rotation: the distance covered by the tip of the arrow is identical to the distance covered by any point of the disk's circumference; the rod follows the disk's rotation.

The rod represents the quantum phase, while the disk is the connection through which the arrow's rotation transforms into rotation for the rod. If this contraption is to provide an accurate representation of a physical particle with spin, it is necessary that it possesses the physical symmetries of spin. This implies in particular that a rotation of the arrow by 360 degrees should not effect any physical change in the total system—*including the rod*. But this is not always possible because of the difference in length between the spin arrow and the disk's radius. One may convince oneself that the rod's axis after a 360-degree rotation of the arrow coincides with the axis before the rotation, only if the ratio of the arrow's length to the disk's radius is an integer multiple of one half. Since the disk's radius equals the Planck's constant, the possible values of the spin arrow are integer multiples of half the value of Planck's constant. It is a usual convention to consider Planck's constant as the unit of spin—one therefore speaks of particles with spin ½, spin 1, spin 2, and so on. We shall conform to this convention from now on.

We now proceed to examine the quantum properties of particles with spin in more detail. The simplest case is the trivial one, the arrow's length being zero; the particle has then no spin. The next spin value is one half times the Planck constant; the electron is such a particle. In this case, the rotation of the arrow by 360 degrees causes the rod to rotate 180 degrees. If one of the rod's endpoints points initially at twelve o'clock, it will point after the rotation at six o'clock. However, we do not distinguish between these two cases: unlike a clock's hand, the rod does not carry an arrow to distinguish between these two directions. The quantum phase is defined by the direction of the rod's axis, which is exactly the same at both six and twelve o'clock.

We need a further rotation of the spin arrow by 360 degrees to bring the rod exactly to its initial position (going from twelve o'clock noon to twelve o'clock midnight). In total the arrow will have rotated 720 degrees. Similar behavior characterizes all particles with spin that takes half-integer values (times Planck's constant) like ½, ³⁄₂, and so on. On the other hand, particles with integer spin behave "normally": a rotation of the arrow by 360 degrees brings the rod to its initial direction.

There is a different example that may help one to visualize the above property of spin ½ particles. We place a glass of water on top of our palm and rotate the arm without spilling the water. After a rotation of 360 degrees the arm is twisted. We need an additional rotation of 360 degrees (moving the arm behind the back) in order to untwist the arm. This behavior is due to the particular set of connections between the bones on the human arm, much like the behavior of the spin particles in

Figure 6.6 **Feynman's demonstration of half-integer spin.** The plastic glass is rotated 360 degrees so that the mark appears forward again, but one such rotation brings the hand to a different configuration. Further rotation by 360 degrees brings the hand to its initial configuration. Reprinted from R. P. Feynman and S. Weinberg, *Elementary Particles and the Laws of Physics: The 1986 Dirac Memorial Lectures* (Cambridge: Cambridge University Press, 1987). Reprinted with the permission of Cambridge University Press.

our mechanical model is a consequence of the properties of the disk that connects the rod to the arrow.

We should recall our discussion in section 4.4 about particles being irreducible systems associated with the Poincaré symmetry. There, I pointed out that every type of particle is characterized by two intrinsic properties: its rest mass and its spin. In the Stern-Gerlach experiment, electrons alone are involved, which have *fixed* values of spin equal to ½. But the results of the experiment were explained in terms of a variable discrete quantity that takes *two* values. What is this quantity? Why did we also call it spin?

To answer this question we recall (sec. 6.1) that the Stern-Gerlach experiment involves a magnetic field, which points to a specific direction in space. The interaction of the particle's spin with the magnetic field depends *not on the length of the spin arrow*, but on the length of *the arrow's projection* along the direction of the magnetic field. In studying this interaction, we should repeat all our previous arguments about spin being discrete, but with reference to *the projection* of the spin arrow. The conclusions are not dramatically different: again the projected arrow should be an integer multiple of half the value of Planck's constant. The only difference is that one also has to include negative values for the length of the projected arrow. (The negative values correspond to the arrow pointing in opposite direction to the magnetic field.) Moreover, we must make sure that the behavior of the projected arrow is compatible with that of the full arrow: when we rotate the full arrow by 360 degrees, the projected arrow should follow accordingly. This can be shown to imply that the spin arrow and its projection have a certain relation: their lengths both have to be multiples of half-integers or both multiples of integers. Eventually, we arrive at the following picture: if the arrow has length $N/2$ (times the Planck constant), where N is an integer, there exist $N + 1$ different values of the spin projection (see fig. 6.7). It is conventional to refer to the different possible values for the spin projection as quantum *spin degrees of freedom*. For electrons the number N equals one (spin ½), hence the spin degrees of freedom are two, exactly what was inferred from the Stern-Gerlach experiment.

The description of spin provided above holds for massive particles. To describe particles with zero rest mass, we should recall that the only possible direction of the spin arrow in that case is either parallel or antiparallel to the particle's direction of motion (see sec. 4.4). The length of the spin arrow is constrained to take discrete values as in the massive case: namely, $N/2$ times the Planck's constant, with N an integer number. Since the spin arrow can only be parallel or antiparallel to the direction of the particle's motion, the spin projection can only take one of two values. It follows that massless particles (like photons) have have only two spin degrees of freedom. When the projected spin is parallel to the direction of motion, we say that the particle possesses positive *helicity*, and when it is antiparallel, we talk about negative

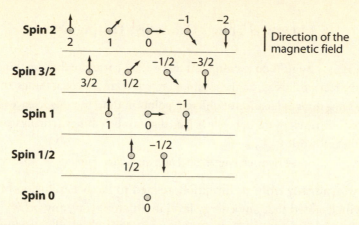

Figure 6.7 **Discrete values for the spin projections.** The interaction of a spinning particle with the magnetic field depends on the value of spin. The result of this interaction is not unique; rather, there are as many different outcomes as possible directions of the spin projection on the axis defined by the magnetic field. For a particle with spin $N/2$, there are $N + 1$ possible directions. Hence a particle with spin ½ has two possible projections (up and down), one with spin 1 has three projections, etc. The particles with spin zero carry no arrow and hence may exist in only one state.

helicity.[10] It is important to emphasize that helicity is an intrinsic feature of massless particles. It cannot change by any external action on the particle.

To summarize, the discreteness of spin appears at two levels:

• First, spin is an intrinsic property of particles, which takes *the same* value in all particles of the same type. The possible values of spin (for different types of particles) are discrete and correspond to integer multiples of half the Planck constant.

• Second, massive particles of the same type may have different discrete values for the projection of the spin in a specific direction. If the spin's value is $N/2$, then there are $N + 1$ possible values for the spin projection in any direction.

6.4 Identical Things Cannot Be Distinguished

Logic too also rests on assumptions that do not correspond to anything in the real world, e.g., on the assumption that there are equal things, that the same thing is identical at different points in time; but this science arose as a result of the opposite belief (that no such things actually exist in the real world).
—Friedrich Nietzsche, *Human, All too Human*

Spin was, already from its inception, related to Pauli's exclusion principle, which stated that any energy level in an atom (i.e., any orbit*) may be filled by at most one electron at a time. This principle accounts for the fact that electrons do not all lie in the orbit of minimum energy. In its absence, the atoms would be much smaller, there would be no external atomic shell to generate the phenomena of chemistry, and matter would be much denser because atoms would be packed much closer together. The exclusion principle is fundamental for the explanation of the macroscopic properties of matter.

Nonetheless, Pauli's principle is not universally valid. It works for electrons, but if one applies it to photons, the results come out completely wrong. The assumption that any number of photons may possess the same quantum numbers works perfectly. One is then entitled to ask, why is the behavior of electrons so different from that of photons?

To answer this question, we must study in more detail the behavior of physical systems that consist of many particles. For this purpose, we consider two particles with identical mass and spin. How do we distinguish between them? In classical mechanics—or in any deterministic theory—different particles are distinguished through their history. If we identify a particle at a particular moment of time, we can follow its time

---

*I mentioned earlier that after the development of mature quantum theory, the term "electron orbits" was gradually superseded by that of "electron energy levels." Both terms refer to the same thing, namely, the possible configurations of a system with a fixed value of energy. However, the image of sharply defined orbits does not fit within the general framework of quantum theory; for example, it does not sit well with the uncertainty principle.

evolution and identify it again at a later moment of time. In a deterministic theory, the identity of a particle is never lost because its motion can be monitored with arbitrarily high accuracy.

However, nature—as seen through quantum mechanics—is not deterministic. Even if we try our utmost to follow a quantum particle's path, we come across the uncertainty relation. There is no way to distinguish between two particles of the same characteristics that have come close to each other. Any (mental) tag that we associate with a particle by virtue of an earlier measurement is completely and irretrievably lost. Hence, the basic features of quantum theory make the distinction of one particle from the other impossible.[11]

We conclude that it is impossible to distinguish between quantum particles that possess the same internal properties. So why should we care? The answer is that the distinguishability of particles affects every calculation that involves probabilities. To see this, we consider a simple example, which involves two boxes A and B and two balls (see fig. 6.8). If we can distinguish between the two balls, we can also give them names. Hence, we call the first ball 1 and the second ball 2. There exist four possible ways to distribute the balls in the boxes: both balls in box A, both balls in box B, ball 1 in box A and ball 2 in box B, ball 2 in box A and ball 1 in box B. If, on the other hand, the balls cannot be distinguished, there exist only three distinct alternatives: both balls in box A, both balls in ball B, and one ball in each box. Hence, the number of physically distinct alternatives is smaller when particles cannot be distinguished. This fact affects any calculation of statistical mechanics, which relies on the counting of possible different states available to a physical system of many particles. In other words, the behavior of large numbers of identical particles differs from that of large numbers of distinguishable particles.

The nondistinguishability of particles in quantum theory has further consequences. The quantum phase is a register of the histories of any physical system, also including systems of many particles. On the other hand, each particle has one such register and its associated connection (corresponding to the rod and the disk of fig. 6.5, respectively). The important point is that, whenever we consider many particles as forming

I. The four alternatives for placing two distinguishable balls in two boxes

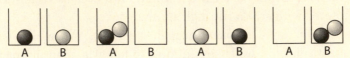

II. The three alternatives for placing two indistinguishable balls in two boxes

III. Only one alternative for placing two indistinguishable balls in two boxes, if the connection is of the Fermi type

Figure 6.8 **The different statistics.** This figure shows the possible rules for placing two balls in two boxes. If the balls can be distinguished (case I) there are four alternatives, if the balls are indistinguishable (case II) there are three alternatives, and if they are indistinguishable but are connected by the Fermi connection (case III) there is only one alternative.

a *single* system, we must have one *single* register. We must therefore find a way to connect the discs of individual particles to a *single rod*, which describes the quantum phase of the *whole system*. We can visualize a system of belts transmitting the motion of each disk into the motion of a master disk that supports the register, as is shown in figure 6.9. The belts are supposed to transmit the motion of either disc to the master rod faithfully. Hence, if the disc of the first spin arrow rotates by 60 degrees and the disc of the second arrow by 40 degrees, the rotation in the master rod will equal 60 + 40 = 100 degrees.

If the particles are identical and cannot be distinguished, an exchange in the configuration of any two particles should not affect the combined system. (If the first arrow points toward two o'clock and the second

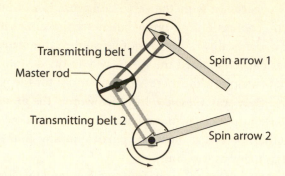

Figure 6.9 **Connecting two quantum systems.** Two belts transmit the motion of both spin arrows into the motion of a single master rod, which represents the quantum phase for the combined system of two particles with spin.

toward six o'clock, an exchange of configuration will bring the first arrow pointing toward six o'clock and the second toward two o'clock.) The connection to the master rod must be such that the master rod does not change its direction when an exchange such as the one above takes place. Like the rod for the single spin, the master rod cannot distinguish between twelve o' clock and six o'clock—its direction is unaffected by a rotation of 180 degrees. It follows that there exist two possible ways that we may construct the connection of the particles' discs to the master rod: either the master rod remains unchanged when we exchange the two particles, or it rotates by 180 degrees. I shall refer to the first connection as being of the Bose type and to the second as being of the Fermi type, and I urge the reader to wait a little before I explain the origin of these names.

To understand the physical significance of the two connections, we consider the special case that the two particles have exactly the same configuration, namely, the case that the two spin arrows point toward the same direction. Since the particles are identical, we do not need to move any of them in order to implement the exchange. The exchange is trivial, and this implies that the rod's axis remains unaffected. Hence, a rotation by 180 degrees is not possible. This implies that the possibility of having two particles with the same configuration is incompatible with the Fermi connection. We immediately see the similarity to Pauli's principle: two identical particles cannot lie on the same state. Hence, we conclude that particles joined the Fermi way (named *fermions*) satisfy the exclusion principle, while there is no constraint for the particles that are joined the Bose way (*bosons*). In other words, the existence of the quantum phase implies that many-particle systems follow one of two behaviors: either there is at most one particle in any configuration, or there can be as many as we want.

Let us now recall the simple system of two balls and two boxes we considered earlier. In systems described by the Bose connection, there exist three possibilities (see fig. 6.8). If the system involves the Fermi connection, the two balls cannot be in the same box. In this case, there exists only one possible alternative. It follows that the exclusion principle reduces significantly the number of possible alternatives in collections of particles. The two different connections, therefore, refer to two different statistical descriptions for many-particle systems. These two descriptions were first identified by Satyendra Nath Bose and Enrico Fermi, respectively, whose names were given to the two types of connection.[12]

We saw earlier that electrons satisfy the Fermi statistics, while photons satisfy the Bose statistics. We next recall that electrons have spin ½ and note that photons have spin 1. This is not an accident. It turns out that all particles with half integer spin have to be described by Fermi statistics, while the ones with whole integer spin are described by Bose statistics. This statement carries the name of *spin-statistics theorem,* and it constitutes one of the most fundamental features of quantum theory. It relates two seemingly unrelated things: the description of many-particle systems and the particle property of spin.

The spin-statistics theorem was very difficult to prove. The first proof (by Pauli[13]) was obtained much later, and it involved concepts of quantum field theory; the description in terms of particles did not suffice. Nonetheless, the theorem can be made plausible by the following remark. Consider the system described in figure 6.5. A rotation of the spin arrow by 360 degrees rotates the rod by 180 degrees in a half-integer spin particle, and by 360 degrees in an integer-spin particle. By

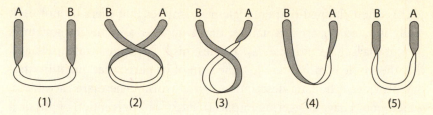

Figure 6.10 Feynman's twisting belts. This is a mechanical example that serves to make plausible the relation between spin and statistics. If we exchange the ends of the belt in the sequence (1) to (5), the belt will necessarily be twisted. However, this twist can be undone by a rotation of one end by 360 degrees. Hence, a rotation of one end by 360 degrees is equivalent to an exchange, exactly as the spin-statistics relation suggests. The drawing is based on an example by R. P. Feynman in Feynman and Weinberg, *Elementary Particles and the Laws of Physics* (Cambridge: Cambridge University Press).

definition, the exchange of two particles rotates the rod by 360 degrees if they are connected the Bose way, and by 180 degrees if they are connected the Fermi way. If the spin-statistics relation is satisfied, the effect of an exchange on the quantum phase can be compensated by a rotation of 360 degrees of a single particle's spin.

Richard Feynman has provided a very elegant visualization of the spin-statistics relation (see fig. 6.10). We assume that the two identical particles are connected through an elastic belt. If the particle's positions are exchanged *without rotating the belt,* the belt will necessarily exhibit a twist. This twist can be undone by rotating one of the particles by 360 degrees.

Of course, particles are not really connected with belts; the belt is yet another mechanical representation of the fact that when *particles are combined to form a single composite system, they are intertwined through their quantum phases.* The total system seems to have certain novel properties in comparison to the sum of its constituents. This intertwining that arises out of the quantum phases carries the name of *quantum entanglement.* This is a generic quantum phenomenon; it is also manifested even if the particles are not identical.

It is often claimed that entanglement implies that there do not exist really isolated systems in nature; that whenever a quantum system is part of a larger system, it always carries this knowledge inside itself, and that the whole is more than just the sum of its parts; that quantum theory provides a holistic description of the world. These are nice ideas, which would have staggering implications for our perception of physical reality if they could be shown to be true. The problem is that these ideas employ a specific interpretation of quantum theory, namely, that the ψ-function—from which the quantum phase is derived—has an objective meaning and that it describes physical systems "by themselves," without reference to measurement. This is not the only possible interpretation of the quantum mechanical formalism, and it is not entirely trouble—free. However, getting into this discussion is a temptation we should avoid, if this book is to keep its coherence. We should remember, though, that in the Copenhagen interpretation, the ψ-function is a semisubjective object (see sec. 5.7), which we employ to keep track of our probability calculations. As such, the ψ-function simply refers to experiments that are carried out in ensembles of physical systems. Consequently, the quantum phase—however exotic and remarkable an object it may seem—serves primarily the same function of statistical bookkeeping. It is not necessary that its behavior implies a fundamental wholeness of nature, or conversely, that the unity of the world finds its physical expression in the basic rules of quantum theory.

7

FORGING THE PERFECT TOOL

THE DEVELOPMENT OF QUANTUM
FIELD THEORY

7.1 Quantum Light

[I]t would depend on the taste of the observer which he wishes to regard
as real, the particle or the guiding field. There is certainly no criterion
for reality if one does not want to say: the real is only the complex of
sense impressions, all the rest are only pictures.
                        —E. Schrödinger, letter to Willy Wien

The synthesis that had been achieved by Bohr, Heisenberg,
Schrödinger, and Born by 1927 was hardly the end of the road in
the development of quantum theory. It did provide a broad framework
for the new physics, but there was still a long way to go before the theory
would account for the wealth of phenomena that were being uncovered
in the atomic world.

One of the most important issues to be addressed was the quantum
mechanical treatment of the electromagnetic field. We should recall (secs.
5.1–5.3) that electromagnetic phenomena had played a pivotal role in
quantum theory from the very beginning: the major breakthroughs had
come from the study of the thermal properties of the electromagnetic

field (in black body radiation) and of its interaction with the atoms (in the photoelectric effect and Bohr's model). It was essential that electromagnetism find its place in the quantum world order.

According to the rules of quantum theory, a quantum description of the electromagnetic field should involve the construction of q-numbers that correspond to the basic physical quantities: the electric field, the magnetic field, and all conserved quantities corresponding to the Poincaré symmetry (energy, angular momentum, etc). This construction is simpler when no electric charges are present because in this case Maxwell's equations only predict the existence of electromagnetic waves that propagate at the speed of light (see sec. 2.8). The quantization of these waves can then be addressed with relative ease. It yields impressive results.[1] The resulting quantum states allow the existence of q-numbers that describe the electric and magnetic fields. But the same states may be interpreted as referring to a system with an arbitrary number of fundamental particles. These particles are massless, have spin equal to 1, and can have either positive or negative helicity. They are nothing but the photons. Hence, *the duality between particles and waves arises as a consequence of the quantization of the electromagnetic field.*

To understand the particle-field duality better, we should first see how physical systems with an arbitrary number of particles are described in quantum theory. The first step in such a description is to identify the physical properties of a single particle: the possible states and the q-numbers that represent the particle's basic physical quantities. Once this is achieved, we can describe the possible states of systems with N particles. For this purpose, we need to know whether the particles will be combined according to the Fermi or the Bose connection. Photons carry spin equal to 1, so according to the spin-statistics theorem (sec. 6.4), the construction proceeds with Bose connections for the quantum phases. If we now glue together the mathematical spaces corresponding to these N-particle systems for all values of N—including the case of no particles ($N = 0$)—we construct a mathematical space that hosts states, which describe an arbitrary number of particles. This space is named after Russian mathematician Vladimir Fock, who first constructed it. It

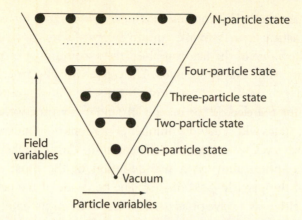

Figure 7.1 **The Fock construction.** A single black ball represents the space of possible configurations for the single particle. One then constructs the space of two, three, etc. particle configurations up to infinity. In quantum theory, identical particles cannot be distinguished, so one has to employ either the Bose or the Fermi connection for the description of many particles. (In our sketch, the lines joining the balls represent the connections.) The Fock space is then the total space containing the zero, one-, two-, three-particle states, up to infinity. A tiny ball represents the zero-particle state, the vacuum. To obtain a physical observable that corresponds to properties of particles, one has to traverse this diagram horizontally, while for an observable that is associated with the field properties, one has to "move" in the vertical direction.

turns out that the Fock space for photons 'carries' q-numbers, which can be interpreted *naturally* in terms of the electric and magnetic fields.

The relation between fields and particles has certain complications. Suppose we can design an experiment in which we prepare the electromagnetic field in a state characterized by specific values of the electric (or magnetic) field variables, in the sense that any measurement of the field variables will always yield the same result. We next attempt to measure the number of photons "contained" in this quantum state. In every run of the experiment we will obtain different values for this number. The differences will not be small (say, measuring 40 photons in half the runs and 41 in the others) but extremely large. We may measure 40 photons in one run and 400 in another.* This conclusion is inevitable, and it is forced on us by the mathematical formalism of quantum theory: if a state is characterized by a definite value of the field variables, then the number of photons it contains cannot be exactly specified. The converse is also true. If we prepare a state with a definite number of photons, we will obtain very different values for the electric (or magnetic) field every time we attempt to measure it. States with a definite number of photons do not describe electromagnetic waves, as we understand them in the classical theory.

The description above is a manifestation of the most important attribute of the particle–field duality. The behavior of the electromagnetic field either as a wave or as particles depends on its state. However, the state refers to the specific procedure we followed in order to prepare the physical system for measurements. It is, therefore, this preparation of an experiment that determines whether we observe the particle or the field aspect of the electromagnetic field. Hence, according to the Copenhagen interpretation, it is our mode of interacting with the quantum world that determines the way this world appears to us.

One, however, may choose to disbelieve the particle-field duality. "What we really have," one would say, "are the fields; these are the fundamental ingredients of nature. The electromagnetic field is characterized by the Poincaré symmetry; it can therefore be *conceptually* analyzed into irreducible components of this symmetry. These irreducible components are the photons. But photons do not have separate physical existence. The field simply transfers its energy in space through some discrete irreducible processes, which we choose to name particles. These 'particles' are really secondary objects, derived concepts, of no fundamental significance. The truly physical object is the field."

One may choose to disbelieve in a different way: "There are only particles in this world. Fields are just mathematical idealizations. They are employed in order to describe the systems of many particles in accordance with the laws of quantum mechanics and relativity. An electromagnetic wave is a collection of photons, and its wave behavior is a

*These numbers are given only for the purpose of demonstration. In realistic experiments, such states may involve a number of photons that is immense; it may easily be on the order of 1,000,000,000,000,000,000.

consequence of the quantum laws of motion—essentially due to the presence of the quantum phase. The reason the number of photons is not definite in some states is that the photons interact with the particles that form our measurement device, and in these interactions photons are either created or absorbed. But at the end of the day, the world consists only of particles, which interact according to the rules of quantum theory."

There are then three distinct attitudes in understanding the wave-particle relation: the acceptance of the mysterious duality as a manifestation of Bohr's complementarity principle, the belief in the supremacy of the field concept, and the belief in the supremacy of particles. One attitude or the other does not affect the predictions of the theory because the Poincaré symmetry always guarantees that any physical system—in this case the field—can be conceptually decomposed into irreducible systems that behave like particles. But the debate is hardly meaningless. Not only does it have a strong philosophical interest, but it also determines the strategy for the study of systems that are not characterized by the Poincaré symmetry, when, for instance, gravitational phenomena have to be taken into account (see sec. 10.1).

For the moment we shall refrain from developing the merits and drawbacks of each of these attitudes. For we have studied only one category of field and seen only one version of the particle-field duality. What about its manifestation in electrons? Is the picture going to be the same, or do other factors need to be taken into account? We now proceed to answer this question.

7.2 Dirac's Sea

> Roll on, thou deep and dark blue ocean, roll!
> Ten thousand fleets sweep over thee in vain;
> Man marks the earth with ruin—his control
> Stops with the shore.
> —George Gordon Noel Byron, *Childe Harold's Pilgrimage*

To describe a physical system that consists of many particles, one first needs to construct a theory that describes a single one. That much had been established from the treatment of the electromagnetic field. It proved, however, a very difficult task to construct a quantum mechanical description of single electrons, in a way that respects the Poincaré symmetry. Schrödinger's description of electrons was valid only in the limit of small velocities and consequently had to be modified. As a matter of fact, Schrödinger had developed such a theory a few months before discovering his famous equation, but this theory did not work in a satisfactory manner. It predicted that each state of constant positive energy for the electron is accompanied by another state, similar in all respects, but with energy taking the opposite (negative) value. Since the number of positive-energy states is infinite, this prediction implies that there exists an infinite number of negative-energy states and no state of minimum energy. Electron systems could not be stable in this theory. As explained in section 4.3, the absence of a minimum energy state in a system allows one to extract indefinitely large amounts of energy from it with practically no effort. Another problem of Schrödinger's theory was that there appeared occasionally negative numbers for probabilities; this contradicts the very definition of the concept of probability. These problems were addressed by a new theory of the electron, which was developed by Paul Dirac.

Dirac was one of the first few scientists to work on the new mechanics—he had been captivated by Heisenberg's ideas even before they had taken their final form. Dirac's contribution to quantum mechanics was significant: he had developed the theory of correspondence between classical and quantum symmetries, and he was responsible for the term "q-number." His main asset as a researcher was his remarkable fluency in the language of mathematics: he could build a solid mathematical edifice around a very simple idea without obscuring it in the slightest way.

Perhaps in no other work of his is the elegance of Dirac's blend of mathematical and physical intuition more aptly demonstrated than in the theory of the electron. Dirac followed a completely different approach to the problem. He did not start from any concepts of classical physics: his arguments revolved around the possible implementations of the relativistic symmetries in quantum theory. These allowed him to construct an equation for a "wave" that generalized Schrödinger's. Dirac's equation turned out to provide a natural description of spin and to fully solve the problem of negative probabilities. However, it did not solve the problem of the negative energies that plagued Schrödinger's theory. The new theory still predicted the existence of one negative-energy state for every positive-energy one. However, the whole construction was too beautiful to be dismissed.

Eventually, Dirac resigned himself to the fact that the nonpositivity of the energy was a generic property of *every* wave equation that respects the Poincaré symmetry. This conclusion could not be accepted without controversy: the existence of states with arbitrarily low (negative) energy is incompatible with the stability of any form of matter. It was imperative that a way out be found. Dirac found one. He had noticed that there exists an energy gap between positive and energy states. The minimum positive energy for any particle equals its rest mass. Because of the symmetry between positive and negative energy states, the maximum negative energy is minus the electron's rest mass. Hence, there exist no states with values of energy lying between the negative and the positive value of the rest mass (see fig. 7.2).

If stability of matter were to be preserved, Dirac thought, electrons should not be able to jump over the energy gap from the positive into the negative energy states. What physical law could be invoked to forbid such jumps?

The key observation is that electrons satisfy Pauli's exclusion principle. Hence, if all the negative energy levels are *already occupied* in our world, no positive energy electron could jump into a negative energy state because two electrons cannot coexist in the same state. On the other hand, electrons should not occupy all the positive energy states. The vast majority of these states must be unoccupied.

An immediate objection can be raised to the idea above. If electrons, which are electrically charged particles, occupy the infinite tower of negative energy states, the total electric charge in the world should be infinite. This would exert a force of infinite magnitude on any charged body. Clearly, this is not the case. This elementary argument would seem to discredit Dirac's idea, but he succeeded in finding an adequate answer. We must assume that the electromagnetic field interacts with electrons only through positive energy states, and it ignores the negative energy ones. Clearly, this is an additional hypothesis, with no justification at that moment, but which could very well be true in a general theory of the quantum interaction between electrons and photons. Such a theory did not exist at that time, but this was not a good reason to reject Dirac's theory out of hand.

Dirac's proposal implied that only the positive energy states correspond to physical particles, the ones we observe and interact with. The negative energy states do not manifest themselves in physical reality. They form a background, a *sea*, or a womb that gives birth to the physical particles in the world. The sea itself does not constitute a thing: it does not interact, and it does not have a material form of any sort. Only something that emerges out of this sea, by crossing the energy gap, has physical existence of its own. It then becomes part of the material world: it acts and it is acted upon by other physical systems, and in this process it arises into our perception. Dirac's conception was very intuitive: it touched upon the old myth about the birth of things out of a formless and ever-present void, the creation of form out of formlessness. Perhaps this was one reason for the huge popularity the image of the Dirac Sea has enjoyed, even after people realized that the theory was in need of substantial improvement.

However, Dirac's theory contained more than the beautifully crafted imagery of the sea. It made a concrete and spectacular prediction: the existence of antiparticles. I shall describe this in the following section.

7.3 Antiparticles

Every destruction is a negative creation; the disappearance of a concrete thing that exists requires a true definite principle, the same as its production if the thing did not originally exist.
—Imannuel Kant, *Essay on the Introduction of the Negative Magnitude in Philosophy*

Dirac's image of the sea implied a sharp distinction between the positive and the negative energy states. The former are mostly unoccupied, and they correspond to the particles we observe in our experiments; the latter are always filled and inaccessible to observation. However, this result holds only for a closed system of electrons, that is, when there is no external intervention. What happens if we transfer to an electron system an amount of energy that can bridge the gap between positive and negative energy states? At least one "electron" with negative energy will jump into positive energies and emerge into spacetime. A negative energy state will be unoccupied. Hence, *a hole will appear in the sea.*

We may better understand the "hole" concept by considering a carton containing a few eggs. To describe the "state" of the carton, we must specify the position of every egg inside. If only one egg is missing, it is easier to specify the position of the missing egg rather than the position of all eggs present. Analogously, it is much simpler to focus on the hole in our discussion of the Dirac Sea. The hole carries the negation of negative energy, hence positive energy. It amounts to an absence of one negative electron charge, hence presence of one positive electron charge. Positivity of energy implies that the hole behaves like a particle, a physical object that has crossed the energy gap. It has the same rest mass as the electron and opposite charge. It will be an *antielectron*, the first example of an *antiparticle*. Therefore, if an electron gets enough energy to jump over the energy gap, it will not appear on its own. It will be accompanied by its antiparticle, which corresponds to the hole that is created in the Dirac Sea.

According to Noether's theorem, energy is the generator of time translations (see sec. 4.3). When we shift our description from the unoccupied negative energy state of the electron to a positive energy state of its antiparticle, we reverse the sign of the energy. This reverses the direction in the time translations, which—we should recall—are generated by energy. Sometimes this is taken to imply that the antiparticle travels backward in time. Such a statement appears often in textbooks and in popular science books but is not literally true. The antiparticles behave ordinarily as far as the direction of time is concerned: their motion fully respects the common idea that the cause should precede the effect. If we encode a message in a beam of antielectrons and send it forward, it will reach an observer in our future (and not one in our past). The observer

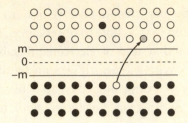

Figure 7.2 **Dirac Sea and the antiparticles.** The circles denote energy levels (states); black circles are filled ones and white empty. According to Dirac's theory, all negative energy levels are filled. Any gap in the negative energy levels is perceived as an antiparticle (gray circle). Only few positive energy levels can be filled at a moment of time. Note that there are no states with values of energy between −m and m (the negative and the positive value of the electron's mass).

will pick up the message, decipher it properly, and follow our instructions. The observer will act exactly the same way he or she would have acted if the message had been encoded in electrons.

The difference between particles and antiparticles lies in an *internal* notion of time,[2] which characterizes the behavior of particles under the transformation of "time inversion," or T-transformation (see sec. 7.7). The name "time inversion" is misleading; the corresponding transformation does not refer to inversion of physical time in nature, that is, clocks running backwards. It is related to the fact that there exist two different ways for energy to generate time translations in a quantum system. The T-transformation switches between these two ways. Energy generates time translations differently in electrons than it does in antielectrons, but in both cases *the translation is forward in time*.

To explain this point in more detail, we consider a yo-yo, which consists of a disk and a string wound around it. There are two different types of yo-yo. In the first, the disk rotates clockwise when the string unwinds; in the second, counter-clockwise. Let us now assume we have a yo-yo of the first type. If we monitor the yo-yo's motion on video and play the tape in reverse (the only way we may get a hint at what actual

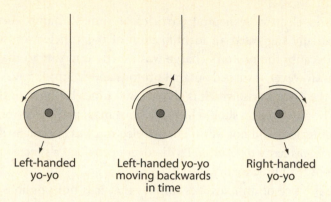

Left-handed
yo-yo

Left-handed yo-yo
moving backwards
in time

Right-handed
yo-yo

Figure 7.3 **Two types of yo-yo.** The first yo-yo in the drawing is left-handed, i.e. its disk rotates anticlockwise, when the string is released. If we reverse its motion in tape, the disk looks as though it moves clockwise. The string is then wound back. In a right-handed yo-yo, the disk also rotates clockwise. However, in this case the string is unwound. If we ignore the strings, it looks as though the inverted in time left-handed yo-yo is identical to a right-handed yo-yo. The only thing that differentiates between them is the string. The direction of the string's winding represents the direction of time in the case of particles. Time only goes forward in nature. We should therefore always consider an unwinding string. The physically correct way to implement the T-symmetry is by transforming left-handed yo-yos to right-handed ones and not to left-handed ones that move backwards in time.

time inversion looks like), the yo-yo disk looks as though it rotates counter-clockwise. However, this does not mean that any counter-clockwise motion of a yo-yo disk is backward in time. The yo-yo may be simply of the *other* type! This analogy accurately describes the distinction between particles and antiparticles—the role of the yo-yo's string being played by a particle's quantum phase.

Do antiparticles exist in nature, or was Dirac's prediction an empty one? To answer this question, I first summarize the progress that had been made toward the understanding of the atomic nucleus. The nucleus was shown to consist of positively charged particles with spin ½ that were named *protons*—they were equal in number to the electrons of the atom—and electrically neutral particles of spin ½ named *neutrons*.[3] Dirac initially suggested an identification of the antielectron with the proton because the proton's charge was exactly opposite to that of the electron. However, it turned out that protons and neutrons have approximately the same mass, which is about 1,800 times larger than the electron's. The antielectron should have the same mass as the electron; hence this identification did not work. The antielectron was eventually detected in 1933; it was named *positron*. Its discovery provided a spectacular verification of the Dirac theory.

The existence of antiparticles implies that it is not possible to construct a fully consistent theory for one single electron. If the electron's energy becomes too high, chances are that a negative energy state will be excited and will cross over the energy gap. Thus an electron–positron pair will appear. Hence, even though we start with a single particle, as the energy increases we may end up with three. It is therefore necessary that we write a theory describing an indefinite number of particles (and antiparticles), much as was the case for the electromagnetic field.

To this purpose, we follow Fock's method again (see sec. 7.1).[4] That is, we construct the spaces that describe the configurations of N particles, for all possible values of the number N, and then glue all of them together (fig. 7.1). However, the Fock construction for electrons is different from that for photons in two respects. First, the theory must describe two different types of particles, namely, electrons and positrons. Second, the spin-statistics theorem implies that electrons are joined together with the Fermi connection because their spin is equal to ½. This procedure leads again to a duality between fields and particles, in the sense that the theory may be equivalently formulated in terms of some q-numbers that represent fields, usually called the *Dirac fields*. In fact, Dirac fields do not describe only electrons and positrons. Different versions of them can describe other massive particles with spin ½. We are going to encounter many particles of this type in later chapters. At this point, it is important to note that the field-particle correspondence in the Dirac field works properly only if we include both particles and antiparticles in the theory; take only one of them and the correspondence is lost.

In the field theory description, the Dirac Sea is incorporated into the definition of a special state, in which all negative energy levels of particles are filled up, but no particle of positive energy exists (whether electron or positron). This state is the field's *vacuum*, the basic reference state. It has the lowest possible value of energy, namely, zero. Any other state can be constructed by adding particles (or antiparticles) to the vacuum and carries, therefore, positive energy.

The vacuum is defined as the state in which the field's energy takes its lowest value: zero. However, energy is not a physical quantity that remains invariant under the Poincaré symmetry. Its value depends on the choice of a reference frame (see sec. 3.5). Does this imply that the choice of vacuum depends on the reference frame? This would be highly undesirable for it would make the presence or absence of particles a frame-dependent property. Happily, this problem does not arise: the vacuum state remains invariant under the Poincaré symmetry. In fact, it is the *only state* that does so (see, however, sec. 7.6).

Dirac's theory provided the basis for a solid understanding of the electron's behavior as a particle, but many questions remained unanswered in relation to its field properties. The natural question to ask was whether it is possible to measure the Dirac fields directly. The answer is an emphatic no. There are two very good reasons for this. First, the Dirac field is not the quantum analogue of any classical field. In this sense, it is very different from the quantum electromagnetic field, which is the quantum analogue of the classical electromagnetic field that is described by Maxwell's equations.[5] In the absence of a classical quantity to account for the q-number representing the Dirac field, it is doubtful that we will ever think of a concrete device that could perform such measurements.

This point can, however, be contested. One could perhaps measure the Dirac field indirectly, by associating its values to some other measurable classical quantities. Such quantities are, for instance, the electric current, or the density of the electric charge, or some novel ones that characterize the Dirac field specifically. This is theoretically possible, so we need to state a stronger reason to explain why the Dirac field cannot ever be measured. The answer is that such a measurement would be *incompatible with the conservation of the electric charge.*

To explain the statement above, we first recall the analogous discussion concerning the electromagnetic field (sec. 7.1). States that are characterized by a sharp value of the field variables have an indefinite number of particles. This is a feature of the Fock construction, which is valid whether the connection is of the Bose or of the Fermi type. It is, then, true also for the Dirac field. Hence, the number of particles in any state, corresponding to a definite profile (i.e., value) for the Dirac field, fluctuates wildly. Different individual measurements would yield wildly different values for the particle number. For photons, this is not a problem because they carry no electric charge. However, electrons are charged particles. This implies that different measurements of the same state would reveal different values of the electric charge. This is physically impossible, since the charge is a conserved quantity. Hence, we conclude that no states can be prepared in which the Dirac field has sharp values. Such states would result from a measurement of the values of the Dirac field because the outcome of an individual measurement always has a specific value. Hence, we conclude that the measurement of the values of the Dirac field is impossible.

The arguments above use the language of the Copenhagen interpretation, in which the act of measurements plays a special role. Nevertheless, other approaches to quantum theory reach the same conclusion: there can be no states in nature characterized by sharp values of the Dirac field. In fact, charge conservation allows us to make a stronger statement. Of all the states mathematically associated with the Dirac field, only the ones with specific value of the electric charge are physically realizable. Any measurement of the electric charge in a given state should always give the same value. However, it does not matter into how many particles this charge is distributed. We can prepare a state that manifests a single electron in one measurement, in another two electrons and one positron, in yet another 1,000 electrons and 999 positrons. This state is physically acceptable because the total value of the charge is the same in all measurements.[6]

Having discussed the properties of the Dirac field, what can we say about the general relation between fields and particles? Let us recall the three main attitudes toward this issue (sec. 7.1): the field-particle duality, the field "fundamentalism," and the particle "fundamentalism." The idea that the fundamental material entity is the field suffers from a severe problem. It is clear from the previous discussion that the Dirac field never manifests itself in experiments as a field but only in the form of particles. We cannot measure it without violating cherished conservation laws. Moreover, there exists no physical state characterized by definite values of the field variables. It is therefore very hard to make the case that the Dirac field exists as a physical entity, at least not in the same way that one would say that the electromagnetic field exists—as a continuous "form of matter" that fills space. These considerations seem to be in favor of the idea that particles are the fundamental ingredients of matter and the field is simply an emergent object. Still, the debate does not end here because, as we shall in the following section, the field concept is indispensable even in a theory phrased solely in terms of particles.

Modern physics perceives the field as material because it is a carrier of energy/inertia. However, one would be mistaken in viewing it as something as crude as, say, a "fluid" on spacetime. Perhaps the most accurate interpretation for the quantum fields is the one that the eighteenth-century dynamists ascribed to force—the principle that is the cause of motion and change. In marked similarity to Leibniz's interpretation of the force, Heisenberg saw the field as a manifestation of the Aristotelian potentia, the inherent principle in matter that allows it to change its appearance and to appear in a multitude of different ways.[7] In the same vein, the emphasis on the field primacy could be viewed as a direct descendent of the theory of Boskovitch and the insights of Faraday, who viewed material particles simply as focal points of the field of force. Still, such identifications can only be partial and incomplete. The formalism of quantum theory is like an oracle: it gives signs and omens, but it does not reveal or endorse which perspective—if any—is a more accurate representation of the world.

7.4 QED and Feynman Rules

Such is the advantage of a well-constructed language that its simplified
notation often becomes the source of profound theories.
—Pierre-Simon de Laplace

I argued earlier that it is very difficult to hang on to the view that fields
are themselves the fundamental ingredients of matter. However, it is
even more difficult to assert that fields are not necessary and that particles are the sole basic ingredients of the world. The reason, as we shall
see, is that particles interact with each other, and this interaction can
only be described in terms of fields.

Our considerations so far involved systems of free electrons (or
positrons) or free photons—the word "free" in this context refers to particles that do not interact with anything else, or with themselves. However, a physical theory yields interesting predictions only if it incorporates a description of particles interacting with each other. There exist
two possible strategies to this end: we may describe the interaction
either on the basis of particles or on the basis of fields. The first description assumes that the interactions are *contact interactions* of particles,
namely, processes of the following form: at a specific spacetime point,
an electron emits a photon, which is later absorbed by another electron
at another spacetime point; an electron approaches a photon, they collide at a specific spacetime point, and then each follows a different path.
Processes such as the above are not different conceptually from collisions of billiard balls, except for the fact that they ought to be described
by quantum rather than classical physics.

The alternative description assumes that particles interact through
local interactions of their corresponding fields. In other words, the interaction depends fundamentally not on the particles' degrees of freedom
(momentum, position, etc.), but on the values of the fields associated
with the particles. Moreover, such interactions are local. Any particle
process that takes place in a specific region of spacetime depends only on
the values of the fields within that region. The values of the fields outside
the specified regions are simply irrelevant to the outcome of the process.

The two procedures sketched above lead in general to different theories. Which one is correct? Luckily, we do not need to wait for the judgment of experiment to settle the issue: mathematical consistency alone provides the answer. If we apply the principles of special relativity and quantum mechanics, the theory of contact interactions is ruled out: it is impossible to construct a theory of contact interactions that respects both the Poincaré symmetry and the structure of quantum theory. Hence, a physically meaningful description of interactions between particles is possible only using local fields. For this reason, many physicists have a strong tendency to view the field description as more fundamental than the particle description, even though the case of the Dirac field argues otherwise. In any case, the conclusion is that the interaction between photons and electrons should be constructed using their corresponding fields—the electromagnetic and the Dirac fields. The theory that describes this interaction is quantum electrodynamics (QED). QED describes in fact the interaction of any spin ½ particle with the electromagnetic field.

A world in which photons, electrons, and positrons do not interact would be very convenient for the physicist, except for the fact that nothing of interest would ever arise out of matter. The mathematical techniques for the description of free particles are very accurate, but unfortunately they cannot be immediately generalized to the physically realistic case of interacting particles. In fact, we have not managed to find any mathematical method that yields exact results for the latter case—our calculations always rely on approximations. The specific approximation that works in QED is *perturbation theory*.

The fundamental concept for perturbation theory is that of a *coupling constant*. A coupling constant is a physical parameter that determines the strength of a specific interaction. For example, the coupling constant of QED is the electron's charge. This charge is a constant of nature, but one may imagine a world in which it takes a different value. If that value were larger than that of our world, the electrons would interact more strongly through the electromagnetic field, while if the charge were zero, they would not interact at all.

If the value of the coupling constant is small, one expects that the interactions cause only small changes to the free particles' states of motion. It is reasonable to assume that we can describe a system of interacting particles by making small corrections to the description of free particles. Perhaps we could incorporate these corrections in a piecewise manner, first adding some large ones that are easy to discern, then some smaller ones that are more difficult to isolate, and so on. We may even obtain a good agreement with experiment if we incorporate a sufficient number of corrections. If that method worked, we would be able to describe the system of interacting particles without performing the technically dreadful task of constructing its states of motion in detail.

It turns out that the method above works: a precise mathematical procedure allows us to implement it. The corrections that have to be incorporated follow a repeatable pattern, which can be represented in a highly intuitive graphical manner without any loss of mathematical rigor. This graphical representation analyzes the interactions of field theory in terms of *contact interactions* between particles. It was initially conceived by the American physicist Richard Feynman.[8] For this reason, the rules of this representation are known as *Feynman rules* and the corresponding diagrams as *Feynman diagrams*. We saw earlier that contact interactions are not really consistent with both quantum theory and relativity. It is then no surprise that an *infinite* number of such interactions are needed, (i.e., an infinite number of Feynman diagrams) to obtain the results of quantum field theory.

The first step toward the construction of a Feynman diagram involves the representation of free particles. For this purpose, we use lines that connect spacetime points: a wavy line for photons, a full one with an arrow for electrons (or positrons). (This represents the fact that in classical physics free particles move in straight lines.) We employ a convention that the direction from past to future corresponds to the direction from left to right on the page. If the arrow of a full line points toward a future direction, the line represents electrons. Otherwise, it represents positrons. This is only a notational convention, and it by no means implies that positrons move backwards in time.

Concerning the endpoints of the lines, there are two possibilities. An endpoint can be free, in which case we visualize the corresponding particle

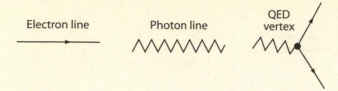

Figure 7.4 **The basic components of Feynman diagrams in QED.** There are three components to the Feynman diagrams in QED: two lines that correspond to the two types of particles (electrons and photons) and one type of interaction vertex.

as either emanating from an external source or ending up in a detector. Alternatively, the endpoint may correspond to a contact interaction. Such contact interactions are represented by the diagram's *vertices*, that is, spacetime points where more than two lines meet. The structure of the vertices, namely, the number of lines of any given type that "meet" in a vertex, is not arbitrary: every theory is *uniquely* characterized by the different types of vertex (usually a few) that can appear in a Feynman diagram. Each type of vertex corresponds to a specific way that particles can interact with each other; physical considerations place a limit on the total number of such ways. QED is one of the simplest of theories, for it admits only a single type of vertex that corresponds to the intersection of two filled lines and a single wavy one. If in a Feynman diagram one sees a vertex, in which, say, two photon and two electron lines meet, then this diagram does not correspond to QED, but to a different theory (probably a bad one).

The experiments that provide information about the microscopic interactions of particles typically involve colliding particle beams. The initial energies of the beams and the angle between them are the *input* parameters of the theory. Their values can be adjusted by a suitable calibration and preparation of the instruments. When the beams collide, the particles scatter to all possible directions possessing very different values of energy, and even new particles may appear. Quantities, such as the energy and the scattered particles' directions of motion, are the *output*

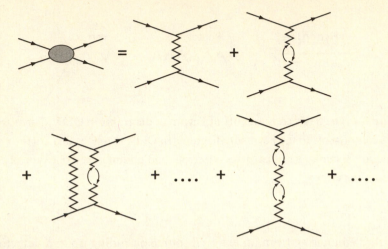

Figure 7.5 **Feynman diagrams for electron-electron scattering.** The blob
that represents the scattering of electrons splits into a sequence of contact
interaction characterized by the vertices of fig. 7.4. The splitting does not stop
here; we can continue including more diagrams with six vertices, diagrams
with eight vertices, ten, etc., up to infinity.

parameters, namely, the quantities that are determined by the experiment.
Since quantum theory is not deterministic, different runs of an experiment yield different results. Each individual collision has a different outcome and quantum field theory only determines the probabilities that
correspond to each outcome.

For example, the product of collision between two electrons could be
another pair of electrons. Perturbation theory describes this process by
splitting the blob of fig. 7.5 (which represents the interaction among
the corresponding *fields*) into a sequence of contact interactions. In the
simplest case, we consider a diagram with only two vertices. One
incoming electron emits a photon, which is then absorbed by the other
electron. The electrons then interact through the exchange of a photon,
and as a result, their energy and momentum changes—as if one had
acted upon the other through a force. However, this is not the only

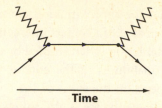

Figure 7.6 **Compton scattering.** This diagram contains two vertices and describes a process by which an electron encounters a photon and absorbs it, and at a later time, another photon (of different energy) is emitted from the electron.

eventuality. At any moment during the interaction, the exchanged photon may split into one electron and one positron, which will recombine to form a photon. As this process does not affect the incoming or outgoing lines, the corresponding Feynman diagram is also acceptable and contributes toward the total probability of the scattering process. In general, any graph that uses the QED vertices contributes to the total probability for the process as long as the external lines remain the same. The external lines are the *defining* ones: they identify the process as corresponding to electron scattering. Figure 7.6 presents the Feynman diagram with the largest contribution to the probability for the process of a photon scattering off an electron (a phenomenon known as Compton scattering).

We may thereby write an infinite number of graphs for any physical process that quantum field theory describes. One such graph represents a mathematical term,* which contributes to the total probability for this process. Clearly, the presence of an infinite number of terms means an infinite amount of calculating work. Does this mean we have gained nothing from this procedure?

*It is important to emphasize that the processes depicted by the graphs are not in any sense real. QED or quantum theory does not endorse any image of particles that come into contact at specific points of spacetime. When we talk about electrons emitting or exchanging photons, we do not refer to physical processes—we only embellish an otherwise dry description of the corresponding lines in the Feynman diagram.

This is the point that the approximation comes in. A single vertex corresponds to a single instant of a contact interaction between particles. The strength of this interaction is determined by the coupling constant of the theory. The probabilities of the corresponding physical processes clearly then depend on the value of the coupling constant. The rule that governs this dependence is the following: for every vertex that appears in a Feynman diagram, we multiply the corresponding mathematical term once by the coupling constant. Hence, the mathematical term corresponding to a diagram with N vertices is proportional to the coupling constant raised to the N-th power. If the coupling constant has a small value, its higher powers have even smaller values.* As a result, a graph with many vertices corresponds to a very small number and thus contributes relatively little to the physical predictions of the theory. This fact suggests the approximation of *truncating* the infinite tower of diagrams that describe a process. For instance, we can choose the approximation of considering only graphs with two vertices. To improve this approximation, we take into account graphs with three vertices, which contribute to the physical predictions by a substantially smaller amount. For further improvement, we consider the graphs with four vertices; they contribute even less. We can proceed further if we want to. The number N of vertices at which we terminate our calculations characterizes the accuracy of the approximation: one speaks of an approximation up to *the N-th order of perturbation theory*.

We emphasize that perturbation theory only works if the corrections become increasingly smaller in the higher orders. For this reason, it is necessary that the coupling constant be small. We are lucky that this is the case in quantum electrodynamics. If, however, the coupling constant takes a large value in some theory, the perturbation expansion gives results of limited accuracy. Higher-order terms are not really corrections any more; they may cause great qualitative differences.

*If the value of the coupling constant is 1/100, its square is 1/10000, clearly a much smaller number.

7.5 The Taming of Infinities

Though this be madness, yet there is method in't.
—William Shakespeare, *Hamlet*

The ideas underlying quantum field theory were put in place by the mid-1930s and have changed little ever since. Still, the universal validity of these ideas had not been realized at that time because quantum field theory found itself face to face with a problem that almost strangled it at birth. The problem was that the "corrections" of perturbation theory turned out to be infinite in magnitude. The predictions of the theory seemed then completely nonsensical.

The cause of this problem was easy to discern. The basic premises of perturbation theory we described in the previous section may be practically sensible, but from a mathematical point of view they are wrong. Perturbation theory, we should recall, starts from the theory of free particles (the lines in a Feynman diagram) and gradually incorporates interactions in the form of small corrections. However, photons and electrons interact through their corresponding fields, and these fields are present at all points of space and time. These interactions are never and nowhere switched off: in effect, the notion of the free particle is negated. In the interacting theory, we can only define "isolated particles," that is, particles that are very far away from any other material objects. These are very different from the idealized free particles postulated in the first step of perturbation theory because they incorporate in their definition the persistent interaction of fields in all points of spacetime.

The difference between free and "isolated" particles is fundamental. It affects all physical predictions of the theory. The problem is that this difference does not simply amount to a numerical correction—say of 10 or 20 or even of 100 percent. When we start our analysis from the fields that describe free particles, we ignore interactions that take place at all points of space and in the tiniest of scales (hence at extremely high energies). It is not surprising that the resulting error is infinite in magnitude.

Do the considerations above imply that the perturbation theory is useless? Not quite. The lowest-order predictions of perturbation theory are not plagued by infinities, and in the case of QED, they provide a substantial agreement with the experiment. Something can be salvaged from the mathematical catastrophe. Is it possible to make physical sense out of it? It took some time before this question was definitely answered, but the answer was affirmative. There *is* a way to tame these infinities, rectifying in the process the error in the basic premises of perturbation theory. This is the scheme of *renormalization*, which was developed for QED in the late 1940's, mainly through the efforts of the American scientists Richard Feynman, Julian Schwinger, and Freeman Dyson, and the Japanese Shin-Ichiro Tomonaga.[9]

Physical particles are characterized by a number of parameters (like rest mass and charge). These parameters enter into the theoretical description of the particle's laws of motion. The theory does not determine their values: one needs information from experiments for this purpose. Strictly speaking, these parameters refer to the complete theory that includes all interactions. However, in perturbation theory we start from the idealization of free particles. Consequently, we place the experimentally determined values of the physical parameters into the approximate free particle description. This would not be a problem, if the free particle description were contained (as a limit) within the full theory. But this is not the case. As explained earlier, the free particle description is mathematically incompatible with the fully interacting theory: we get all these nasty infinities when we try to connect them. This implies that the parameters appearing in the mathematical description of perturbation theory are unphysical: they are not related to the values that are actually determined from experiment. Still, one cannot do without them. These parameters refer to fundamental properties of the particles, without which no theoretical description is possible. In other words, the mathematically convenient description that forms the basis of perturbation theory cannot accommodate the observed physical parameters. We find ourselves at a loss: we cannot compare the theoretical predictions with the experimental outcomes. What are we then to do?

A practical person would say: "If the definitions of your theory do not work properly, then change them in the most convenient possible way. Do not care very much about mathematical or logical rigor, but try to find an adequate redefinition that might just do the job you require. You only want to get a practical rule that will relate the mathematical quantities that appear in perturbation theory to the physical parameters that are determined by experiment. With a little bit of luck, you might side-step all problems related to the infinities, instead of having to face them head on."

However opportunistic and unprincipled such an attitude may sound, it turns out to work well. We can conceal the infinities of perturbation theory in a suitable redefinition of the theory's physical parameters. From a mathematical point of view, the redefined physical parameters are also infinite in magnitude, and hence useless for any calculation. However, at this point, practicality takes the lead. We forget how we constructed the redefined parameters in the first place, and we pretend that they are well defined. We then assert that these redefined parameters are the true ones, that is, the ones relevant to the experimental results. Since all infinities have been absorbed in the parameters' redefinition, any calculation involving the new parameters will be perfectly consistent. Hence, a little bit of mathematical cheating compensates for the fact that perturbation theory is at its basis mathematically wrong. This act of cheating, when orderly and methodically executed, forms the basis of the theory of renormalization.

The key postulate of renormalization is that the infinities of quantum field theory are not arbitrary, but they refer to the erroneous translation of the physical parameters of the fully interacting theory into the parameters of the free particle approximation that is employed in the first step of perturbation theory. If this assumption is true, we need only to construct a "dictionary" that will provide a proper translation.

In general, physical theories contain only a small number of parameters, so there should be only a small number of types of infinity in perturbation theory.[10] This turns out to be the case in quantum electrodynamics. All infinities can be absorbed into a redefinition of the electron mass and of the electric charge. No matter how high an order of perturbation theory we use in our calculations—dealing consequently with huge numbers of diagrams—such a redefinition is always possible. The reason is that only a few types of infinity exist in QED. They all appear at the lowest orders of perturbation theory (in simple diagrams that contain only a few vertices), and they are reproduced in higher orders without further complications. Hence, the theory's physical predictions are perfectly well defined when they are phrased in terms of the redefined quantities. Moreover, they are in excellent agreement with the experimental results.

Theories like QED, in which all infinities of perturbation theory can be removed through a redefinition, are called *renormalizable*. They are hardly the norm among mathematically conceivable theories. In most theories, more and more types of infinite terms appear as one considers higher and higher orders of perturbation theory. Eventually one runs out of physical parameters that can absorb them. Terms of infinite magnitude survive, and these spoil any possibility of making accurate physical predictions. A nonrenormalizable theory is a bad theory because it makes ambiguous predictions. It is therefore reasonable to demand that any quantum field theory be renormalizable.

However, renormalization does not come without a price. The incorporation of the infinities into a redefinition of a theory's physical parameters is not a procedure that is provided naturally by the mathematical formalism. A degree of arbitrariness is always involved in the handling of mathematical infinities. When we want to relate the redefined physical constants with numbers that can be determined in experiments, it becomes necessary to introduce certain conventions. The situation is similar to the choice of units in ordinary measurements. For instance, to measure the length of an object, we first choose a reference rod that defines our unit of length. We then proceed to determine the ratio of the object's length to the rod's length. This ratio determines the object's length in units of the reference rod.

In quantum field theory, we study physical processes rather than static properties (like length), and for this reason our "units" must also be processes. In other words, to provide a reference "scale" for the physical parameters measured in experiments, we choose a *reference process*, and we relate the physical predictions of our theory to the parameters characterizing this process. This procedure is much more complicated than the determination of ratios involved in length measurements, and it involves a host of mathematical tricks. It is nonetheless well defined and unambiguous. One usually chooses reference processes that involve the collision of two particle beams at a specific energy. The numerical value of the relevant physical parameters (a particle's mass or electric charge) relative to the reference process will, therefore, depend on the chosen energy scale.[11] The choice of a reference process is not fully arbitrary, though. It must be made in accordance with the experiment under consideration. For example, if we perform an experiment that involves collisions of particles at very high energies, it is of little help to consider a reference process at low energy. The chosen process should be one realized in our experiment, one also involving high energy of the particle beams.

To summarize, the removal of the infinities through a redefinition of the physical parameters of the theory involves an ambiguity that can only be removed by the arbitrary choice of a reference process. Hence, the experimental determination of the physical parameter partly depends on our choice of reference "scale." As a result, experiments carried out at different energies—and hence having to employ different reference processes—lead to different values for the theory's physical parameters. To denote this dependence of the measured parameters on energy, we place the adjective "running" in front of their name: we talk about the *running mass* of a particle, the *running coupling constant* of an interaction, and so on.

It turns out that the running coupling constant of QED increases with energy. This implies that experiments in high energies will reveal a larger value for the coupling constant than those in low energies, provided that we use the same procedure for its determination in terms of the reference process.

We emphasize that the comparison of the running parameters at different energy scales is possible only because QED is a renormalizable theory, and as such it can be described by a finite number of physical parameters. There is thus no problem in devising unambiguous rules for the determination of these parameters in terms of experimental data. This is in sharp contrast to nonrenormalizable theories. These are characterized by an infinite number of types of infinity, which can be

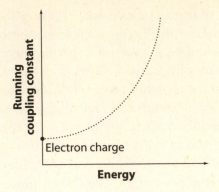

Figure 7.7 **Running coupling constant in QED.** The running coupling constant in QED increases with energy. As energy goes to infinity, the coupling constant increases indefinitely. This implies that two electrons may not come very close together because the energies grow at small distances and the repulsive forces become very strong. Note that at small energies the coupling constant is equal in value to the electron's charge.

absorbed only by an infinite number of physical parameters. Clearly, there is no way to devise any meaningful rules for the determination of an infinite number of (independent) parameters.[12]

One may doubt the universality of renormalizability as a physical principle because it seems rather artificial: it refers to the way we manipulate the mathematical formulas that describe a physical system and not to the properties of the system itself.[13] However, as things stand at this moment, renormalizability is an essential demand from any quantum field theory that relies on perturbation theory. It is after all necessary for a physical theory to make unambiguous predictions. A theory with ambiguities may be useful occasionally—nonrenormalizable theories often are—but it cannot really be fundamental because it fails to provide a guide to future experiments or act as a basis for theoretical generalization. Perhaps the importance of renormalizability will diminish if we find a mathematical procedure of dealing with quantum fields without perturbation theory. At present, alternative methods cannot provide sharp predictions (see, however, sec. 9.4). Perturbation theory is much more reliable within its domain of applicability. For QED, in particular, its predictions are remarkably accurate: in some instances it provides an agreement with experiment of the order of 0.00000001 percent.

Still, the success of the renormalization scheme did not remove the uneasiness felt by many physicists about the existence of infinities in the first place. Renormalization, however useful, reliable, and convenient it may have proved, involves in its essence nothing less but downright mathematical fraud. Shin-Ichiro Tomonaga—one of the pioneers of the renormalization scheme—readily acknowledged that "our method by no means gives the real solution of the fundamental difficulty of quantum electrodynamics but an unambiguous and consistent manner to treat the field reaction problem without touching the fundamental difficulty,"[14] while Dirac was more openly critical: "The rules of renormalization give surprisingly, excessively good agreement with experiments. Most physicists say that these working rules are, therefore, correct. I feel that is not an adequate reason. Just because the results happen to be in agreement with observation does not prove that one's theory is correct."[15]

In retrospect, one can say that the development of the theory of renormalization was a turning point in the history of twentieth-century physics. Following Planck's postulate of the quantum of energy, the first decades of the century witnessed a dramatic overturn of principles that scientists believed fundamental for the understanding of the physical world. A generation of physicists grew up who witnessed the foundations of their science changing at a rapid pace: the challenge to established ideas had almost become a tool of the trade. Nothing is more telling of this fact than the eagerness of most participants at the fifth Solvay Conference to embrace the revolutionary interpretation of the Copenhagen school.

When the issue of the finiteness of quantum field theory came up during the 1930s, a majority of scientists had the inclination to suggest very radical solutions. Radicalism had worked only a few years ago; why not again? The renormalization theory was perhaps the most conservative approach. It did not involve the breakdown of any of the newly acquired physical principles. Its remarkable success signaled a change of perspective. The days of revolution were over, and a new age was about to start in physics. The new theory had been completed and consolidated: it had survived its first severe consistency test. Even though its foundations were not crystal clear, quantum theory had proved itself a very powerful tool for the study of the world. Physicists gradually perceived that the task ahead had become different: they had to apply the momentous work left by their predecessors to the explanation of the myriad of phenomena of the microscopic world.[16]

7.6 The Basic Principles of Quantum Field Theory

> Invention is not the product of logical thought, even though the
> final product is always tied to a logical structure.
> —Albert Einstein, *Helle Zeit, Dunkle Zeit*

As explained in section 7.4, the description of quantum fields through perturbation theory is an approximation. This would not pose a problem if the only aim of a physical theory were to account for the results of experiments. Approximations suffice for that purpose. But we usually demand more from a physical theory: we want to understand the relationships between the fundamental physical concepts. A good theory should allow us to demonstrate that one specific property of a physical system is a consequence of some other property, so that when the latter is satisfied, the former may be inferred. Identifying such relationships is often more important for our understanding of nature than mere predictions of experimental results.

Relations between concepts are by necessity sharp: *they either hold or they do not*. Approximations are not relevant in this case. One may prove for instance that the spin-statistics theorem holds up to the fourth order of perturbation theory in QED; this does not imply that it is true in an absolute sense. A calculation at a higher order of perturbation theory may prove this statement wrong. No approximation method can make conclusive proofs about relations between concepts. The problem is that in quantum field theory, approximation methods are the only ones we have available.

The solution to this problem required an approach of a spirit different from that of renormalization theory. To deal with the problem of infinities, one had to abandon the ideal of mathematical rigor and adopt a "cut and paste" method specific to each problem. Here, one has to follow the exactly opposite direction, namely, that of overwhelming mathematical abstraction. Rather than concentrating on the description of specific physical processes, one should try to isolate the general features that characterize all possible quantum field theories and then build an axiomatic system out of them. In other words, the physical theories were to be defined through the *axiomatic method*, so dear to mathematical purists. This procedure involves the following steps. We first identify some basic postulates, which express properties that are common to a large class of theories (in our case all conceivable quantum field theories). These postulates must provide a correspondence of physical concepts to mathematical objects and of physical properties to mathematical statements. We then treat the postulates as axioms, and we use sharp and rigorous mathematical reasoning to derive their consequences.[17]

The axiomatization of physical theories is very much an architect's job. One tries to design a building that will house not just one theory describing a specific physical system, but a large class of such theories. The architect has to deal with the constraints arising from both the building's desired function (i.e., the accommodation of the physical concepts) and the available building material (i.e., mathematical techniques). The axioms are the supporting pillars of this construction; they have to be laid down with great care or the whole foundation will collapse. They must be physically intuitive and mathematically simple; otherwise they darken rather than enlighten our understanding. They must allow the derivation of any physically interesting property of the studied systems but must also be restrictive, so that they do not lead to oversimplified or unphysical properties. A fine balance must be treaded here, for if the axioms are too restrictive, there will be no physical theory that satisfies them; the axiomatization will then be devoid of physical content.

The identification of an adequate set of axioms for quantum field theory was first completed by the American mathematical physicist Arthur Wightman in the 1950s.[18] Even though his axioms were stated in a difficult mathematical language, the ideas they express are very simple and intuitive.

Axiom 1: *Quantum field theory is a quantum theory.*

This implies that the physical concepts and mathematical techniques of quantum theory can be employed in the description of quantum field theory.

- Physical quantities are represented by q-numbers.
- The preparation of an experiment is reflected in a mathematical object that incorporates all information related to the probabilities for all possible outcomes of the experiment, that is, a quantum state.
- The time evolution of these states should be given by a version of Schrödinger's equation.
- The information we obtain after carrying out an experiment is incorporated in a change of the system's state.

Axiom 2: *Quantum field theory is a theory of relativistic fields in the spacetime of Minkowski.*

Axiom 2 expresses two things. First, the basic mathematical quantities of the theory are fields on Minkowski spacetime. Such fields have typically different values at different spacetime points. This value might be a single number, or a four-vector, or some generalized mathematical object. Second, the theory should satisfy the Poincaré symmetry of transformations between different inertial frames on Minkowski spacetime. This means that the q-numbers that represent the fields should be constructed in a way compatible with the Poincaré symmetry.

Axiom 3: *All states have energy that is either positive or zero.*

This axiom incorporates the principle of energy positivity, which is necessary in order to render the field into a thermodynamically stable physical system (see sec. 4.3). In the Dirac Sea language, this postulate implies that all negative energy levels are filled.

> **Axiom 4:** *There exists a unique state that remains invariant under the Poincaré transformations.*

The state distinguished by axiom 4 is the *vacuum*. The invariance of the vacuum under Poincaré transformations implies that observers in different inertial reference frames will agree in its definition. Furthermore, this invariance implies that the vacuum's energy is zero, for otherwise it would take different values in different reference frames: it would then appear different to different observers. Axiom 3 implies that there exists no state with energy lower than the vacuum, so the vacuum is the unique lowest energy state of the field.

However, there exist certain physically important field theories in which this axiom has to be relaxed. These theories form a necessary ingredient of the Standard Model of fundamental interactions; they will be described in section 9.3.

> **Axiom 5:** *All physical quantities in the theory can be expressed in terms of fields.*

This means that any quantity that is described by the theory can be written in terms of the q-numbers that represent the fields. The axiomatic formulation of quantum field theory is based on the working assumption that fields are the fundamental objects and that all other physical concepts—whether they are particles or any weird object that might conceivably appear—can be expressed in terms of fields.

> **Axiom 6:** *The principle of local causality is satisfied.*

Any measurement of (or action upon) the field is by necessity local. By our nature as finite beings, we can interact with a physical system only in a *finite* region of spacetime: a laboratory must be located somewhere, and every experiment has a beginning and an end in time. We also know from relativity that the speed of light is the highest in nature, and that there can be no instantaneous communication or transmission of information. This implies that two experiments that measure properties of the same field should not influence each other, if they take place in locations so far apart that no light signal can travel from one to the other during the execution of the experiments. It is necessary that we constrain our theories to respect this fundamental condition. For this purpose, the principle of *local causality* is introduced: *two measurements of the field that take place in spacelike separated regions cannot influence each other.*[19]

The local causality postulate is a special case of a fundamental principle of modern physics that has a very long history. We should recall the unease that many of Newton's contemporaries felt about the action at a distance that was implied by the gravitational force (sec. 1.7). Even people wary of the mechanistic worldview had a hard time accepting it. The field was then introduced, to provide a mediator of the forces. The success of the field theoretic description suggested a law of nature that no force can be exerted instantaneously at a distance. This law is the principle of *locality*.[20] The theories of relativity provided further support for this principle. An instantaneous transmission of force was ruled out because the speed of light sets an upper limit on how fast a physical effect can propagate on space. This chain of ideas led toward the postulate of local causality as a fundamental physical principle that has to be satisfied by any quantum field theory.

The principle of locality is a direct descendant of the basic intuitions that led to the introduction of mechanism in the physical sciences. The mechanistic ideal may have been put to rest in modern physics, but it died content that its most beloved child survived it. Quantum field theory completely renounces any mechanistic interpretation, but in the principle of local causality it incorporates one of the most fundamental insights of the mechanistic worldview: motion can be directly transmitted from one "machine part" to another, only if these parts are in direct contact.

One important consequence of the axioms above is the spin-statistics theorem, which we already discussed in section 6.4. Fields that correspond to particles with half-integer spin are constructed through Fermi connections, while fields that correspond to particles with integer spin are constructed through Bose connections. It is important to remark that this theorem, even though it ultimately refers to a property of particles, cannot be proved without the introduction of quantum fields. The key piece that is missing from the particle description is axiom 6, the postulate of local causality. It plays a key role in the proof of the theorem, and there is no way it can be phrased in a language that employs only particle concepts.[21]

It can be shown that the set of axioms above is not empty: it is satisfied by the field theories that correspond to free particles. However, any attempt to rigorously construct a quantum field theory that describes realistic particle interactions has failed.[22] The reason is that the only way we know that allows us to describe interactions is perturbation theory, and this treads on a mathematical ground so shaky that it is impossible to reconcile it with the strict and unyielding mathematical language of the axiomatic description.[23] One may joke that this is another instance of the complementarity principle, which is manifested not at the level of our descriptions of physical reality (the theory of quantum fields), but at the level of descriptions of these descriptions (perturbation theory vs. axiomatic method). If one wants rigor and conclusions of a general validity, one should use the axiomatic method. But then one would have little to say about specific physical processes. To obtain concrete physical predictions, we have to abandon all expectations of precise mathematical demonstration and learn to appreciate the good fortune that such an ill-defined theory can be made not only to give meaningful results of high accuracy, but to also provide good insight for the construction of new theories in regimes that have not yet been tested.

However, unlike the quantum complementarity, which is believed to be a fact of nature, the "complementarity" in the formulations of quantum field theory is only a historical accident. We have to live with it now, as we had to live with it for the past half-century or so, but it is not here to stay. Either through more effort on the foundations of quantum field theory, or through a new mathematical language, or through a yet unsuspected physical principle, we hope that this paradoxical situation will be resolved, preferably sooner rather than later.

7.7 Three Elegant Symmetries: P, T, and C

> Since you know you cannot see yourself
> So well as by reflection, I, your glass,
> Will modestly discover to yourself
> That of yourself which yet you know not of.
> —William Shakespeare, *Julius Caesar*

The previous section described the basic properties of quantum field theory, which were expressed in the form of axioms. We now proceed to examine one of the most important consequences of these axioms: the so-called *CPT theorem*. To this purpose, we will first introduce some important new concepts, namely, the P-, C-, and T-transformations.

The P-symmetry

The P-transformation or *parity transformation* is also known as space inversion transformation. To provide a proper definition, we choose an inertial reference frame on Minkowski spacetime. We then ignore the time axis and focus only on the spatial coordinates. The P-transformation inverts the sign of all spatial coordinates of a point; hence if one coordinate of a point is 5, by spatial inversion it becomes -5. The time coordinate, however, is not affected.

Every physical quantity is affected by spatial inversion. Velocity is defined as the rate of change of the spatial coordinate over time. The P-transformation takes the spatial coordinates to their opposites and does not affect the time coordinate: the velocity then goes to its opposite.

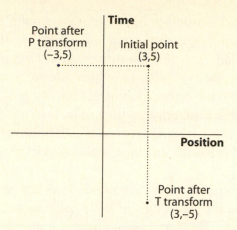

Figure 7.8 **Space and time inversion transformations.** The parity (space inversion) transformation takes one spacetime point into another spacetime point with the opposite value of the spatial coordinate and the same time coordinate. The time inversion transformation takes one spacetime point into another spacetime point with the same value of the spatial coordinate and the opposite value of the time coordinate.

The momentum is proportional to the velocity, so it also goes to its opposite. The energy, however, remains unchanged because it is related to time translations, which are not affected by spatial inversions. Similarly, spin is a degree of freedom that has no relation to location in space. Hence the direction of the spin arrow remains unaffected by space inversions. Eventually the effect of space inversion can be studied in all physical quantities: they either remain unchanged or take opposite values. Note that even though we employed a reference frame for the definition of the P-transformation, our conclusions do not depend on the choice of the frame.

Spin in massless particles is either parallel or antiparallel to momentum. This relative orientation of the spin is invariant under the Poincaré transformations and cannot change. It is a defining feature of the particle, and it is usually referred to as *helicity* (see secs. 4.4 and 6.3). Momentum reverses sign under parity transformations, while spin does not. Hence a massless particle with spin parallel to its motion is transformed through space inversion into a particle with spin antiparallel to its motion. The parity transformation changes the particle's helicity. It takes one type of massless particle to another, much like reflection on a mirror makes a left hand look like a right one and vice versa. In fact, this comparison with hands has served as a basis for naming the two types of helicity: particles with positive helicity (parallel spin) are called *left-handed*, and those with negative helicity (antiparallel) *right-handed*.

So far, our description of the P-symmetry has lain entirely within the domain of classical physics. We can easily translate it into the quantum language: the behavior of a physical quantity under the P-transformations is the same in both the classical and the quantum theory. The only additional object in quantum theory is the quantum phase, which can be shown to remain unaffected by the P-transformation.

As far as quantum field theory is concerned, the most important question is to determine whether its physical predictions are invariant under space inversion. Since quantum field theory refers to interactions between particles, such invariance will appear in particle collisions. If the theory is invariant under the P-transformation, the probabilities of all possible outcomes in a scattering experiment should be the same for processes that are related by a P-transformation.

As an example, we consider an electron beam directed toward a static target (consisting, for example, of heavy atoms). The electrons then scatter upon each other. If we place detectors all around the site of collision, we will register a pattern of outgoing electrons varying in their directions and energies. We can implement the transformation of spatial inversion by considering a different experiment, identical in all respects to the original, except for the fact that the electron beam comes from exactly the opposite direction. If the interaction is invariant under the P-transformation, the monitored pattern of outcoming particles should remain the same. Hence, a detector placed at a specific point will measure the same number of particles (and with the same energies) in both the initial and the "inverted" experiment. The forces responsible for the scattering of electrons off atoms are the electromagnetic ones. The theory of QED, which describes these forces, turns out to be invariant under space translation. The P-transformation is then a symmetry of the electromagnetic interactions.

Space inversion is so natural a symmetry that almost all physicists expected it to be of universal validity. They turned out to be wrong—they had discounted the possibilities afforded by the unconventional behavior of massless particles. The P-transformation takes left-handed to right-handed particles. In QED, the two types of photon are treated symmetrically, and for this reason the P-transformation leaves the theory invariant. If, however, a physical theory treats left-handed particles differently from right-handed ones, that is, if the interactions depend on the particle's helicity, then the theory cannot have the symmetry of spatial inversion. This proved to be the case in the so-called weak interactions, which we shall examine in detail in the next chapter. Hence to the great astonishment of the physics community, space inversion turned out not to be a universal symmetry in nature.

The T-symmetry

We already encountered the T-transformation (or time inversion transformation) in our discussion of Dirac's hole theory and the physical interpretation of the antiparticles. The T-transformation is complementary to the P-transformation. It refers to the inversion of the time axis in a spacetime coordinate system: the time coordinate goes to its opposite, while the spatial ones remain unchanged. The T-transformation affects all physical quantities: velocity, momentum, and spin all go to their opposites under its action.

Since energy is the generator of time translations, one would expect that it would also go to its opposite under the action of time inversion. However, this would imply that the T-transformation takes particles of positive energy to ones with negative energy. Hence it cannot be a symmetry of any theory satisfying the principle of positive energy (like quantum field theory) because such theories do not have room for negative energy states. However, there exists an alternative account of the

T-symmetry, namely, that it changes the way energy generates the time translations. We recall from section 6.2 that the quantum phase is a register that keeps track of the history of the quantum system, and that we have represented it by a rod rotating on a plane. The rod's rotations may be either clockwise or counter-clockwise. The T-transformation can be seen as causing the reversal of the rod's rotation during the time evolution of the system rather than an inversion of time. The energy of the physical system remains positive; hence this interpretation of the T-transformation is compatible with quantum field theory. Our earlier analogy with the yo-yo (fig. 7.3) provides another representation of this idea. The conclusion is that in quantum theory we can have both invariance under the T-transformation and positivity of energy, as long as the clockwise or counter-clockwise motion of the quantum phase does not affect the physical predictions of the theory.

The experimental determination of the P-symmetry is relatively easy: one only needs to invert the direction of the incoming particles to realize a P-transformed experimental configuration. This is not the case for the T-symmetry.

How can we construct a T-inverted experimental configuration? Surely, one should not have to perform experiments that run backwards in time! To answer this question, we consider a particle process of the type

$$A + B \rightarrow C + D,$$

in which the particles C and D arise from the collision of particles A and B.

If the theory that describes this process is invariant under the T-transformation, the opposite process

$$C + D \rightarrow A + B \text{ (momenta and spin reversed)}$$

takes place with the same probability. How can we verify this statement? A typical experiment involves the careful preparation of beams with specific energy and direction, and the subsequent monitoring of the produced particles, which are scattered in all possible directions. The experiment corresponding to the process $A + B \rightarrow C + D$ proceeds from a very controlled and ordered state of the particles A and B (the colliding beams) to a really chaotic state for the particles C and D. If we perform a similar experiment with C and D contained in the controlled incoming beams, then the particles A and B appearing at the end will lie in a chaotic state. The two experiments are not really the time-inverse of the other.

The time-inverse of the process

$$A + B \, (\text{as beams}) \rightarrow C + D \, (\text{chaotically})$$

is the process

$$C + D \, (\text{chaotically}) \rightarrow A + B \, (\text{beams}) \, (\text{momenta and spin inverted}),$$

that is, we go from a disordered state to an ordered one. However, there exists really no way we can do this in a laboratory; order is always followed by disorder, and the opposite never happens. A test of the T-symmetry would involve picking very specific final configurations of C and D in the process $A + B \rightarrow C + D$. We should then reproduce these configurations as initial states (with momenta and spins reversed) in the process $C + D \rightarrow A + B$, and finally we should study the behavior of the products in this reaction that correspond to the configuration of A and B in the initial reaction. This is really a very complex task, much more so than that involved in the determination of the P-symmetry violation.

Admittedly, the discussion above is oversimplified, and it ignores certain indirect ways that can be used to test the T-symmetry. Its aim is to convey a feeling for the enormous difficulties involved in a direct determination of the T-symmetry. Scattering experiments do not have a single product. Hence, to make a controlled study of the time-reverse reactions involves a very sensitive determination of the outcoming particles' properties. For this reason, the first claim for a direct observation of the T-violation came only in the late 1990s.[24]

The C-symmetry

The last transformation we examine is the C-transformation, also known as charge conjugation. Its effect is to substitute a particle by its antiparticle. Unlike the P- and T-symmetries, charge conjugation is not defined in terms of spacetime geometry, but its definition involves a purely quantum concept, that of the antiparticle.

A particle and its antiparticle share the same values for mass and spin but have *opposite values of the electric charge*. Some particles may be their own antiparticles, or in other words, they may have no antiparticle. The photon is such a particle. If a particle is electrically charged (like the electron), its antiparticle must have the opposite value of charge and will therefore be distinct. It is, however, possible that an electrically neutral particle has an antiparticle (e.g., the neutron)—in that case particle and antiparticle are distinguished by the values of *charges* other than the electric one, which are introduced to account for forces other than the electromagnetic ones (see sec. 8.2).

In quantum theory, the C-transformation reverses the direction of rotation of the quantum phase, like the T-transformation does. This is to be expected because antiparticles can be viewed as holes in the Dirac Sea. Their description involves a "jump" from negative to positive energies. To preserve energy positivity, this has to be nullified by inverting the rotation of the quantum phase. In quantum field theory, charge conjugation is implemented by inverting the role of particle and antiparticle in the relation between field q-numbers and the corresponding particles.*

*The reader may have noticed an ambiguity in the use of the word "particle." On one hand, the term particle refers to every pointlike body that moves in spacetime. It is in this context that we talk about the field-particle duality. On the other hand, we talk about the distinction of particle and antiparticle. Here the word "particle" is used only as a designation of one element of the pair particle–antiparticle. Hence when we talk about electrons and positrons and we refer to the electron as a particle, we will refer to the positron as its corresponding anti-particle and vice versa. Both electrons and positrons are particles in the former sense. When we talk about the field-particle duality for electrons, we mean the relation between the Dirac field, on one hand, and *both* electrons and positrons, on the other. The q-numbers of the Dirac field are specific functions of the physical quantities that describe the electrons and the positrons. The C-transformation acting upon the Dirac field changes these functions, by inverting the role played by the electron and the positron in their definition.

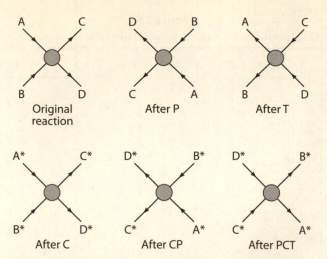

Figure 7.9 **The action of the P-, T-, and C-transformations on the reaction**
$A + B \rightarrow C + D$. Here A, B, C, D can be any particles; A^*, B^*, C^*, D^* are their
corresponding antiparticles.

An interaction respects the C-symmetry, if particles and their
antiparticles participate in it symmetrically. For instance, we consider
the process of one photon with high energy splitting into an electron-
positron pair. The electron's motion should be a mirror image of the
positron's motion with respect to the point of their creation. This is the
case in QED: its physical predictions are invariant under the C-transformation. However, nature does not respect the C-symmetry. Again,
massless particles are responsible. The antiparticle of a left-handed particle is necessarily a right-handed one. Hence, a theory that treats left-
handed particles differently from right-handed ones violates the charge
conjugation symmetry. In particular, the theory that describes the weak
interactions lies in this category.

The C- and the P-symmetries both fail because weak interactions
treat left-handed massless particles differently from right-handed ones.
In fact, the C-transformation cancels out the effects of the P-transformation in the particles' helicity. Hence, even if the theory is asymmetric

TABLE 7.1
The action of the P-, T-, C-transformations on physical quantities

Quantity	P	T	C	CP	CPT
Position	−	+	+	−	−
Velocity	−	−	+	−	+
Momentum	−	−	+	−	+
Energy	+	+	+	+	+
Spin	+	−	+	+	−
Helicity	−	+	−	+	+
Quantum phase	+	−	−	−	+
Electric charge	+	+	−	−	−

Note: If a physical quantity remains invariant under a transformation, we denote it by a plus (+); if it goes to its opposite value, we employ a minus (−).

between left-handed and right-handed particles, the combined transformation of C and P (referred to as the CP-transformation) may very well keep being a symmetry. Nonetheless, CP turned out not to be an exact symmetry of nature. We shall postpone the explanation of the CP violation because the physical mechanism that is responsible for it is very different from anything we have encountered so far. It can be understood only within the context of the theory of spontaneous symmetry breaking, which we examine in section 9.3.

Having described the P-, C-, and T-symmetries, we now state the CPT theorem:

CPT theorem: *Any quantum field theory that satisfies axioms 1–6 of section 7.6 is invariant under the combined action of the P-, C-, and T-transformations.*

The most interesting thing about the CPT theorem is that it follows only from the fundamental axioms of quantum field theory. Its proof

involves no assumptions about the type of particles appearing in the theory or their modes of interaction. Consequently, if a violation of the CPT-symmetry is ever determined (so far, all tests have been negative), one would have to question the fundamental physical principles of quantum field theory! Theoretical conclusions of such universality, which allow a single experimental result to cause the reappraisal of important physical principles, are very rare in physics. For this reason, the CPT theorem is one of the most important theoretical predictions of quantum field theory.

8

PIECES OF A PUZZLE

THE PHYSICS OF ELEMENTARY PARTICLES

8.1 Radioactivity and Forces

One wraps a Lumière photographic plate with a bromide emulsion in two sheets of very thick black paper, such that the plate does not become clouded upon being exposed to the sun for a day. One places on the sheet of paper, on the outside, a slab of the phosphorescent substance, and one exposes the whole to the sun for several hours. When one then develops the photographic plate, one recognizes that the silhouette of the phosphorescent substance appears in black on the negative. If one places between the phosphorescent substance and the paper a piece of money or a metal screen pierced with a cut-out design, one sees the image of these objects appear on the negative. . . . One must conclude from these experiments that the phosphorescent substance in question emits rays, which pass through the opaque paper and reduce silver salts.
—Antoine Henri Bequerel, announcing the discovery of radioactivity

The main conclusion of the previous chapter was that interactions between particles can be described consistently only in the language of quantum field theory. To construct a theory that describes a specific interaction, we need to identify which particles participate in this interaction, so that we can construct the corresponding fields. It is therefore important that we gain a secure knowledge about the number and properties of particles that exist in nature. We thus enter the realm of particle physics, which is also called high-energy physics because it is at high energies that new particles appear.

The nucleus is an essential part of the atomic structure. It is not an elementary physical system. It consists of two types of particles: protons, which have positive charge equal in absolute value with that of the electron, and neutrons, which are neutrally charged particles. The number of protons in the nucleus equals the number of electrons in the atom, so that the atom is overall neutral. Both nuclear particles proved to be quite heavy: their mass is about 1,800 times larger than that of the electron (which is the reference energy for particle physics), the neutron being slightly heavier than the proton.

The introduction of the nucleus concept was almost from the beginning accompanied by severe problems. A nucleus consisting of positively charged particles cannot be stable because these particles repel each other. The understanding that neutrons are also part of the nucleus does not change the nature of the problem. Protons repel each other, whether neutrons exist or not.

The constituents of the nucleus can stay together only if a new type of *attractive* force exists. This force should be stronger than the electromagnetic one, at least in the length-scales characterizing the nucleus. This force should also vanish far away from the nucleus because otherwise a strong attractive force between nuclei of different atoms would appear. This would spoil all our understanding of chemical activity, which is supposed to arise solely from the electromagnetic interactions of the electrons. The newly introduced attractive force was called (predictably) *nuclear*. It affects both protons and neutrons, but not electrons. If electrons interacted through the nuclear forces, they would be easily captured by the nuclei, and the atoms would be completely unstable.*

---

*Note that in the present context the word "force" is used only for convenience. The Newtonian force does not survive as a fundamental concept in quantum physics, where we talk only about interactions of particles through their corresponding fields. Excepting gravity, all forces of classical physics (pushing and pulling, friction, etc) have been reduced to electromagnetic interactions between the particles (molecules) that compose macroscopic material bodies.

The nuclear forces provided an explanation of certain phenomena that had been observed in the 1890s and were lumped under the collective name of radioactivity. Certain substances (the element radium in particular) emit a kind of ray, which can pass through opaque materials and cause the blackening of photographic plates. These rays are not of a single type. When the rays coming from a radiating material pass through a large electromagnet, some of them are left unaffected by the magnetic field (they were named gamma rays), others bend toward one direction (alpha rays), and yet others bend toward the opposite direction (beta rays). This implies that the alpha and beta rays, at least, consist of particles that have different types of electric charge (positive/negative). It was also realized that the alpha particles must be much heavier than the beta particles because they do not penetrate matter as efficiently as the latter do.*

Subsequent experimentation revealed that the alpha rays consist of positively charged particles formed by two protons and two neutrons and coincide with the nuclei of the element helium; the beta rays were eventually identified with electrons, while the gamma rays proved to be nothing but photons of very high energy. As to the reason for their emission, classical physics had nothing to say. A proper explanation had to await the advent of the quantum theory.

I explained in section 4.3 that physical systems lose energy to their environment, and they therefore tend to fall into their state of minimum energy. According to quantum theory, this process is a random phenomenon whose duration is not fixed. Even if we prepare two wholly similar systems in the same state, each system will complete this transition at a different moment of time. There is, however, an average time for such transitions, and this time may be predicted by the rules of quantum theory.

To see how quantum theory explains the emission of the alpha rays, we consider a physical system consisting of, say, 50 protons and 60 neutrons. It is not difficult to enumerate all possible nuclear configurations

---

*A heavy particle is slower than a light particle of the same energy. Hence, if a light and a heavy particle lose energy at similar rates, the heavy one will stop earlier.

that can be formed from these particles. For example, we can form one nucleus with 50 protons and 60 neutrons, or 1 nucleus with 30 protons and 35 neutrons and another with 20 protons and 25 neutrons, or 2 nuclei with 10 protons and 12 neutrons each together with a nucleus of 30 protons and 36 neutrons, and so on. On the other extreme, another possible configuration is the one with all 110 particles being free, that is, being so far away from each other that they do not interact. Having identified all possible configurations, we use quantum mechanics to calculate the energy characterizing each one of them. This is in principle possible as long as we know the properties of the nuclear forces. In most configurations, the energy will be quite high, and the corresponding nuclei will be unstable: they will rapidly decay into configurations of lower energy. For this reason, configurations of high energy rarely appear in nature. On the other hand, low-energy configurations are relatively stable and persist long enough to be observed.

It may happen that the configuration of the single, large nucleus (consisting of 50 protons and 60 neutrons in our example) corresponds to the state of lowest energy, in which case the nucleus is stable in an absolute sense. This case is representative of the stable nuclei that exist in nature.

A common alternative is that the configuration of the single nucleus has slightly more energy than the lowest-energy configuration. Let us assume that in our example the lowest-energy configuration involves a nucleus with 48 protons and 58 neutrons and 1 alpha particle. The large (50, 60) nucleus will then split into the daughter (48, 58) nucleus and the alpha particle. The resulting alpha particle will typically move away from the daughter nucleus, and by detecting it, we may infer what nuclear process took place. Whether a nucleus can split by emitting an alpha particle depends solely on the energy of the two different configurations, which itself depends on the detailed properties of the nuclear forces. For some nuclei, the alpha emission is energetically favorable; for others, it is not.

One may ask at this point, why are alpha particles emitted and not some other light nucleus, say, one consisting of 3 protons and 4 neutrons. The explanation involves a detailed knowledge of the nuclear forces and extensive calculations, which show that the alpha emission is usually energetically favored and for this reason occurs more often in nature. Other light nuclei may be emitted, but the probability is so incredibly smaller as to be practically zero.

However, there exist unstable nuclei, for which the energetically favored transition does not correspond to an emission of an alpha particle. The initial nucleus may split into two smaller ones of roughly the same size. Usually some neutrons are also released in this process. This phenomenon is called nuclear *fission*. Since the final state of two nuclei has less energy than the initial one, the energy difference is converted into motional energy of the daughter nuclei and the neutrons. If many nuclei split at the same time, a large amount of this energy is transferred to the environment. As a result, the environment will be heated and reach extreme temperatures. This is what happens in the explosion of a nuclear bomb.

Nuclear fission may cause great devastation, but viewed as a physical phenomenon it is only a secondary consequence of the nuclear forces. It may be interesting to understand how nuclei split, or even more to learn how we can make nuclei split, but this brings very little benefit to the fundamental understanding of the nuclear forces. The understanding of the nuclei's stability is much more important for this purpose than that of their spectacular breakup.

It is clear from the above that emission of alpha particles is related to the nuclear forces. It is not so with the beta particles. The measurement of their mass and charge showed that they were nothing but electrons. However, electrons cannot interact through the nuclear forces without jeopardizing the stability of atoms. The nucleus would rapidly absorb the surrounding electrons, and the atomic structure would disintegrate. It follows that the forces responsible for the beta rays are different in nature from the ones responsible for the alpha rays. Further study determined that the emission of the beta particle is associated with the process of a nuclear neutron turning into a proton and an electron. The proton typically remains within the nucleus, while the electron carries sufficient energy to escape. The energy carried by these electrons is typically much smaller than that associated with the alpha particles.

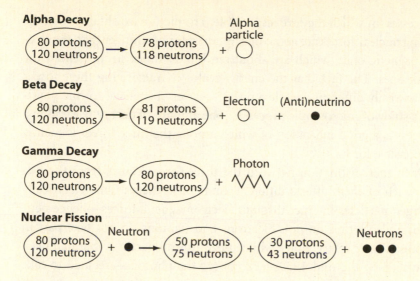

Figure 8.1 **Nuclear processes.** The numbers of protons and neutrons in the drawing are provided only for purposes of demonstration and do not correspond to real nuclei.

This suggests that the corresponding forces are weaker. For this reason, they were called weak nuclear forces or simply *weak* forces, while the forces responsible for the nucleus's constitution were renamed as strong nuclear forces.

The gamma particles turned out to be nothing but photons of very high energy. They are typically emitted from energetically excited nuclei: since photons are related to the electromagnetic forces, it is natural to assume that these forces are responsible for the processes that lead to the emission of gamma rays.

The study of the alpha, beta, and gamma rays—whose discovery predated the identification of the atomic nucleus—established the existence of three different types of forces in nature, namely, the (strong) nuclear, the electromagnetic, and the weak forces. These forces are responsible for different processes and have distinct properties. We can estimate the relative strength of these forces by measuring the energy released in the processes for which they are responsible. The picture we obtain is that the strong nuclear forces are about one hundred times stronger than the electromagnetic ones, which are about ten billion times stronger than the weak ones. The fact that the energy scales characterizing these forces are so vastly different—while a puzzle—is very beneficial for the physicist's study of microscopic processes. The amount of energy released in a process is a good indication of which among the three forces is mainly responsible for it.

A further distinction between the three forces comes from the fact that each of them affects different kinds of particles. Only electrically charged particles interact through electromagnetic forces—hence neutrons and photons do not. Electrons and photons do not take part in strong nuclear interactions, while apparently all particles that had been identified by the mid-1930s—except for the photons—seemed to take part in the weak interactions.

8.2 The Hunt for Symmetries

How can we divide an elementary particle? Certainly only by using
extreme forces and very sharp tools. The only tools available are other
elementary particles. Therefore, collisions between particles of extremely
high energy would be the only processes by which the particles could be
eventually divided.
—W. Heisenberg, *Physics and Philosophy*

A great surprise lay in store for the physicists who studied the emission of beta particles (electrons) from nuclei. The principle of energy conservation implies that the beta particle would carry energy equal to the difference in energy between the initial and the final state of the nucleus. The simplest assumption was that for a given type of nucleus the energy of these states would be constant. Hence, the energy of the emitted electrons would also be a constant. However, it was soon found (the first hints had appeared already in 1914) that this is not the case. The energy of the beta particles emitted from a given type of nucleus takes any value within a continuous range. This phenomenon could be interpreted as a violation of energy conservation, the first such recorded instance in physics. However, only with great reluctance would any twentieth-century physicist ever agree to abandon the principle of energy conservation. All understanding achieved by modern physics was based on this principle. To abandon it would be a literal intellectual suicide. For this reason, most physicists preferred not to think of the problem at all.

It took some time before an alternative explanation was proposed. The person responsible was Wolfgang Pauli. Having been invited to a workshop at Tübingen (in 1930) that he could not attend, he sent his colleagues a rather strange letter with the sketch of a solution.

Dear radioactive ladies and gentlemen,

I beg you to most favourably listen to the carrier of this letter. He will tell you that . . . I have hit upon a desperate remedy to save . . . the law of conservation of energy. This is the possibility that electrically neutral particles exist which I will call neutrons, which exist in nuclei, which have a spin ½ and obey the exclusion principle, and which differ from the photons also in that they do not move with the velocity of light. The mass of the neutrons should be of the same order as those of the electrons and should in no case exceed 0.01 proton masses. The continuous beta spectrum would then be understandable if one assumes that during beta decay with each electron a neutron is emitted in such a way that the sum of the energies of neutron and electron is constant. . . . I admit that my remedy may seem incredible, because one should have seen these neutrons long ago if they really exist. But only he who dares can win. . . . Henceforth every possible solution must be discussed. So, dear radioactive people, examine and judge . . .

Your humble servant,
W. Pauli[1]

Pauli's suggestion was right in its essentials, but the details needed corrections. In 1932 the neutron was discovered. It was much heavier than the electron, so it did not correspond to the particle imagined by Pauli. To distinguish the latter from the former, he accepted a suggestion by Enrico Fermi to name his particle *neutrino*, namely, light neutron. Fermi himself went on to develop a theory of weak interactions, in which the neutrino played a fundamental role.[2]

Further study revealed that the neutrino must be much lighter than Pauli had initially supposed. Its mass would have to be much smaller than that of the electron, and it seemed quite possible that it could be zero. Apparently, neutrinos would be very hard to detect. Since they are not electrically charged, they cannot interact with other matter through electromagnetic forces. Moreover, they are not affected by strong nuclear forces—in that, they were very different from neutrons. Hence, the only way they can interact with other matter is through the weak forces, which are literally very weak. A rough calculation shows that on average a neutrino can cross through Earth without colliding even once with another particle. The neutrino was eventually detected in 1953. Detection had to await the availability of gigantic neutrino sources (these were nothing but ordinary nuclear plants).[3] Nonetheless, the majority of physicists had adopted the new particle already since the 1930s. This was partly because Fermi's theory of weak interactions was moderately successful, but mainly because the neutrino provided a rescue from the terrible challenge to the principle of energy conservation.

The neutrino also has an antiparticle, the antineutrino. A convention was chosen later to refer to the particle appearing in the beta decay as the antineutrino, the name neutrino being reserved for its antiparticle. With this naming, the neutrino was determined as a left-handed particle and the antineutrino a right-handed one (assuming, of course, that they are indeed massless).

The story of the neutrino emphasizes the importance of conservation laws to the understanding of the microscopic structure of matter. Our trust in fundamental symmetries turns out to be a much stronger factor for the acceptance of a new theory than any requirement of direct experimental verification. This is, in fact, the only way to proceed, in order to understand the structure of the micro world: we have to hunt for symmetries.

Typical experiments in high-energy physics involve collisions between particle beams, which lead to the realization of many different processes: two particles may just scatter off, a pair of particle-antiparticle may appear in a highly energetic collision, particles may be created or destroyed, and so on. Unless we have a thorough theoretical understanding of the forces involved, there is no way to predict what will happen and how often. The important thing is, however, that any particle that arises in such processes can in principle be detected because it leaves a track in suitably prepared material that surrounds the locus of collisions. This material may be a photographic plate, a bubbly fluid, or a solid sensitive to the passage of charged particles. The selection of this material is important for the design and execution of any experiment, but at the fundamental level we care only that it does the job required: to record of the tracks left by the particles created in the collisions. From these records one tries to determine as many of the particles' properties as possible.

A carefully designed experiment provides an adequate picture of all processes that take place during the collision of the particle beams and allows one to determine how often each of them is realized. However, more important than the processes that are realized are the ones that are not. A fundamental maxim in physics is that *anything can happen as long as it is not forbidden*, and the only prohibitions that are strongly enforced in nature are the ones associated with conservation laws. A process that violates a conservation law cannot happen. A careful study and comparison of allowed vs. forbidden processes provides strong clues about the *definition* of conserved quantities for the elementary particles.

I next describe in detail a simple but important example, which demonstrates how the procedure sketched above works. Some attention is demanded from the reader in order to follow the arguments. Perhaps a pen and paper would be helpful, but the effort needed is no more than that needed to solve a brainteaser.

The particles are represented by small letters. In particular:

- the electron is represented by e^- and the positron by e^+
- the proton by p
- the neutron by n
- the neutrino by the Greek letter ν and the antineutrino by ν^*

Figure 8.2 **Particle tracks in a bubble chamber.** A bubble chamber contains a fluid at a temperature higher than its boiling point. When a high-energy charged particle crosses through the fluid, it removes electrons from the fluid atoms. Bubbles of vapor then form around the positively charged atoms, thus marking the particle's path. The bubble chamber was developed by D. A Glaser and L. Alvarez in 1952. This picture shows the annihilation of a proton and an antiproton, yielding a large number of particles known as mesons. Reprinted with permission from S. Goldhaber, G. Goldhaber, W. M. Powell, and R. Silberberg, *Phys. Rev.* 121, 1525 (1961). Copyright 1962, American Physical Society.

I mentioned earlier that the beta emission corresponds to the process of a neutron turning into a proton, an electron, and an antineutrino, namely,

$$n \rightarrow p + e^- + v^* \qquad\qquad\qquad \text{(A)}$$

There are certain associated processes that are never observed, even though they would violate none of the well-established conservation laws of energy, momentum, or electric charge. The following are examples of nonobserved processes.

$$n \rightarrow p + e^- \qquad\qquad\qquad \text{(B)}$$
$$n \rightarrow e^- + e^+ \qquad\qquad\qquad \text{(C)}$$
$$n \rightarrow p + e^- + v \qquad\qquad\qquad \text{(D)}$$
$$n \rightarrow v, n \rightarrow v + v, \ldots \qquad\qquad\qquad \text{(E)}$$
$$n \rightarrow v^*, n \rightarrow v^* + v^* \qquad\qquad\qquad \text{(F)}$$

Keeping in mind the principle that the only reason that a process may not be realized is because it violates a conservation law, we proceed to a comparison of the realized process (A) with the forbidden ones (B) and (D). We immediately see that the neutrino and the antineutrino must possess a property that is preserved in these interactions. We call this property the L-charge (I shall explain this name shortly) and choose units so that the neutrino's value of L-charge is +1. Since antiparticles carry the opposite value of all charges, the antineutrino will have L-charge equal to −1.

The processes (E) and (F) are forbidden; in fact, a neutron cannot split into *any* number of neutrinos or antineutrinos. This implies that the neutron has zero value of L-charge.[4]

The fact that (A) takes place and that the process

$$p \rightarrow n + e^+ + v \qquad\qquad\qquad \text{(G)}$$
(a kind of beta decay with positrons)

TABLE 8.1

Baryon and lepton numbers

	Proton	Neutron	Electron	Neutrino	Positron	Antineutrino
B-charge	1	1	0	0	0	0
L-charge	0	0	1	1	-1	-1

is also observed (within nuclei) implies that the L-charge of the electron is also $+1$, and that the L-charge of the proton is zero.*

The absence of the process (C) suggests the existence of another conserved quantity, which we will call B-charge. The neutron carries this quantity since in a pair of electron and positron the total value of all charges is zero. One may choose the units so that the B-charge of the neutron is $+1$. The absence of (E) implies that neutrinos have no B-charge, while the realization of (A) and (G) implies that the B-charges of the proton and of the electron are, respectively, $+1$ and 0.

Hence, the comparison of the observed processes (A) and (G) with the unobserved ones (B)–(F), leads to the definition of two conserved charges. We can make a table (table 8.1) that contains the value of these charges for every particle.

The example above is oversimplified, but it provides a representative demonstration of the procedure involved in the identification and definition of new conserved quantities. This procedure, however, does not stop here. The two conserved quantities were identified through inspection of a small number of realized and unrealized interactions. We have to check that these charges are conserved in *all* known interactions. Moreover, when new particles are discovered, we have to find what value of these charges they carry. Then we have to check again whether these charges are conserved in all processes that involve the new particles.

---

*Note that the process G is never observed in isolated protons because it would violate energy conservation: the mass of the proton is smaller than the total mass of the particles in the right-hand side of G.

I mentioned in the previous section that the energy scales characterizing the three fundamental forces are separated. This fact is very convenient: it allows us to isolate processes for which only one of the forces is responsible and to study the conservation laws for such processes alone. This way we identify quantities that are conserved in one type of interaction but not in another. Thus, we can distinguish between the different types of symmetry that characterize each of these forces.

An example of a quantity that is conserved in the strong interactions but is violated in the electromagnetic and weak ones is the so-called *isospin*. The introduction of isospin was motivated by the fact that the neutron and the proton behave in an identical manner, when they interact through the strong nuclear forces. Therefore, it is convenient for the purpose of mathematical bookkeeping to treat the neutron and the proton as two states of the same particle. This particle is conventionally referred to as the *nucleon*.* The two states of the nucleon are distinguished only by the introduction of a new physical quantity: isospin. The name denotes that the new physical quantity behaves mathematically in a way similar to spin. The two states (neutron, proton) of the nuclear particle are mathematically similar to the two possible values of the spin projection characterizing a particle with spin $1/2$. For this reason, it is convenient to state that the pair neutron-proton has isospin equal to $1/2$ and that the proton corresponds to isospin projection $+1/2$ and neutron to isospin projection $-1/2$.[5]

---

*This is a very common mathematical trick in theoretical high-energy physics. We effectively represent a qualitative difference by a numerical one through the introduction of a new generalized mathematical object. In principle, we can do so for all qualitative differences and treat all particles as different states of a single particle. This is not of much help because the corresponding mathematical object would be too unwieldy as it has to incorporate all distinctions between particles (mass, spin, all kinds of charges). We gain nothing from this process: we just use numbers to represent qualities. Still, when we come across a symmetry (even an approximate one), such a translation is immensely helpful in our effort to understand the underlying laws of motion. The introduction of isospin works in the study of strong interactions because they are much stronger than any other forces, and to a first approximation we may forget that the proton and the neutron behave very differently when electromagnetic effects are taken into account. The reader is reminded that the ancestor of this technique is the eighteenth-century realization that the two types of electric charge can be described by a simple number that can also take negative values (sec. 2.6).

Isospin is not an exact symmetry: substituting a neutron with a proton makes a lot of difference. These particles have different electric charge to start with, and the neutron is a bit heavier. However, these differences are not important for the strong interactions: the laws of motion they induce are invariant under an exchange of isospin. Hence, isospin is preserved in the strong interactions.

Unlike isospin, the L-charge and the B-charge defined earlier are conserved in all known interactions. They are among the most important conserved quantities of particle physics. The L-charge is known as the *lepton number*, and the particles that carry it are called *leptons* (i.e., light). The B-charge is known as the *baryon number*, and the corresponding particles are called *baryons* (heavy). These names were chosen because, during the early days of high-energy physics, the known baryons were much heavier than the known leptons. Nowadays, these names are used only conventionally since we have discovered leptons that are heavier than some baryons.

The technology has changed dramatically since the early 1930s, when the first experiments involving high-energy collisions were designed and executed. The basic method of research, however, has remained the same. We accelerate beams of particles, bring them into collision, and meticulously monitor the resulting particles. Our most important objective is to identify their intrinsic properties: spin, mass, electric charge, other charges. But there is a large amount of data we collect besides that: particle orbits, distributions of particle energies, frequencies of specific processes, really a huge amount of information, chaotic and seemingly senseless. What makes order out of chaos and creates coherent theories out of formless data is the search for symmetry. To discern the pattern underlying the particle interactions, we look for conserved quantities, and from them we infer the symmetries of the laws of motion.

The second basic principle of high-energy physics is the particle-wave duality. To see deeply into the structure of matter, that is, to probe into very short distances, we have to employ particles with very high energies. To transfer energy to a particle, we have to increase its speed. We achieve this by placing them in an electric field, provided of course that the particles carry electric charge. The transferred energy increases with the time the particle is accelerated and with the electric field strength. To obtain a very high energy, it is more convenient to increase the former parameter. In some accelerators this is achieved by forcing the particles to move in circular orbits, so that they may pass many times from the area in which the electric field is situated. This way they will spend a longer time being acted upon by the electric field, even if the accelerator is relatively small. To force particles to move in circular orbits, we usually employ magnetic fields. It follows that a suitable combination of electric and magnetic fields can be used to accelerate particles to very high energies.

Particles that move in a circle lose energy because circular motion involves acceleration and accelerated charged particles always emit electromagnetic waves. This implies that we have to tread carefully in our design of particle accelerators. The particles cannot stay in the accelerator too long because sooner or later the energy loss due to electromagnetic radiation may cancel all gains. Hence for every experimental configuration there exists a maximum value of energy that can be transferred to the accelerated particles. It is a matter of engineering and intelligent design to construct an experimental setup that will maximize this value of energy for a given technological level and budget.

The basic rule of thumb in the engineering of particle accelerators is that *size is important*. The particles follow their circular orbit in suitably constructed tubes. The larger the radius of the orbit, the smaller the force that needs to be exerted on a particle to keep it there. Hence the energy loss due to radiation (which increases with this force) is also smaller. For this reason, the size of accelerators is becoming increasingly larger: from the first accelerators in the 1930s, which could be housed in the basement of a building (the smallest could even be hand-held), we have now the Large Hadron Collider, which will be activated in Geneva in 2008 and involves a ring of twenty-seven kilometers circumference.

The latter accelerator is a part of the European Council for Nuclear Research (CERN). CERN and other great centers of high-energy physics involve thousands of workers, theoretical or experimental physicists, engineers, technical personnel, and administrators, of diverse nationalities.

The experimental data are distributed among many research groups in countries all over the world. The distribution of work and the global involvement are similar to that of a multinational corporation: the only difference is on the issue of profit. The centers for high-energy physics are as yet nonprofit organizations, and their survival depends on the fluctuating commitment of national governments to invest in basic research.

The construction of particle accelerators that can probe to higher and higher energies is widely considered as the safest bet for the discovery of new fundamental physics. The study of energy scales that have not been probed before is similar to the exploration of an uncharted territory. It may be a New World, full of unexpected wonders and challenges; or again it may turn out to be nothing but a familiar place, subject to the known rules and limitations. We shall next see that, in the last sixty years, high-energy physics has been a mixture of both. Significant new discoveries have been made, and they led to a significant reappraisal of our understanding of matter, but on the other hand, all these discoveries turned out to be very respectful of the fundamental rules of modern physics: those of quantum mechanics and special relativity.

8.3 The Breakdown of Simplicity

The aim of science is to seek the simplest explanation of complex facts. We are apt to fall into the error of thinking that the facts are simple because simplicity is the goal of our quest. The guiding motto in the life of every natural philosopher should be, "Seek simplicity and distrust it."
—Alfred North Whitehead, *The Concept of Nature*

During the 1930s, the physics community tried to digest the consequences of the newly born quantum theory. The principle of particle-field duality, in particular, had significant implications for the understanding of the fundamental forces. In classical physics, the field played the role of a mediator for forces. Quantum theory indicated that any quantum field could be described in terms of fundamental particles. Indeed, the photon is the mediator of the quantum electromagnetic forces. It is therefore natural to expect that there will be particles mediating the two types of nuclear force.

The properties of the mediating particles determine to a large extent the features of the corresponding forces. The most important property is the mediator's mass. If this is large, then it is more difficult for the mediating particles to be created. As a result, the force will not be sustained at long distances. It turns out that the interaction of two particles through such force becomes vanishingly small if their distance is larger than a specific length, which is known as the force's *range*. At low energy, there is a very simple relation between the mediator's rest mass and the range of the corresponding force: one is inversely proportional to the other. If, however, the mediator is massless, the corresponding interaction persists even at long distances; the force is then said to be *long range*. Since the photon is massless, the electromagnetic forces can be exerted even at macroscopic distances—this is why they have been known for such a long time.

In contrast, the weak and strong nuclear forces are *short range*; they do not persist away from the nucleus. This suggested that the corresponding mediating particles are massive. Weak forces in particular are characterized by such a short range that their mediators would have to be extremely heavy. They turned out to have a rest mass about 180,000 times that of the electron, some hundreds of times higher than the energies that could be achieved by particle accelerators in the 1930s and 1940s.

The strong nuclear forces had a relatively longer range; the corresponding particles should be lighter, and they could perhaps be detected with the technology of the times. The relevant theory was developed by the physicist Hideki Yukawa in 1935.[6] Yukawa realized that it is necessary to postulate three particles as mediators of the strong nuclear force. The reason is that the strong interactions do not distinguish between protons and neutrons. Hence, there has to be one mediator for each possible interaction, namely, proton with proton, neutron with neutron, and proton with neutron. The first two cases need mediators of zero electric charge because the charge of the interacting particles is not changed during the interaction The third case, however, may need particles with charge equaling that of the electron; in fact two of them, a particle and its antiparticle

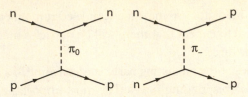

Figure 8.3 **Neutron-proton interactions by means of pion exchange.** These diagrams describe two possible ways that protons and neutrons interact by pion exchange. In the first diagram, a neutron emits a neutral pion, which is later absorbed by a proton, thus transferring energy between the two particles. In the second diagram, a neutron emits a negatively charged pion, thus becoming a proton. The pion is absorbed by a proton, which then becomes a neutron. This process results in an exchange between the proton and the neutron.

(fig. 8.3). These particles were collectively known as the *pions* because they were represented by the Greek letter π. The neutral pion was denoted as π^0; the pair of charged ones, as π^+ and π^-. The spin of all three pions had to be equal to zero.

One particle with some of the predicted features of π^- was soon discovered. Its mass was approximately two hundred times that of the electron, very close to the original prediction of Yukawa. But the enthusiasm for this discovery was short-lived. The newly discovered particle did not participate in the strong nuclear interactions. Moreover, its spin turned out to be ½ rather than zero, which was the value Yukawa's theory predicted for the pions. In fact, the new particle was in all respects (but mass) identical to the electron. It was called mu-particle or *muon*. It is denoted as μ^-, and its antiparticle is denoted as μ^+.

The existence of the muon was something of a puzzle, and in a sense it remains so. We do not understand why nature needs a bigger, heavier brother to the electron. It has seemingly no role to play in the economy of the material world.

Muons do not live long. They rapidly decay into electrons with a simultaneous emission of neutrinos. It was later realized that some of the neutrinos that appear in the muon decay were different from the neutrinos people were familiar with.[7] It turned out that there is a new type of neutrino (and antineutrino) that has the same relation to the muon as the original neutrino has to the electron. In other words, for every process involving electrons and neutrinos of the old type, there is a corresponding process with muons and neutrinos of the new type.[8] As a result of this discovery, the original neutrino was renamed *electron neutrino* (ν_e), and the new neutrino was called *muon neutrino* (ν_μ).

The pions were eventually discovered a few years later (1947).[9] The charged pions had a mass about 270 times larger than the electron, while the neutral one was slightly lighter (about 260 electron masses). Further study showed that pions participated in all three of the interactions, and that their lepton and baryon numbers were equal to zero. This last property set them apart from any other type of particle (but photons).

The pions were the first representatives of a new class of particles, the *mesons*, which were given this name (meaning the middle) because they lie in the middle between baryons and leptons and could not be identified with either of them (their leptonic and baryonic numbers both being zero).

TABLE 8.2

Particles that had been discovered by the end of the 1940s

Name	Symbol	Charge	Mass	Spin	Antiparticle	Interactions
Electron	e	−1	1	½	exists	weak, EM
Proton	p	+1	1836	½	exists	weak, EM, strong
Neutron	n	0	1842	½	exists	weak, strong
Neutrino	ν	0	0	½	exists	weak
+ pion	π^+	+1	267	0	exists	weak, EM, strong
Neutral pion	π^0	0	260	0	none	weak, strong
Muon	μ	−1	207	½	exists	weak, EM
Photon	γ	0	0	1	none	EM

Note: This table contains the particles thought of as fundamental by the late 1940s. The charge is in units of the electron charge, and the mass in units of electron mass. The last column describes the interactions in which these particles were known at that time to participate; EM stands for electromagnetic.

By the middle of the twentieth century, physicists had either discovered or predicted with a high degree of certainty a relatively small number of fundamental particles, almost all of which seemed to have an important physical function. The atoms of chemistry were composed of electrons together with protons and neutrons in the nucleus. Electromagnetic interactions were mediated by the photon, and strong nuclear forces by the three pions. The (electron) neutrinos participated in the beta decay. Only the muons seemed to play no role in the greater scheme of things.

In less than half a century, one moved from a description of the world in terms of about one hundred elementary substances (the different atoms of chemistry) to one in terms of about ten fundamental particles. This was a dramatic improvement, but it was not destined to last long. The new, more powerful accelerators that were developed in the 1950s unleashed a huge number of new particles, all vying for the prestigious characterization of "elementary."

In 1953 the first very high-energy accelerator, the Cosmotron, started working at Brookhaven National Laboratory in the United States. A new breed of particles was discovered there. They were called *strange* because their behavior seemed strange to physicists at the time. The family of strange particles consisted of a few members—which were either mesons or baryons—that interacted mostly through the strong nuclear forces. Processes through which strange particles turned into other strange particles were much more common than processes through which strange particles turned into nonstrange ones. This suggested that the latter processes were due to *weak interactions*, and the former due to strong or electromagnetic ones. In a world with no weak forces, strange particles would never decay into nonstrange ones.

I explained in section 8.2 that the absence of seemingly reasonable processes implies the existence of symmetries and conserved quantities. The behavior of strange particles suggested that they were characterized by a new physical quantity that was preserved in the strong and electromagnetic interactions, but violated in the weak ones. Since this quantity characterizes strange particles and vanishes for all others, it was named *strangeness*.

TABLE 8.3
Strange particles and their properties

Particle	Charge	Mass	Spin	Strangeness	Isospin	Isospin projection	Baryon number
K$^+$	1	966	0	1	½	½	0
K^0	0	974	0	1	½	−½	0
Λ^0	0	2183	½	−1	0	0	1
Σ$^+$	1	2328	½	−1	1	1	1
Σ^0	0	2334	½	−1	1	0	1
Σ$^-$	−1	2343	½	−1	1	−1	1
Ξ^0	0	2573	½	−2	½	½	1
Ξ$^-$	−1	2586	½	−2	½	−½	1
Ω$^-$	−1	3272	½	−3	0	0	1

Note: The charge is measured in units of electron charge, and the mass in units of the electron mass. Note that we have only written the particles and not the antiparticles. The antiparticles are all denoted by an overbar, except for the antiparticle of K$^+$, which is denoted conventionally as K$^-$.

The strange particles were only the tip of the iceberg. A huge number of new particles (baryons and mesons) were discovered in the following years. These particles were short-lived and decayed rapidly by means of strong interactions to other, more stable particles. Because of their short lifetime, they were rather difficult to detect, but they were nonetheless identified.

In the mid-1950s the table of "elementary" particles contained about 150 entries; many of them corresponded to particles with very short lifetimes. The expectation that particle physics would provide a simple picture of the basic constituents of matter seemed to be falling apart. The situation seemed as hopeless as that of nineteenth-century chemistry trying to guess the structure of the atoms. To add to the sense of bewilderment, certain symmetries that had initially been thought of as natural turned out to be violated in the microscopic

world. We refer to the P-, C-, and CP-symmetries, which were introduced in section 7.7.

The breakdown of the P-symmetry was discovered first. It originated
from a rather innocuous problem. Two particles had been discovered
that were in all respect identical: mass, spin, electric charge. Their only
difference was that they decayed into particle configurations characterized by different value of the conserved quantity associated to the
P-symmetry (named *parity* for short).

This problem was addressed by two young physicists working at the
Institute for Advanced Study in Princeton: Tsung-Dao Lee and Chen-
Ning Yang. They realized that there are two alternatives: we either have
two distinct particles that differ only in their behavior under the
P-symmetry or we have a single particle whose interactions violate the
P-symmetry. The decays in question were due to the weak forces,
hence in the latter alternative one would have to assume that the weak
interactions violate parity. But how could this be? The P-symmetry
may not have been a cherished principle of physics, but it was so natural and there had been no indication of its violation in the past.

The issue of parity violation was being discussed at that time, but
more in the form of a highly unlikely theoretical possibility. Lee and
Yang realized that while parity conservation had been verified with a
high degree of accuracy in the electromagnetic and strong interactions,
"past experiments on the weak interactions had actually no bearing on
the question of parity conservation. . . . The fact that parity conservation in the weak interactions was believed for so long without experimental support was very startling. But what was more startling was the
prospect that a space-time symmetry law which the physicists have
learned so well may be violated."[10] They then suggested a number of
experiments that could test whether the P-transformation is indeed a
symmetry of the weak interaction.[11]

Their idea was well received, but with a lack of enthusiasm. The idea
that parity may be violated seemed too remote to warrant immediate
attention. However, Lee and Yang persuaded Chien-Shiung Wu, a col-
league of theirs at Columbia University in New York, who proceeded to

perform one of the proposed experiments. The associated problems were enormous, but they were eventually overcome.[12] The experiment was a success, and it gave the physics community a shock: parity was not conserved in the weak interactions. The world did not treat left- and right-handed particles the same way. The simple symmetry that had been taken for granted for so long was not a universal one.

Once the violation of the P-symmetry was established, it was not far-fetched to assume that other symmetries were violated in the weak interactions. This was immediately evident for the C-symmetry of charge conjugation. As discussed in section 7.7, the most likely reason for the violation of the P- and C-symmetry is that the underlying theory treats left-handed and right-handed massless particles asymmetrically. Since neutrinos participate in the weak interactions, and they were thought to be massless particles, it was natural to assume that weak interactions distinguished between left-handed and right-handed neutrinos and that this distinction lies at the root of the C and P violation.

Even if C and P are violated independently, their combination, CP, may still be a valid symmetry of the weak interactions. We may recall from section 7.7 that the violations of the C and P symmetries in theories with massless particles have a common cause, and for this reason the violation of the C-symmetry cancels out that of the P-symmetry. Nonetheless, things turned out differently. In 1964 it was shown that some rare decays of certain strange particles (known as the *neutral kaons*) violate the CP-symmetry.[13] I will explain the reason for this violation in the next chapter. For now, I will only state that it is not directly related to that of the C or the P violation.

Summarizing, the situation in particle physics was transformed dramatically in the early 1950s because of the new particles discovered in the new generation of particle accelerators. Things were much more complicated than people had been expecting. At first sight, the structure of matter at the subatomic level appeared chaotic. It was necessary to look deeper into the chaos and try to discern a principle that would bring back some order.

8.4 Some Simplicity Restored

In all chaos there is a Cosmos, in all disorder a secret order.
—Carl Gustav Jung[14]

During a short time period (between 1913 and 1930), physicists believed that it was possible to explain all microscopic phenomena in terms of the electrons and the nucleus, which consisted of protons and neutrons. However, as the energies one could access experimentally grew, more and more particles started making their appearance, and by the early 1960s a very large number of them had been accumulated.

The huge number of particles was annoying. It killed off the desire to identify a simple ultimate substratum of matter, which is the deepest motivation of particle physics. It would be very uneconomical—not to say downright ugly—if nature worked with as many as 150 elementary particles. There had to be a hidden structure underneath. The particles that appear to us should be composed of some elementary ingredients. The developments were analogous to those at the beginnings of the nineteenth century, when the exploration of the new world of chemical elements had just started. There were many elements—more than fifty at the time—and the most speculative chemists were trying to find a simpler underlying structure that would explain their properties. The atomic theory and the periodic table were the results of this endeavor. Mendeleev succeeded in the classification of the elements because he identified the periodically repeated pattern in their chemical properties: he recognized in effect a symmetry in the chemical properties. History provided a clear hint that the only way out of the chaos of elementary particles would be to look for a kind of "periodic table" for the properties of the subatomic particles.

It took about ten years from the moment the host of new particles was discovered until the formation of a theory that would identify their underlying structure. Many people provided significant contributions in that direction, but the most important ones came from the American physicist Murray Gell-Mann, who was working at the time at the California Institute of Technology. In 1961 he suggested (together but independently with the Israeli physicist Yuval Ne'eman) a system of classification for the lighter mesons and baryons.[15] This system was based on the number 8—or rather on a symmetry that was based on the number 8. It became known as the Eightfold Way. The name was apparently a pun on the diffusion of Buddhist philosophy in California at that time. The classification scheme above was very successful: it even predicted the existence of a particle that was only detected a couple of years later.

However, the eightfold classification was mainly descriptive since there was no underlying justification. It was difficult to find one because there was as yet no reliable theory for either strong or weak interactions. Still, in 1964 Gell-Mann demonstrated that the Eightfold Way could be explained by the assumption that baryons and mesons consist of three fundamental particles, the quarks.[16] The name "quark" was a playful loan from James Joyce's *Finnegans Wake* and had no deeper connotation.[17] George Zweig came up with a similar theory at about the same time (starting from a different direction), only in his scheme the elementary constituents were four.[18] He called them aces after the four aces in a deck of playing cards. Eventually, Gell-Mann's choice of name stuck to the newly postulated particles.

There are three types of quark, or, as Gell-Mann stated, quarks come in three *flavors*. The choice of the word "flavor" is arbitrary. It is only a name for the property that distinguishes between different types of quark. This naming is remarkably appropriate. It highlights the fact that in modern high-energy physics, one still employs qualitative distinctions. Particles are distinguished one of the other not only by their characteristics that relate to the geometry of spacetime, namely, mass and spin, but also by the possession of different qualities like charge or flavor. Modern physicists prefer the name internal property instead of quality, but irrespective of how we choose to call them, these characteristics carry the fundamental intuition that certain aspects of matter cannot be analyzed in terms of motion and shape.

The three quarks are named *up* (denoted u), *down* (d), and *strange* (s), and each of them has an antiparticle. The first two are constituents of ordinary particles, and they form an isospin doublet—just as the proton and the neutron are described as an isospin doublet. The strange quark is contained in all strange particles. Except for the leptons, which are seemingly fundamental, all other known particles (mesons and baryons) consist of quarks.* The following rules should be satisfied for the theory to provide an adequate explanation for the properties of the observed particles.

- A baryon consists of three quarks, while a meson consists of one quark and one antiquark.
- Quarks have electric charges, which are fractions of the electron's charge. In particular, the u quark has charge equal to ⅔ of the electron charge,* while the d and s quarks have charge equal to −⅓.
- Since no particles with the charge of the quarks has ever been observed, we are forced to postulate that quarks can never be found free, that is, without being constituents of either a baryon or a meson. Initially, there was no physical explanation for this behavior.

One may construct ten possible triplets out of the three quarks, such that the total charge of each is a multiple of the electron charge. Each triplet corresponds to a baryon. For instance, the proton consists of two

TABLE 8.4
Gell-Mann's quarks and their properties

Quark	Baryon number	Charge	Mass	Strangeness	Spin	Isospin	Isospin projection
up (u)	1/3	2/3	10	0	½	½	½
down (d)	1/3	−1/3	18	0	½	½	−½
strange (s)	1/3	−1/3	250	−1	½	0	0

Note: These are the properties of the three quarks of the Gell-Mann model. The units for charge and mass are the electron charge and mass, respectively. The quarks' masses can only be roughly estimated because they do not appear free in nature.

*Mesons and baryons, being composite, are collectively denoted as *hadrons* (i.e., coarse particles).
*When we employ the electron's charge as a unit of charge, we refer to its absolute value. In this convention the electron has charge equal to −1.

u quarks and one d quark, while the neutron consists of two d quarks and one u quark. One then may write the lightest baryons as quark triplets. To be precise, the lightest baryons correspond to the lowest energy state of a quark triplet. The triplet may very well possess states of higher energy. In that case, some of its properties are different, and the triplet behaves like a different particle of higher rest mass (see table 8.6). However, these particles have a very short lifetime, and they decay into lighter ones, as the triplet tends toward its state of lowest energy.

TABLE 8.5
Baryons constructed from quarks

Quark triplet	Symbol	Charge	Spin	Isospin	Isospin projection	Strangeness	Mass
uud	p	1	1/2	1/2	1/2	0	1836
udd	n	0	1/2	1/2	−1/2	0	1839
uuu	Δ^{++}	2	3/2	3/2	3/2	0	2407
uud(2)	Δ^+	1	3/2	3/2	1/2	0	2410
udd(2)	Δ^0	0	3/2	3/2	−3/2	0	2412
ddd	Δ^-	−1	3/2	3/2	−3/2	0	2415
uus	Σ^+	1	½	1	1	−1	2328
uds (2)	Σ^0	0	½	1	0	−1	2334
dds	Σ^-	−1	½	1	−1	−1	2343
uds	Λ^0	0	½	0	0	−1	2183
uss	Ξ^0	0	½	1/2	1/2	−2	2573
dss	Ξ^-	−1	½	1/2	−1/2	−2	2586
sss	Ω^-	−1	3/2	0	0	−3	3272

Note: The table provides the structure of some of the lightest baryons in terms of quark triplets. Note that the baryons Δ^+ and Δ^0 have the same quark triplet as the proton and neutron, respectively, but they have different spin and isospin. These baryons correspond to a higher energy state of the triplets uud and udd, respectively, while the proton and the neutron correspond to the ground state of those triplets. Similarly, the particle Σ^0 can be viewed as an energetically excited state of the Λ^0.

TABLE 8.6
Mesons constructed from quarks

Quark	Symbol	Charge	Spin	Isospin	Isospin projection	Strangeness	Mass
ud*	π^+	1	0	1	1	0	280
uu* − dd*	π^0	0	0	1	0	0	270
uu* − dd* − 2ss*	η	0	0	0	0	0	1098
us*	K^+	1	0	½	½	1	998
ds*	K^-	0	0	½	−½	1	996

Note: The table describes the quark structure of some light mesons. Note that certain mesons involve the "mixing" of different types of quark-antiquark pairs. The mass unit is always the electron's mass. The antiquarks are denoted as u*, d*, and s*.

Mesons involve one quark and one antiquark. In table 8.6, we see the structure of some light mesons.

It is important to remark that all interactions between baryons we have discussed so far can be expressed in terms of quark processes. For instance, the beta decay process corresponds to the decay of the quark d to the quark u with an emission of an electron and an antineutrino (see fig. 8.4).

Similarly, the strong interactions between nuclear particles are a consequence of the forces between quarks. When the interaction between

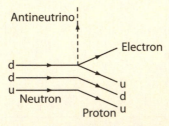

Figure 8.4 **Quarks and weak interactions.** In the beta decay, a neutron turns into a proton with an emission of an electron and an antineutrino. According to the quark model, the beta decay is due to one d quark in the neutron decaying into a u quark with the emission of an electron and an antineutrino.

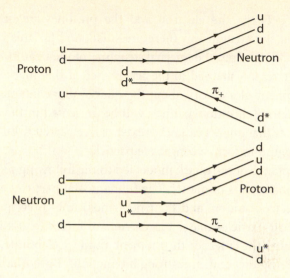

Figure 8.5 Quarks and strong interactions. The diagram describes the role played by the pions in the strong interaction. A triplet of quarks corresponds to either a neutron or a proton. At some moment, the excess energy creates a quark-antiquark pair. The created antiquark joins a quark of the baryon and creates a pion—a quark-antiquark pair. This way the processes $p \rightarrow n + \pi^+$ or $n \rightarrow p + \pi^-$ may take place (the former occurs only within nuclei). The produced pions are absorbed by other nucleons through a similar process, thus giving rise to the strong nuclear forces.

two quarks in a nucleon becomes very strong, a quark–antiquark pair may appear. The antiquark is then combined with another quark of the nucleon, and together they may move away. This quark–antiquark pair is a pion; it can be absorbed by a different nucleon and be destroyed in the interactions with its quarks.

The quark model brought a dramatic simplification to particle physics. Nevertheless, there were reservations about its validity, mainly because nobody had ever observed free quarks. The balance was eventually tipped in its favor for two reasons. First, there were strong indications that the proton was a composite particle. In experiments where a beam of electrons scattered off protons, the behavior of the electrons was different from what one would expect if the proton were a point particle. In effect, the photons that transmitted the electromagnetic interaction between the electron and the proton were of very high energy. Their wavelength was therefore small, and for this reason they could probe deeply into the internal structure of the proton.

The other reason that led to the acceptance of the quark model was the development of a theory about the quark–quark interactions. This theory, quantum chromodynamics, will be presented in the next chapter. It predicts that quarks can never exist as free particles, but they must always be bound to form a composite particle.

The acceptance of the quark model led to the following picture about the basic constituents of matter. The fundamental particles are the four leptons (electron, muon, and their neutrinos), the three quarks (u, d, and s), and all particles that act as carriers of forces (e.g., the photon). We shall ignore the latter for the moment, until we elaborate on the theories describing the weak and strong interactions. Both quarks and leptons are Fermi particles, with spin ½. However, there are three of the former and four of the latter. This does not look nice. However, the combined effort of theory and experiment soon removed this asymmetry. A new quark had been theoretically predicted (for a rather different reason[19]), and a meson containing this quark was discovered in 1974. The new quark was named *charmed* (c). It defined a new physical quantity named charm. Charm was in all respects similar to strangeness: it was preserved in the strong interactions, but not in the weak interactions. Except for the quality of charm, the c quark seemed to be nothing but a heavier version of the u quark, much as the s quark was in relation to the d quark and the muon in relation to the electron.

The matching between leptons and quarks was then perfect. We had two doublets of leptons (or two families, as they came to be called), electron and its neutrino, muon and its neutrino, corresponding to two families of quarks (u and d are the first, s and c the second).

The correspondence between quarks and leptons persisted even after the discovery of a new family of particles. A new rather heavy lepton, the tau (τ),[20] was identified in 1976. The tau together with its neutrino formed a third family of leptons. The corresponding quark family consists of the bottom quark b and the top quark t. The former was discovered in 1977 and the latter in 1999. In fact, the trust in the lepton-quark symmetry was so strong that after the discovery of the b quark, the existence of the t quark was taken for granted long before its actual detection.

It seems, then, that there are three families of elementary particles. Each family contains two leptons and two quarks. Families are in all respects identical, except for one thing. Each particle of the second family is heavier than its corresponding particle of the first family and each particle of the third family is heavier than its corresponding particle of the second family. The particles of the first family, being the lightest, are the most stable. The reason is that particles typically decay into lighter

TABLE 8.7
The three fermion families

Family	Type	Name	Symbol	Charge	Spin	Mass
	lepton	electron	e	-1	$\frac{1}{2}$	1
	lepton	electron neutrino	ν_e	0	$\frac{1}{2}$	<0.000014
1	quark	up	u	2/3	$\frac{1}{2}$	~10
	quark	down	d	$-1/3$	$\frac{1}{2}$	~18
	lepton	muon	μ	-1	$\frac{1}{2}$	207
	lepton	muon neutrino	ν_μ	0	$\frac{1}{2}$	<0.00053
2	quark	strange	s	$-1/3$	$\frac{1}{2}$	~350
	quark	charmed	c	2/3	$\frac{1}{2}$	~2,700
	lepton	tau	τ	-1	$\frac{1}{2}$	3,483
	lepton	tau neutrino	ν_τ	0	$\frac{1}{2}$	<0.06
3	quark	bottom	b	$-1/3$	$\frac{1}{2}$	~9,000
	quark	top	t	2/3	$\frac{1}{2}$	~360,000

Note: The particles in this table are at present thought to be elementary. All families follow the same pattern, but the energies involved increase about 100–200 times as we go from one family to the next. The first family consists of the lightest fermions, which are the ingredients of ordinary matter—nuclei and atoms. The tilde indicates an estimated value of mass for the quarks. Recent experiments suggest that the neutrinos have nonzero masses, even though these are very small (see sec. 9.3).

ones, if the symmetry of the theory allows it. Hence, while particles of the second and third families (with the exception of the neutrinos) decay into the ones of the first through weak interactions, the converse does not happen because particles of the first generation are the lightest. It follows that the *stable* formations of matter (nuclei and atoms) exclusively consist of members of the first family.

The reason for the existence of three families of fundamental particles remains a great mystery of high-energy physics. One can imagine a world where only particles of the first family exist. This world would be almost identical to ours—at least as far as the facts of everyday life are concerned. Chemistry and biology would be the same. Our bodies and our metabolism would not differ, and neither would the behavior of animals and plants or their nutritional value. The properties of minerals and all earthly phenomena—whether on Earth's surface or in Earth's core—would not change. The function of technological artefacts, from Stone Age tools to computer chips, would remain unaffected. Stars—including the Sun—would also behave the same way. Only in the energies that correspond to the extreme temperatures in the interior of the Sun is the contribution of the higher families to the physical processes (through their neutrinos) substantial, but even then the distinction among the three types of neutrinos does not seem to be crucial for the star's evolution.

We do not know the reason for the multiplicity of families. It seems to play no role in the economy of the physical world. This picture may change when we obtain a better and deeper knowledge of subatomic matter. For all we know, the three families may be the first step in a repeated pattern: more and more families being discovered in high energies, until we have again an overabundance of fundamental particles.[21] Then someone would have to do again the work of Mendeleev and Gell-Mann and look for a new symmetry pattern among the particles we now view as fundamental.

9

REACHING THE LIMITS

THE GAUGE PRINCIPLE
AND THE STANDARD MODEL

9.1 The Birth of the Gauge Principle

Every line of force, therefore, at whatever distance it may be taken from
the magnet, must be considered as a closed circuit, passing in some part
of its course through the magnet, and having an equal amount of force
in every part of its course.

—M. Faraday

The knowledge of the fundamental particles is not by itself sufficient
to provide a complete description of the structure of matter.
Unless we also know how the particles interact with each other, our information is not very useful because the basic properties of an elementary
particle refer mainly to the way it evolves and interacts with other particles. The electromagnetic forces are accurately described by quantum
electrodynamics; the strong and weak interactions are the ones we need
to account for.

It turned out that the discovery of a new principle was necessary for
the attempts to describe these interactions to succeed. This was the principle of gauge symmetry or *gauge principle*. Its roots lie in the work of

the German mathematician Hermann Weyl, but also in an elaboration by Noether of her theorem on symmetry in physics.[1] These theories had been formulated in the late 1910s—before the identification of the strong and weak forces and before the advent of quantum mechanics.

It was 1915 and Einstein had just presented his general theory of relativity. The mathematical beauty and elegance of the new theory captivated Hermann Weyl, a thirty-year-old mathematician at the Federal Institute of Technology in Zurich who was to become one of the most influential mathematicians of the twentieth century. Weyl's aim was to extend general relativity in such a way that it would easily incorporate electromagnetism. Maxwell's equations would then partake some of the geometrical character of Einstein's theory—that was his hope. He failed in this endeavor, but in the process he identified what turned out to be an intrinsic fundamental feature of electromagnetism: its gauge symmetry.

> Weyl's idea was to introduce a scale, or gauge, that varied from point to point and whose variation round a closed path in space-time would encapsulate the electromagnetic force. Almost immediately (in fact in an appendix to Weyl's paper) Einstein criticised the idea on physical grounds. If Weyl was right, then the size of a particle would depend on its past history, whereas experiments showed that all atoms of hydrogen, say, had identical properties. One might have thought that such a telling criticism from someone of Einstein's standing would have discouraged Weyl and that he might have withdrawn his paper. It is a tribute to his mathematical insight and self-confidence that he went ahead. The idea was too beautiful to discard, and Maxwell's equations came out like magic.[2]

To see how Weyl's idea translates into the language of modern quantum field theory, we first recall (sec. 7.3) that fermionic particles of spin ½—such as quarks or leptons—are described by Dirac fields. The Dirac field at a spacetime point contains information about the momentum and spin of any associated particle that may potentially appear at that point. It also contains information about the *quantum phase* of a particle at that spacetime point. Unlike spin and momentum, the quantum phase is not itself an unambiguous physical quantity: only differences in quantum phase are measurable (see sec. 6.2). If one changes the quantum

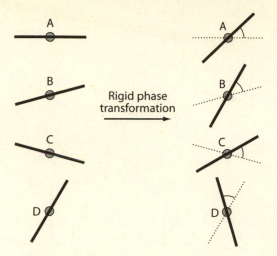

Figure 9.1 **Rigid transformation of the quantum phase.** A Dirac field associates a configuration for the quantum phase (corresponding to the rod's axis) at each of four spacetime points, A, B, C, and D. A rigid transformation rotates the phase by the same amount at every point.

phase of all potential particles *rigidly*, namely, by the *same amount in every spacetime point*, the relative quantum phase between any two points remains the same. One expects that the physical quantities and the laws of motion will not be affected by such a transformation. This is indeed born out by a precise mathematical analysis.

It follows that the rigid change of phase is a symmetry for the free particles of spin ½. From Noether's theorem (sec. 4.2), we infer that an associated conserved quantity exists. This quantity turns out to correspond to the total number of particles (e.g., electrons) minus the number of antiparticles (e.g., positrons) that are present at any moment of time. If the particles are charged, the conserved quantity is proportional to the total electric charge. This is a hint that the electromagnetic interaction is somehow related to the transformations of the quantum phase.

A rigid transformation of the quantum phase is *not* the most natural symmetry to impose on a physical system. We learned from relativity that there is no way to transmit any information between two spacelike

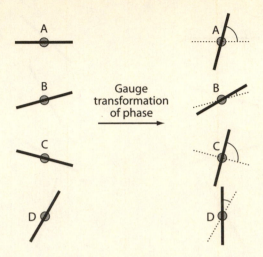

Figure 9.2 **Gauge transformations.** The quantum phase changes by a different amount on each spacetime point, A, B, C, and D. This transformation is a symmetry of particles interacting with the electromagnetic field.

separated points (sec. 3.4). For this reason, the idea that one may impose the same change of phase in two points arbitrarily far away from each other, while mathematically possible, is physically suspect. A symmetry that corresponds to a change of the quantum phase by *different* amounts at different spacetime points seems more natural. An observer at a specific spacetime point would then decide on his own how much to rotate the quantum phase. Weyl had called this transformation a *gauge* transformation because he compared it to changes of a measurement scale (similar to the gauge in a mechanical scale for weights). The name persists until today, even though (at least in the present context) it would be more appropriate to denote these transformations as phase transformations.

It turns out that the gauge transformations do not leave invariant the laws of motion for the free Dirac field. Hence, it seems as though they do not define a physical symmetry. But here a remarkable thing happens, essentially what Weyl had seen when he was captivated by the charm of the gauge principle. If we assume that the fermions interact with the electromagnetic field—as is the case in quantum electrodynamics—then the gauge transformations leave the laws of motion of the total system (fermions + electromagnetic field) invariant, if the electromagnetic field is also transformed in an appropriate way. As a result, the theory invariant under gauge transformations is not one of free particles, but of particles interacting through electromagnetic forces.

The role of the electromagnetic field is to *transport* the information about the quantum phases between different spacetime points. It corresponds to a *rule* that determines the change of quantum phase *along any path* we choose to follow as we move from one spacetime point to another. This rule of transport is neither unique nor a priori specified in a physical system. In fact, every possible state (profile) of the electromagnetic field corresponds to a different rule of transport. The equation

$$\text{profile of the electromagnetic field} = \text{rule of transport for the quantum phase}$$

is the most fundamental statement pertaining to gauge symmetry. Since the electromagnetic field is a fully dynamical object, the corresponding rule of transport will also be dynamical: it will be subject to physical change through interactions with matter.

The relation of gauge symmetry to the electromagnetic interactions implies that any discussion about the quantum phases at different spacetime points must necessarily take into account the rules governing the phases' transport. The realization of gauge symmetry in nature involves two types of field: one that carries the quantum phase and another that transports the phase in spacetime. The carrier of the phase is the field that describes charged particles (the Dirac field); the rule of transport is the mediator of the "force" between the charged particles, namely, the electromagnetic field.

To keep the laws of motion invariant under gauge transformations, it is necessary that every change in one field be compensated by a suitable change in the other. This is not easy to achieve: most conceivable laws of motion are simply incompatible with gauge symmetry. For electromagnetism in particular, the requirement of gauge symmetry turns out to be so stringent, that the laws of motion are determined uniquely; they

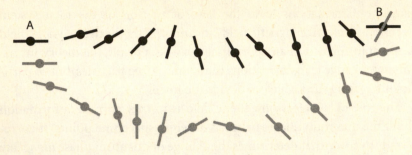

Figure 9.3 **Phase transport by the electromagnetic field.** The electromagnetic field provides a rule for the change of phase as we travel along a path. The diagram shows two different paths (black and gray) connecting two spacetime points, A and B. The circles represent the intermediate spacetime points of each path. The quantum phase changes along the path, as shown by the rotation of the corresponding rods. The final value of the phase at point B depends on the chosen path.

coincide with those that form the basis of quantum electrodynamics. It follows that electromagnetism arises as the *unique* physical theory that describes quantum phases and their transport in a way compatible with Weyl's gauge principle.

9.2 Yang-Mills Theories

> An idea is always a generalization and generalization is a property of thinking. To generalize means to think.
> —G.W.F. Hegel, *The Philosophy of Right*

Initially, most physicists perceived the gauge symmetry of the electromagnetic field as a mere mathematical curiosity. Admittedly, it was an elegant curiosity, but still not good enough to play the role of a fundamental principle, even at a time when electromagnetism was the only known microscopic force. By the 1950s, however, physicists had become familiar with two more types of force, for which there existed no satisfactory description. The fact that QED provided the only successful quantum field theory up to that date suggested that one should look at the electromagnetic theory for new ideas about the treatment of the other forces.

Chen-Ning Yang—whom we already encountered in the discovery of parity violation (sec. 8.3)—and Robert L. Mills took the crucial step in this direction in 1954.[3] They suggested that the other microscopic forces were also characterized by a gauge symmetry of a different type, and that this symmetry could be used to determine their laws of motion. Electromagnetism is based on the freedom to rotate the quantum phase by a different amount in every spacetime point. Perhaps the other forces corresponded to different (more general) types of "rotation."

The gauge symmetry in electromagnetism refers to the change of phase of a single field—and hence of a single type of particle. Could we perhaps obtain more symmetries if we took more fields into account? After all, this situation is closer to reality because microscopic forces involve interactions between many different types of particles.

The simplest possible generalization would be to consider two Dirac fields that correspond to two different types of particle with spin $\frac{1}{2}$. Let us call the first field "black" and the second one "white." The names are chosen arbitrarily, to illustrate the fact that the only difference between the fields is an intrinsic quality that has nothing to do with the basic properties of the corresponding particles (mass and spin). Each field is characterized by a "rod" that represents its quantum phase: hence, we have a black rod and a white rod.

A crucial observation is that quantum theory allows two fields that correspond to particles of same mass and spin to *mix*. This means that one may "construct" a new field by mixing the black and white fields according to some proportion. We need to stress that the field "mixing" is a mathematical prescription and *not* a physical process. For this reason, the word "mix" should be taken with a grain of salt. Fields obtained through mixing are not in any sense composite. The mixing of two mixed fields may very well lead to the original black and white fields, something that can never happen when we mix actual fluids. Mixing is a name for a very specific mathematical operation within the formalism of quantum mechanics that allows one to combine two or more fields

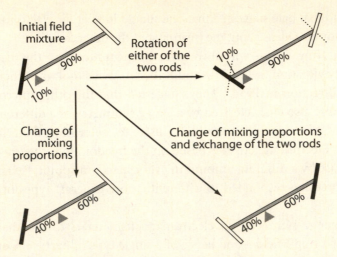

Figure 9.4 **Mixing of two fields.** A mechanical analogue for the mixing of the two fields is provided by the two rods (representing the quantum phases) being joined by a connecting bar. The bar is split into two parts according to the proportions of the mixing. There exist four different ways to change the field mixing. We can rotate either of the two quantum phases; we can change the proportions of the mixing; and—perhaps unexpectedly—we can change the proportions of mixing while performing a flip between the red and green phase.

and obtain a new one. There is no physical reason why fields obtained from mixing should not appear in nature.

I next describe a mechanical model that captures some of the basic features of field mixing (see fig. 9.4). A field that is obtained from the mixing of the black and white fields is represented by a bar connecting the black and white rods in a way that allows them to rotate around their midpoints. The bar splits into two pieces that denote the proportions of the black and white field in the "mixture." The quantum phases play an important role in the mixing—two mixed fields with the same proportions of black and white but with different directions of the quantum phases are physically distinct.

We next study the possible ways to transform the fields obtained from mixing.

- We may rotate the black rod.
- We may rotate the white rod.
- We may change the proportion of black and white fields in the mixture.
- We may change the mixing proportions and at the same time exchange the black with the white rod. The mathematical description of field mixing suggests that it does make a difference whether the black phase is on the left-hand side or on the right-hand-side of the connecting bar.

The most general transformation for the mixed fields is obtained by a combination of four elementary transformations above. It follows that four numbers determine such a transformation: the angle of rotation of the black phase, the angle of rotation of the white phase, the change of the mixing proportions without a flip, and the change of proportions with a flip.

Yang and Mills then suggested that one should describe the black and white fields in terms of laws of motion that remain invariant when different mixing transformations are applied at different spacetime points. As in electromagnetism, the only way to guarantee this symmetry is through the introduction of a mediating field, which will transport information about the mixing between different spacetime points. In the electromagnetic case, there is one quantum phase at each spacetime point and we need only one parameter, its angle of rotation, to describe the phase's change. In the present case, the system is more complex. We may have a different mixture of black and white field at each spacetime point and the corresponding transformations involve four different parameters. For this reason, we have to introduce *four* mediating fields, one for each of the four possible transformations depicted in figure 9.4. These fields are similar to the electromagnetic fields: they correspond to mediating particles with zero mass and spin equal to 1; these particles can be either left-handed or right-handed.

It is truly remarkable that the requirement of gauge symmetry uniquely selects the laws of motion for the combined system of the black and white fields and the four mediating fields. Once we identify the gauge symmetry, we obtain automatically the mediating fields and their dynamical behavior. Is it perhaps possible that the weak and strong interactions are also described this way, just as electromagnetism was?

This hope was eventually fulfilled—all the more a tribute to Yang's and Mills's insight—but fulfillment came with a twist or two. The gauge principle involves a severe constraint, which causes it to fail in the description of the microscopic forces. All mediating fields related to a gauge symmetry (*gauge fields* for short) must correspond to *massless particles*. This is the case for the field mediating electromagnetic forces, and for the gauge fields of the Yang-Mills type. However, both strong and weak interactions are short-range forces. This seems to imply that their mediators are massive particles. What is even worse, the spin ½ particles that carry the quantum phase *have to be massless themselves* in a Yang-Mills theory that describes interactions that violate the P-symmetry (the weak interactions). It follows that the gauge theories of Yang and Mills fail to make contact with reality, since—with the possible exception of neutrinos—all known particles with spin ½ have nonzero mass.

The gauge principle appeared then as another brilliant idea that had been condemned to the dustbin of history by a reality check. Most people had abandoned it by the end of the 1950s. Gauge theory survived only because a new mechanism was conceived that turned the tables on the relation between gauge symmetry and mediating particles. We will now turn to the study of this mechanism.

9.3 Symmetry Is Broken "Spontaneously"

> You boil it in sawdust: you salt it in glue:
> You condense it with locusts and tape:
> Still keeping one principal object in view—
> To preserve its symmetrical shape.
> —Lewis Carroll, *The Hunting of the Snark*

The reconciliation of the gauge principle with experiment demanded a deeper understanding of the manifestations of symmetry in physics. I emphasized in section 4.1 that symmetry refers to the laws of motion, not

to the observed configurations of physical systems. In quantum field theories, however, we have a singular circumstance: the laws of motion for
the interactions between particles are expressed in terms of fields. Consequently, all symmetries must refer to fields. However, the observed quantities usually refer to particles alone. Hence, we must translate any talk
about symmetry from the language of fields to the language of particles.

As explained in section 4.1, there is no reason to expect the observed
behavior of a physical system to bring out the symmetry of its laws of
motion: the observed behavior depends on the system's initial conditions, which are in general chaotic. However, in most physical systems
there are special states that exhibit at least part of the underlying symmetry. Quantum field theories, in particular, possess a very distinguished state, the state of minimum energy: the vacuum. The vacuum
provides a guide for the translation between the language of fields and
that of particles. According to Fock's prescription (see sec. 7.1), the field
states are constructed by adding particles to the vacuum state. Hence, if
we know how the vacuum state behaves under symmetry transformations, we go a long way toward achieving a translation of the field's symmetry into the particle language.

A very important consequence of the basic rules of quantum theory is
that any transformation that leaves the laws of motion invariant transforms a state of definite energy* into another state of the same energy. This
is natural after all: if a transformation changes the value of energy, it would
fail to be a symmetry of the system. Since in a quantum field theory the
vacuum state is the unique state of zero energy (see axiom 4 in sec. 7.6), a
symmetry transformation takes the vacuum onto itself. This implies that
the vacuum is invariant under any symmetry transformation of the fields.

The property above has a remarkable consequence. Recall from
Fock's construction (sec. 7.1) that the states of the field arise from the
"addition" of particles to the vacuum state. A symmetry transformation
that acts on such a state affects only its particle content because the transformation leaves the vacuum invariant. We consider the special case of a

---

*A state of definite energy is one in which all measurements of energy always yield the same result.
For example, any energy measurement on the vacuum always gives the result zero.

state describing a single particle (either elementary or composite) of specific rest mass and spin. A symmetry transformation takes this state into another state, again describing a single particle. The important point is that the particle in the new state must have the same values of mass and spin as the original one. The reason is that mass and spin remain invariant under the Poincaré transformations (sec. 4.4), and any symmetry must be compatible with the Poincaré symmetry because the latter is forced to all physical systems by the structure of spacetime. The presence of a symmetry in the fields' laws of motion suggests *the existence of different particles with the same values of mass and spin*. Conversely, if we observe two different particles with the same mass and spin, we can infer that the corresponding quantum field theory is characterized by a symmetry in its laws of motion.[4]

However, the invariance of the vacuum under symmetry transformations is guaranteed only if a *unique* lowest-energy state exists (i.e., if the fourth axiom for quantum field theories in sec. 7.6 is satisfied). If several distinct states of zero energy exist, the symmetry transformations may very well transform one such state into another. In this case, the lowest-energy states do not remain invariant under the symmetry transformations. Consequently, the symmetry of the fields will not appear in the physical properties of their corresponding particles.

The uniqueness of the vacuum seems a very natural principle to require from a quantum field theory, but it is not inviolate. In the late 1950s some mathematical models appeared in the physics of solids that seemed to describe physical systems with more than one vacuum states. These models were later translated into the context of particle physics. In a system characterized by more than one state of minimum energy, the physical properties of the observed particles will not reflect the symmetries of the laws of motion. This loss of the observable manifestation of symmetry was named *spontaneous symmetry breaking*.[5]

We saw in the previous chapter that all fundamental particles (quarks and leptons) possess spin ½. It turns out that it is impossible to formulate a theory with a nonunique vacuum state for the fields that describe such particles because it would fail to be renormalizable (sec. 7.5). The only possible type of field that could possess a nonunique vacuum

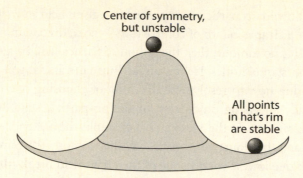

Center of symmetry,
but unstable

All points
in hat's rim
are stable

Figure 9.5 **The Mexican hat.** The top of the hat is the point of highest symmetry. It is, however, unstable. A ball placed at this point will roll and eventually will stop on a point on the hat's rim. These points represent the several states of minimum energy that characterize the Higgs field.

corresponds to particles with zero spin. Such fields were named *Higgs* fields (and the corresponding particles Higgs particles), after the British physicist Peter W. Higgs, who first studied them in relation to symmetry breaking in a gauge theory.[6]

To study the basic features of quantum field theories with spontaneous symmetry breaking, it is useful to consider the so-called *Mexican hat* model. The top of a Mexican hat is its point of highest symmetry: the hat looks the same if we rotate it around a vertical axis passing through this point. However, it is not a stable point. If we place a small ball on the top, it will soon roll over and will eventually rest at a point on the hat's rim. It is, however, impossible to predict which point this will be because there is nothing to distinguish one rim point from another. One may think of the rim points as analogous to the states of lowest energy in a quantum field theory. In this analogy, the Mexican hat provides a simple demonstration of a system with many lowest-energy states that are mutually indistinguishable.

The quantum theory of the Higgs field is in many respects similar to the Mexican hat example. There are many states of lowest energy, but *only one* of them can play the role of the physical vacuum—the state of no-particles. The principle of wave-particle duality forces this upon us. To determine how many particles are present in a given field state, it is necessary to have a clear standard of comparison—the state containing no particles. If this is nonunique, then the notion of particle will be ambiguous. Nature abhors ambiguities. Hence, unlike Buridan's ass,[7] it will not stand still pondering how to resolve the dilemma of choosing one out of the many equally appealing candidates for the no-particle state. It will just choose one, any one. In the Mexican hat example, the ball ends up on one rim point and, by doing so, "breaks" the hat's symmetry. Similarly, the selection of one lowest-energy state of the Higgs field breaks the theory's *apparent* symmetry. It is impossible to determine beforehand which of all lowest-energy states of the Higgs field will be selected. This is the reason that the symmetry breaking is referred to as "spontaneous."

Before proceeding further, I will simplify the language. I will call a field corresponding to a specific particle by the name of the particle with the word "field" appended. Instead of saying "the field corresponding to the u quark," I will say the "u field," and instead of "the field corresponding to the electron," I will say "the electron field." In talking about the fields corresponding to massless particles, it will be necessary to distinguish them by the helicity: therefore I will say "the left-handed neutrino field," meaning the "field corresponding to the left-handed neutrino." I will also use generic designations, such as "quark field," meaning "any field corresponding to a quark"; "left-handed field," meaning "any field corresponding to a left-handed massless particle"; and similarly for "lepton field," "fermion field," and so on. There will be no change in the naming of fermionic particles: when I say "u quark," "d quark," or "electron" without the word "field" appended, I always refer to particles. Finally, I will refer to the particles that correspond to the gauge fields as "gauge particles" or "mediators," and to the particles that correspond to the Higgs field as "Higgs particles."

With the naming conventions above, I will proceed to examine the situation that arises when one or more fields of the Higgs type are present in a theory characterized by gauge symmetry. The Higgs fields will then interact with both the fermionic fields and the gauge fields. As I explained toward the end of section 9.2, the gauge symmetry requires that both fermions and gauge particles be massless. The consequences of spontaneous symmetry breaking in such theories are nothing less than spectacular.

First, some of the Higgs fields are "swallowed" by the gauge fields. This results in the gauge particles becoming massive! Second, left-handed and right-handed (massless) fermionic fields intertwine through their interaction with the Higgs field, and they "condense" to form *massive* particles of spin ½!

Clearly, the effects described above are so remarkable and unexpected that they require careful explanation. To do this, we have to recall the basic features of the field-particle duality. We saw in section 7.1 that certain states that arise out of the quantization of the field are readily interpreted in terms of particles. Any particle is characterized by specific values of mass and spin. These values are related to basic properties of the corresponding field: its geometrical structure (which corresponds to kinematics) and its laws of motion (dynamics). Small changes in the dynamics do not usually cause dramatic changes to the field-particle relation: the properties of the particles do not change when one incorporates additional interactions into the field's laws of motion. However, the interaction of any quantum field with a Higgs field is very special because of the presence of the many different states of lowest energy. Spontaneous symmetry breaking really causes a shake-up in the field's basic structure. This is reflected in the properties of the physically realized particles.

It is important to emphasize that the changes of particle properties by the interaction with the Higgs field are not arbitrary. They have to be compatible with the general structure of the theory, and for this reason they are subject to rules. To explain these rules, we consider an idealized case of a general quantum field (say A) coexisting with (perhaps many different) Higgs fields, but not interacting with them. In this case, we would observe the particles corresponding to the field A coexisting with Higgs particles. The former may have any value of mass and spin, while the latter all have zero spin.

Suppose now that we switch on the interaction between the field A and the Higgs field, with the result that symmetry breaking strongly affects the field A. How do the states of the combined system change? The main requirement is that no (mathematical) information should be lost. In other words, if we use a number of parameters to describe a field state prior to switching on the interaction, we shall use the same number of parameters afterwards. What, then, are the relevant parameters? The rest masses of the corresponding particles are the first ones that come to mind, but it turns out they are not really so important. Rest mass is just another form of energy, and as such, it can change if the energy transforms into another form. What is really important is the particles' spin.

Recall that spin is an intrinsic property of particles that takes discrete values in quantum theory (see sec. 4.3). For the present discussion, the key fact is that the value of spin determines the number of possible values for the projection of the spin arrow onto a particular direction (see fig. 6.7). In particular, for a massive particle with spin $\frac{1}{2}$, there exist two distinct states that are almost identical: they differ only in the value of the spin projection. For a massive particle with spin 1, there are three distinct states, and so on. Field-particle duality implies that the structure of the corresponding fields reflects this distinction. The fields should consist of as many different components as the number of possible spin projections for their corresponding particles.

The key observation is that the number of field components is *additive*. If a quantum field theory involves two fields, we can add the number of components of both fields and thus obtain the number of components of the combined field system. This number characterizes the amount of information needed in the specification of any field state.* It does not depend on the details of the theory's laws of motion. Therefore, it should be the same whether the interaction of the field A with the Higgs field has been switched on or off.

A solid principle guides the change in the states of a field due to spontaneous symmetry breaking. I will apply it to the case of interest, namely, gauge theories.

First, consider the interaction between Higgs and gauge fields. Higgs particles are spinless; their fields have one component each. Due to the presence of the gauge symmetry, the gauge particles are massless, they have spin equal to one, and they may be either left-handed or right-

---

*It is often referred to as the *number of degrees of freedom* of the quantum field theory.

handed. One gauge field will therefore have one component for each value of the particle's helicity—two in total.* On the other hand, a massive particle of spin 1 has three different values for the spin projection. Gauge particles can become massive, only if they acquire an additional spin degree of freedom. The Higgs particles have exactly what is needed. If the Higgs field and the gauge field work together, they may create one *massive* gauge particle. One type of Higgs particle will consequently disappear, so that the total number of field components stays the same. Therefore, a suitable interaction between gauge and Higgs fields results in the latter being "absorbed" by the former. The mediating particles thereby become massive, and an equal number of Higgs particles disappears completely.

However, it is not necessary that the above will happen. The interaction between gauge and Higgs fields may be of a form that will not allow all gauge particles to absorb the Higgs particles. Some gauge particles will therefore remain massless.

We next consider the effect of the Higgs fields' interaction with two fields that correspond to massless fermions of spin $\frac{1}{2}$. The fermionic fields have one component each. On the other hand, a *massive* particle with spin $\frac{1}{2}$ has two possible values for the spin projection—the corresponding field will thereby have two components. It is conceivable that the Higgs fields can induce the two fields corresponding to the massless fermions to coalesce, *thereby forming a single massive particle with spin* $\frac{1}{2}$. This process would clearly preserve the total number of field components. Indeed, explicit mathematical calculations verify the guess above. (In all known physical situations, the two massless particles have opposite helicity. In general, this is not necessary.)

We therefore conclude that even if the gauge symmetry requires the fermionic fields to correspond to massless particles, the introduction of the Higgs field with spontaneous symmetry breaking allows us to relax

---

*Gauge particles of different helicities participate equally in all interactions compatible with the gauge symmetry. For this reason, it is customary to treat the two massless particles as corresponding to two different spin states of a single particle. Hence we refer to the photon as a single particle, while strictly speaking there are two: a left-handed and a right-handed one.

TABLE 9.1

The effect of spontaneous symmetry breaking on particle properties

	Gauge fields	Fermion fields
Particle content before symmetry breaking	1 left-handed + 1 right-handed massless mediators (2 components) + 1 Higgs particle (1 component)	1 left-handed particle of spin ½ (1 component) + 1 right-handed particle of spin ½ (1 component)
Particle content after symmetry breaking	1 massive particle of spin 1 (3 components)	1 massive particle of spin ½ (2 components)

this restriction, without compromising in any sense the gauge symmetry of the system's laws of motion.

To summarize, it appears that the particles we observe in nature cannot be described by a theory with gauge symmetry. However, the study of spontaneous symmetry breaking via Higgs fields suggests that appearances can be deceptive. A theory may be fully compatible with gauge symmetry at the level of the laws of motion and still fail to reproduce the supposed trademarks of this symmetry at the level of observation, namely, the existence of massless particles—both mediators and fermions. One should therefore try all possible combinations of gauge symmetries and Higgs fields to see if the patterns they exhibit after spontaneous symmetry breaking correspond to the observed properties of particles.

For the weak interactions the best working combination was identified in the late 1960s by Stephen Weinberg of the University of California at Berkeley and Abdus Salam of Imperial College, London, who had been working independently.[8] Their model (expanded by later work) possesses the following features.

- The theory is characterized by gauge symmetry. This implies that the gauge and fermionic particles it describes have to be massless—at least before symmetry breaking. Since we know that symmetry breaking results in a pair of massless spin ½ particles forming a single massive

spin ½ particle, we shall introduce in the model two fields (one for
left-handed and one for right-handed massless particles) for each
fundamental (massive) lepton that is observed in nature. There are 12
fundamental leptons; hence, we have 2×12 such fields.

- The C- and P-symmetries are not conserved in the weak interactions.
This suggests (see sec. 7.7) that the theory *does* distinguish between
left-handed and right-handed massless particles. Since the gauge
symmetry guarantees that the mediating particles of different helicity
are treated symmetrically, the origin of the asymmetry must lie in the
fermions of the theory (quarks and leptons) and not the mediators.

- The *left-handed* fermion fields are grouped in *doublets*: the (left-handed)
electron and electron neutrino field form one doublet, the (left-handed)
muon and the muon neutrino fields another, the (left-handed) tau and
the tau neutrino field another. Similarly, the (left-handed) u and d
fields form a doublet, and so do the (left-handed) c and s fields, and the
(left-handed) t and b fields. Each doublet is similar to the pair of black
and white fields we considered earlier (see fig. 9.4). This implies that the
left-handed electron "mixes" with the left-handed electron neutrino, the
left-handed u with the left-handed d quark, and so on. The laws of
motion of each doublet remain invariant under gauge transformations
similar to those of figure 9.4, which involve four parameters in their
description. Four gauge fields are therefore necessary to transport the
information about the mixing of the doublet's fields. (Note that *the same*
gauge fields transfer the phase information of all doublets).

- The right-handed fermion fields interact with only one of the four
gauge fields that we introduced for the description of the left-handed
fields' symmetry. The corresponding gauge symmetry is similar to that
of electromagnetism (sec. 9.1): it corresponds to a rotation of the
quantum phase for each of the right-handed particles. The gauge
symmetry of the right-handed particles is therefore very different
from that of the left-handed ones. This is the source for the violation
of the C- and P-symmetries.

- It turns out that we need to introduce at least four Higgs fields for the
model to work. Three of these fields are swallowed after symmetry
breaking by three of the gauge fields. Hence, three of the gauge

particles become massive. These particles were named the W^+, W^-, and Z^0 bosons. The first two are electrically charged, and they form a particle-antiparticle pair. However, Z^0 is electrically neutral. The prediction of the Z^0 boson was a surprise because it corresponds to processes that had not been observed at the time the theory was stated—they were only observed a few weeks later. Both particles were detected in the early 1980s—they were rather heavy, with a rest mass more than 180,000 times that of the electron.

- One Higgs field is not absorbed after symmetry breaking. The corresponding particle is known as *the* Higgs particle. It has not been observed yet, but there are high hopes that in the near future experiments at the Large Hadron Collider will find it. Its detection will constitute the final confirmation for this model of the weak interactions.

- Since one Higgs field is not eaten up, one mediating particle remains massless. This is nothing but the photon: the mediator of the electromagnetic force. This feature is truly impressive because it implies that the model provides a unification of the weak and electromagnetic forces as different aspect of a single interaction, conveniently named *electroweak*. However, we should note that this unification is not fully satisfactory. When electricity and magnetism were unified in Maxwell's theory, all phenomena were explained in terms of a single magnitude: the electric charge. In the electroweak unification, one has to postulate different "charges" (coupling constants in the field theory language) for the electromagnetic and weak parts of the interaction. At present, we do not know any relation between them, and for this reason, we must accept that the electroweak unification is at this moment incomplete.

- The final part of the model involves the acquisition of mass by quarks and leptons through spontaneous symmetry breaking. The left-handed fermion fields combine with the right-handed ones to form massive particles of spin ½. Hence this model is also compatible with the fact that observed fermions are massive.

The part of the model that involves the acquisition of mass by the fermions has a very interesting physical consequence. There is no evident a priori reason to believe that the interaction of the Higgs field with

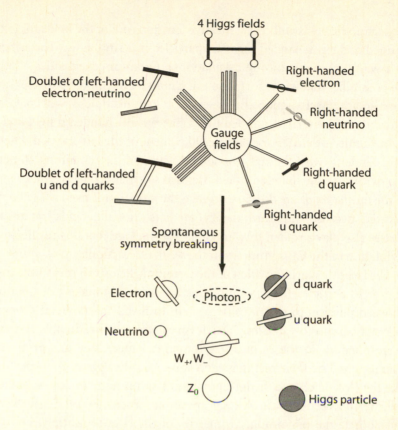

4 Higgs fields

Doublet of left-handed electron-neutrino

Right-handed electron

Gauge fields

Right-handed neutrino

Doublet of left-handed u and d quarks

Right-handed d quark

Spontaneous symmetry breaking

Right-handed u quark

Electron

Photon

d quark

Neutrino

u quark

W_+, W_-

Z_0

Higgs particle

Figure 9.6 **Spontaneous symmetry breaking within the first fermion family.** The figure is a diagrammatic description of the modern theory of electroweak interactions, restricted within the first family of particles. Prior to spontaneous symmetry breaking, all fields correspond to massless particles. Left-handed particles form doublets, and they interact with four components of the gauge field (corresponding to transformations of fig. 9.4). Right-handed particles remain single and interact with only one component of the gauge field (as in fig. 9.2). Four Higgs fields are also introduced in order to cause symmetry breaking. The results of symmetry breaking appear in the bottom half of the diagram. All fermion particles acquire mass, and so do three out of the four gauge particles. One gauge particle (the photon) remains massless. Only one out of the four components of the Higgs field forms a particle. The white rod corresponds to particles that carry electric charge and consequently interact through photons. The weak forces are mediated by the heavy gauge particles. Note that the neutrino is represented by a smaller ball because its mass is much smaller than those of the other fermions.

the fermionic fields will create massive fermions out of the fields for left-handed and right-handed massless particles in a simple way (or rather, in a way convenient to our calculations). Quantum mechanics allows the possibility that in the formation of a massive fermionic particle (say, the d-quark), a left-handed field (say, a left-handed d field) will coalesce with *two* different right-handed fields (say, a right-handed d field and a right-handed s field). For example, 90 percent of the left-handed d field may condense with the right-handed d field and the remaining 10 percent with the right-handed s field; we then say that *the fields d and s mix in the formation of the d-quark*. In fact, the situation is even more complicated because quantum theory predicts that the fields' quantum phases also play a role in this "mixing." We shall see that this results into a violation of the CP-symmetry in the weak interactions.

The issue then arises, which of the fermionic fields can mix in the formation of particles? To determine this, we must first make a distinction between physical properties that pertain to fields and physical properties that pertain to particles. Particles are defined as irreducible systems corresponding to the Poincaré symmetry; hence, they are primarily characterized by their rest mass and spin. All other physical properties, like the electric charge or the lepton and baryon number (the so-called internal properties), are strictly speaking properties of fields. They incorporate the remaining symmetries of the laws of motion (i.e., the ones that do not have a spacetime origin). And we know that the laws of motion are always expressed in terms of fields.

If the fields never mixed in the formation of particles, the distinction above would not be interesting because one type of particle would always correspond to one and only type of field.* Field mixing destroys this correspondence.

---

*This allows us to name the field after the corresponding particle. Hence we talk about the u quark field, the electron field, or even make finer distinctions and talk about the left-handed neutrino field, etc. In presence of mixing, things are not so straightforward because three different fields may be responsible for the laws of motion of a single particle. The usual convention is that we name a field after the particle in which it is "contained" with the highest proportion. For example, the d quark field is defined as the field that participates with the highest proportion in the formation of the d quark particle.

The internal properties refer to the symmetries of the laws of motion. By Noether's theorem, they must be conserved in all physical processes. If, however, the formation of a particle involves the mixing of two fields that differ in the value of some internal property, the particle will sometimes manifest the one value and sometimes the other. Hence, the conservation law associated with the symmetry will be violated. This implies that fields that differ even in a single internal property cannot mix. It follows that quark fields cannot mix with leptonic fields (because they differ in baryon and lepton number). Similarly, fields that correspond to particles of different electric charge cannot mix.

Keeping the rules above in mind, a quick look at table 8.7 will convince the reader that fields of the same particle family cannot mix in the formation of particles. This leaves open the possibility of mixing between fields that correspond to particles of different families. Hence, mixing can take place between the

- u, c, and t fields
- d, s, and b fields
- electron, muon, and tau fields
- three neutrino fields

However, the mixings above are not independent of each other. The theory's laws of motion are invariant under gauge transformations that mix the two fields that constitute each of the particle doublets (sec. 9.2). This mixing is different from the one referred to earlier because it involves fermions of the same doublet (hence of the same family), and it does not refer to the formation of particles. However, the mathematical descriptions are similar. For this reason, the gauge symmetry unravels some of the mixings between fermions of different families. In effect, the u, c, and t fields mix in the formation of particles, the d, s, and b also combine, but the gauge transformations are employed to cancel one of the two mixings. They cannot cancel both, though. The usual convention is that the mixing of the fields corresponding to the positively charged quarks is canceled, but that the fields of negatively charged particles (d, s, and c) are allowed to mix. Something similar holds for leptons: gauge transformations can cancel one of the two possible mixings.

TABLE 9.2

Properties pertaining to particle and properties pertaining to field

Particle Properties	Field Properties
• Defining characteristics of particles that are related to the Poincaré symmetry of spacetime. • Origin is geometrical (*mass, spin, helicity*).	• Refer to symmetries of the laws of motion. • Origin is dynamical (*electric charge, baryon and lepton number, color, etc.*).

Note: The mixing of different fields in the formation of particles highlights a distinction between properties pertaining to particles and properties pertaining to fields that would be otherwise too fine to be interesting. The properties specific to particles are the ones that refer to their allowed motions in Minkowski spacetime—they are essentially kinematical. Fields refer to dynamics; they determine the particles' laws of motion. For this reason, any property related to the laws of motion (via symmetries and conservation laws) refers primarily to fields. These properties are traditionally labeled as internal.

The usual convention is that the surviving mixture corresponds to the three neutrinos.

Field mixing introduces a rather remarkable complication in the field-particle relation. In experiments, we observe only the particles. However, the particles' laws of motion are expressed in terms of the corresponding fields. Because of mixing, the interaction between two particles of the same family (say, a d quark and a u quark) involves an interaction of fields corresponding to different families (the u field will interact with the d field, but also with the s and the b field). The converse also holds. This implies that in a process in which only the d field is relevant, we will see not only the d quark but also the s and b quarks.

The mixing of quark fields is well attested to in experiments. In fact, it had been observed before the discovery of fermion families,[9] and it provided theoretical motivation for the introduction of the c quark. In leptons, the situation is rather different. The neutrinos have very small mass, very close to zero. If their mass is precisely zero, then spontaneous symmetry breaking has not affected their fields at all. In this case, field mixing cannot take place in leptons, since the mixing of the electron, muon, and tau fields is unraveled by gauge transformations. However, in recent experiments the mixing of neutrinos has been observed.

Unlike quarks, the neutrinos exist free in nature, and their ability to mix implies that a beam of neutrinos arising out of a specific process (one that involves one single neutrino field) may contain any of the three neutrino particles. In fact, the mean number for each type of neutrino in the beam is a periodic function of time—the frequency increasing with the mass differences between neutrinos. This phenomenon is known as *neutrino oscillations*, and its observation in 1998 provides a strong—if indirect—indication that at least some of the neutrinos have nonzero mass.

Another important consequence of the field mixing in the formation of particles is that the CP-transformation fails to define a universal symmetry of nature. We saw earlier that a theory that treats left-handed particles differently from the right-handed ones violates the C-symmetry and the P-symmetry, but not necessarily their combination, CP. Field mixing changes this picture (see fig. 9.7). The quantum phases of fields remain unaffected in the formation of particles, with one exception. In one of the particles that are formed by mixing, the phase of a single field must be rotated by a specific angle.[10] The magnitude of this angle is determined by the quark fields' interaction with the Higgs field, and it constitutes a fundamental parameter of the electroweak theory.

The change of the quantum phase above affects the way quarks interact with each other in the weak interactions. The angle of rotation is *external* to the fermion fields because it is provided by the structure of the Higgs field. Since fermion fields incorporate the description of both particles and antiparticles, this change of angle *is the same for both quarks and antiquarks*. On the other hand, quantum phases rotate in opposite directions in quarks and antiquarks, so that their overall effect cancels out. At least this is the case if the C-symmetry is preserved. *But there can be no cancellation if the symmetry breaking angle is incorporated.* For example, if the symmetry breaking angle corresponds to a rotation of one degree, a rotation of the quark's phase by 51 degrees clockwise will be accompanied by a rotation of the antiquark's phase by 49 degrees counter-clockwise. The exact correspondence between quark and antiquark will be lost.

Quantum Phases for the Three Quark Fields

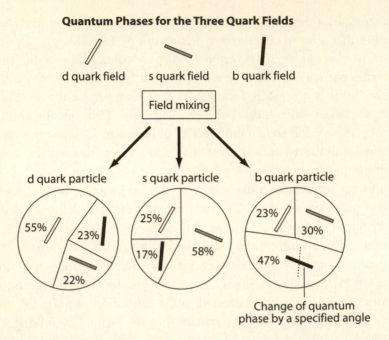

Figure 9.7 **Quark field mixing in the formation of physical particles.** This drawing shows the formation of quark particles from the mixing of the quark fields. Each particle contains a specified proportion of each of the three fields (the numbers provided are not realistic). The black, white, and gray rods represent the fields' quantum phases. The orientation of the quantum phases is preserved under mixing (as can be seen in the diagram), with one exception. In one component of one out of the three particles, the phase must rotate by a specific angle. This is a necessary consequence of the interaction with the Higgs field: the value of this angle is a parameter of the theory of electroweak interactions. It can only be determined from experiment. It is a matter of convention, where one will effect the change of phase. Here we chose to do it in the b field's contribution to the b quark. This phase affects the way the particles interact through the weak forces and is responsible for the violation of the CP-symmetry.

TABLE 9.3
Bookkeeping for the symmetry breaking angle

	Symmetry Breaking Angle	Quark Rotation	Antiquark Rotation	Total
No C violation	0	+50 degrees	−50 degrees	0
C violation	+1 degree	+50 + 1 = + 51 degrees	−50 + 1 = − 49 degrees	+2

Note: This is a simple example of how a symmetry breaking angle creates an overall imbalance in the rotations of the quantum phases for quarks and antiquarks. We assume that clockwise rotation is positive and counter-clockwise negative.

Hence a new source for the violation of the C-symmetry appears. This one cannot be compensated by any changes in the P-symmetry. It follows that the CP-symmetry is violated in the weak interactions, something that had been observed earlier in the interactions of strange particles (see sec. 8.3).

The description above of the weak interactions and their consequences involves a number of parameters (like the angle responsible for CP violation), whose values are not known a priori and have to be determined by recourse to experiment. Once this is done, one may derive precise predictions for all processes that correspond to the electromagnetic and weak interactions. The fact that the weak interactions are indeed weak is a great help in that regard because it implies that the results of perturbation theory are quite trustworthy. It turns out that all predictions of this theory are in good agreement with every experiment that has been carried out so far.

9.4 The Force That Binds

It seems probable to me that God in the Beginning form'd Matter in solid, massy, hard, impenetrable, moveable particles of such Sizes and Figures and with such Properties, and in such Proportion to Space, as most conduced to the end for which he form'd them . . . no ordinary

Power being able to divide what God himself made One in the first creation.

—Sir Isaac Newton, *Opticks*

The previous section showed that the weak interactions are successfully described by a model that involves (1) an asymmetry of left-handed and right-handed spin ½ particles, (2) a specific gauge symmetry (corresponding roughly to the one in fig. 9.4), and (3) spontaneous symmetry breaking. The latter guarantees that the mediating particles are massive. As explained in section 8.2, this implies that the weak interactions are short-range.

The strong interactions posed a different problem altogether, in some aspects easier and in others more difficult than the one posed by the weak interactions. The easy part was that the strong forces do not violate the C- and P-symmetries. This fact suggests that left-handed and right-handed particles would have to be treated in equal footing. On the other hand, a theory of strong forces must be based on the properties of quarks, which are not directly observable. Hence, the only experimental tests of such a theory would come from the prediction of the properties of the observed composite particles (hadrons). This is a very difficult task and much grander in scope—it is comparable to the achievement of the early atomic physicists, which explained the chemical properties of atoms in terms of Rutherford's model and quantum theory.

There were many attempts to find a theory of the strong interactions during the 1960s. Some of them were partially successful, but none was without its problems. For some time, it was widely believed that the problem lay with quantum field theory, and that one should scrap it altogether and introduce different frameworks for the description of microscopic processes. Indeed, at that time quantum field theory seemed to have reached a dead end. We had learned a great deal from it, some people were saying, but it is about time to discard it "as an old, but rather friendly, mistress who one would be willing to recognize on the street if one encounters it again."[11] Alternatively, one could keep quantum field theory but only as an intermediate step, which would provide some insight about abstract relations between physical quantities relevant to the strong interactions.

One should then focus on these abstract relations, clarify and enlarge them, and then abandon quantum field theory. One could compare this process "to a method sometimes employed in French cuisine: a piece of pheasant meat is cooked between two slices of veal, which are then discarded."[12]

Hence, there was no guarantee that the gauge principle (which was riding on quantum field theory's bandwagon) would emerge victorious. What eventually tipped the balance in its favor was a mathematical result. In 1972 two Dutch physicists, Martinus Veltman and Gerard 't Hooft, proved that the quantum field theories constructed from the gauge principle are renormalizable in perturbation theory—renormalizability being an indirect but crucial consequence of the gauge symmetry.[13] This result implies that the gauge theories for particle interactions have one advantage over their opponents: they provide unambiguous predictions for all physical phenomena in their domain. This fact not only strengthened the belief that a gauge theory could describe the strong forces, but also guaranteed the acceptance of the gauge theory describing the weak interactions.

The first task toward the construction of a theory for the strong interactions, characterized by gauge symmetry, would be to identify the symmetry involved. This required a careful look at the experimental data, looking for repeated patterns and, more importantly, for forbidden processes. The quark structure of hadrons provided the conclusive argument. We observe either particles composed of three quarks (the baryons) or pairs of quark and antiquark (the mesons). We never observe particles composed of four or five quarks, or two quarks and one antiquark. This strongly suggests that the underlying symmetry must be based on the number three. It should therefore involve field *triplets*, the same way that the symmetry of the weak interactions involved field *doublets*. Therefore, to describe the strong interactions, we must assume that the quark fields possess a quality that appears in three different forms. The name given to this quality was *color*[14]—of course, it has nothing to do with the colors we see with our eyes. The choice of name was arbitrary, like that of flavor. The corresponding theory was naturally named quantum chromodynamics (i.e., color dynamics), or QCD for short.

According to QCD, every quark field comes in three colors—conveniently labeled red, green, and blue. This implies that there are three

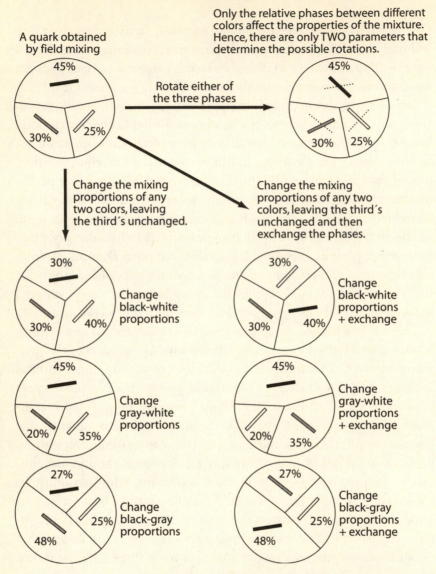

Figure 9.8 **Color mixing and gauge transformations.** Quark fields of different colors (but of the same flavor) mix freely in QCD. Mixing is determined by the relative proportions of each color in the mixture, but also by the quantum phases of each of the colored quark fields. As usual, a rod represents the direction of each quantum phase. In distinction to the mixing described in

different versions of every quark field: a red u, a blue u, a green u, a red d, a blue d, a green d, and so on. Hence, there are altogether $3 \times 6 = 18$ quark fields differing both in flavor and in color. One should also take into account the antiparticles' colors: antired, antiblue, antigreen. We note that anticolors cannot be expressed in terms of colors; they are entirely different. An analogy with real colors would be very misleading at this point: antired does not coincide with the optically opposite color of red, namely, green.

The three colors can freely mix through their corresponding fields— figure 9.8 contains a mechanical representation of such mixing. The freedom to change the color mixing is elevated into the fundamental postulate of the theory. QCD is defined by the requirement that the laws of motion remain invariant under any transformation that changes the color mixture at every spacetime point.* Figure 9.8 demonstrates that

---

Figure 9.8 (continued)

fig. 9.4, only the relative orientations of the phases affect the properties of the mixture. Hence, we may consider one of the rods as unmoving and allow for the rotation of the other two. This implies that there are only two physically distinct ways of rotating the rods. The other transformations that can be applied are the change of the relative proportions between any two colors (three different ways), and the change of relative proportions between any two colors together with an exchange of their "positioning" (also three possible ways). A general transformation is a combination of the eight transformations above, and for this reason it involves the specification of eight parameters. This implies that one needs eight gauge fields to guarantee the invariance of the theory's laws of motion under the corresponding gauge transformations.

*The reader may wonder about the degree of arbitrariness involved in this procedure. Can we really come up with any symmetry we want and then define a gauge theory out of it? The answer is yes, but of course the theory will have to be judged by experiment. Nonetheless, symmetries are not arbitrary mathematical constructions. They follow a pattern. The symmetries involved in gauge theories had been exhaustively studied and classified in a stunning tour de force by the French mathematician Elie Cartan in the 1890s. There are four families of symmetries, and the symmetries in each family conform to a specific pattern. (There are also five weird symmetries that do not belong to any of the four families.) Hence, the arbitrariness in choosing a symmetry for the gauge theory is much smaller than what may naïvely appear.

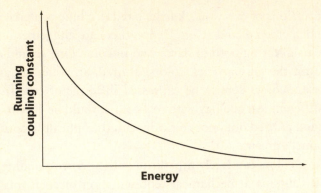

Figure 9.9 **The running coupling constant for QCD.** Unlike the case of QED, the running coupling constant of QCD decreases to zero at high energies, so that at very high energy the quarks do not interact with each other. On the other hand, it becomes very large at low energies, which implies that quarks interact more and more strongly as the distance between them increases. Hence, they are tightly bound to each other and tend to form composite particles.

eight parameters are necessary in order to define the most general transformation that mixes the colors of quark fields. It follows that one must introduce eight gauge fields as mediators of the strong force. The corresponding particles are named *gluons* because their main function is to keep the quarks together.

The most important property of the gluons is that they are themselves colored. For this reason, they interact with each other as much as they interact with the quarks. This is in marked contrast to photons, which are electrically neutral particles and do not interact with each other, but only with the particles carrying electric charge. The fact that gluons are colored turns out to be responsible for a fundamental feature of quantum chromodynamics, which is known as *asymptotic freedom*.[15] As energy increases, the strong interactions between particles become weaker. The language of the running coupling constant (see sec. 7.5) is particularly useful for the description of this effect. At high energy, the running coupling constant of QCD (the effective strength of the interaction) becomes extremely small, so that the quarks behave effectively like free particles: they do not interact with each other. At low energy, the exact opposite happens: the interac-

Figure 9.10 **Baryon described as a bag.** A simple way of looking at the internal structure of baryons is to assume that the gluons form a bag impenetrable from the inside. The quarks move freely inside the bag, but they can never puncture it. If their energy increases, the strength of the bag's walls also increases by the same proportion, so as to keep them contained.

tions between gluons become very strong, with the result that very strong attractive forces act upon the quarks. In a sense, gluons at low energies behave like springs that connect the quarks and do not allow them to move away from each another. This is believed to be the reason that quarks are lumped together in particles and never appear free.

The "springs" corresponding to the gluons are rather weird. When quarks come close to each other (and the property of asymptotic freedom is manifested), the springs disappear into thin air: they exert no force any more. This suggests a simple picture for the visualization of the composite particles formed by quarks. A baryon is similar to a bag containing three quarks. Gluons form the bag's wall, which is elastic but impenetrable. No quark can puncture the wall and escape. In the bag's interior, quarks move freely. They do not interact with each other (in particular they do not collide). If, however, they attempt to leave the bag, the gluon wall pushes them back.

It is almost universally accepted today that quantum chromodynamics is the correct theory to describe the strong interactions. However, it is not entirely trouble free. The running coupling constant of QCD takes very large values at low energies. This implies that the usual treatment through perturbation theory is inadequate at these energies. This is a big problem because the quarks' behavior at low energies determines the way composite particles like the proton, the neutron, and the mesons are formed. In other words, perturbation theory cannot determine the properties of the confining wall in the bag analogy of figure 9.10.

The inadequacy of perturbation theory implies that we lack a precise mathematical procedure that will enable us to derive the basic properties of the hadrons from first principles. One has to rely on other methods, such as, the simulation of the QCD dynamics on a computer. This approach is called *lattice QCD*, and it constitutes today one of the largest areas of research in high-energy physics. In the last few years, lattice QCD has made dramatic progress in achieving accurate predictions: some of them are within an order of 2 percent from the experimental results. While this is good enough precision for some experiments that involve hadrons of high mass, it is still far from the precision of QED, which for some quantities is of the order of one part in a billionth. This fact highlights the enormous mathematical difficulties involved in the study of QCD.

In spite of the increasing accuracy obtained from computational methods, our mathematical control over QCD is still rather weak. We cannot even provide a theoretical proof for the most immediate of observations: that quarks form particles in threes and do not, for instance, coalesce in their billions to form large undifferentiated blobs of nuclear matter. For this reason, the development of new nonperturbative techniques for QCD is considered as a very important area of research in modern theoretical particle physics.

The predictions of QCD have been repeatedly proved consistent with observation. For this reason, present-day physicists believe strongly that it provides a correct description of the strong interactions. However, it is impossible to refute conclusively the (nowadays extremely rare) criticism that QCD is not really a sharply tested theory, and that surprises may be hidden in the yet inaccessible details.[16] If, however, QCD turns out not to be the correct theory of the strong interactions, the surprise will be enormous; it will feel as though nature played a joke on the people who study it—a joke too crude in its subtlety to be gracefully appreciated.

9.5 The Standard Model

> Neither a work of nature nor one of art we get to know when they have
> been finished; we must surprise them in the process of being created so
> as to understand them to some degree.
> —Johann Wolfgang Von Goethe, Letter to Karl Friedrich Zelter

Once the relation between particles and fields had been established
and the machinery of quantum field theory put in place, physicists
had a conceptual and technical framework by which to describe the
behavior of matter at the shortest scales. This point marks the maturity of high-energy physics because the existence of such a framework
allows the translation of ideas into precise estimations about the results
of experiments. However, a general framework alone does not suffice
for the description of the micro world. Two more ingredients are necessary. First, one has to identify the fundamental particles and their
defining properties. Second, one needs to write the particles' law of
motion, specifying essentially how they interact with each other. At
present, we believe that we know these ingredients, at least up to the
energy scale that is accessible by our experiments. Our basic knowledge has not changed much since the mid-1970s; only fine details have
been added.

The gauge principle turned out to be the most important building
block of our present theories. Besides severely constraining our choices
for the particles' laws of motion, it also provides a classification of elementary particles in three types, according to the role they play in the
manifestation of the corresponding symmetry. First, we have the particles with spin ½, which are the carriers of the quantum phases. Leptons
and quarks fall in this category. Second, we have the mediators of the
gauge symmetry. These are particles with spin equal to 1: the photon,
the three particles mediating the weak interactions, and the eight types
of gluons. Third, standing alone, is the Higgs particle, of spin zero,
whose field is responsible of the breaking of the symmetry.

The Higgs particle is the only one predicted by the present theories
that has not yet been detected. There are some rather indirect arguments

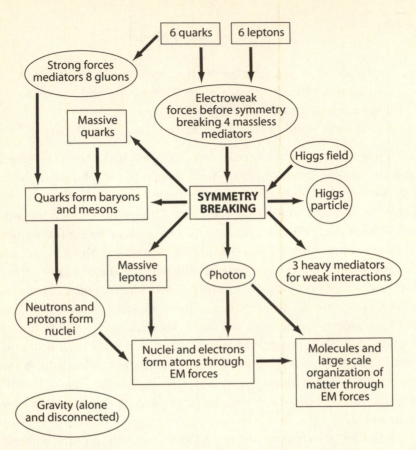

Figure 9.11 **The structure of matter according to the Standard Model.**

that its mass is smaller than one million electron masses, an energy range that will be reached at the Large Hadron Collider in the coming years. At present, the consensus is overwhelming that the Higgs particle exists. The Higgs field is necessary as the cause of spontaneous symmetry breaking, which is necessary in order to render the observed particle masses compatible with gauge symmetry. The resulting theory predicts with a high accuracy the results of high-energy experiments. If the Higgs particle turns out to be missing, our understanding of the basic laws of motion in the micro world will be shaken. Still, the shock will not be that strong—after all, we will still be able to employ the basic ideas of quantum field theory and to rely on the gauge principle—but physicists would have to search anew for a physical process that does what the Higgs field is supposed to do.

Our knowledge of the fundamental particles together with the gauge symmetries corresponding to the electromagnetic, weak, and strong interactions and the mechanism of spontaneous symmetry breaking provide the basis for the construction of a quantum field theory that incorporates our present state of knowledge of the micro world. This theory is known as the *Standard Model*, and it is remarkably successful in the comparison of its predictions to experiment.

The Standard Model was put in place about thirty years ago, and it has not changed since—a rare thing for models in high-energy physics. This also has to do with the fact that groundbreaking experimental results have been sparser in the last twenty years or so: the energy ranges that promise new physical phenomena have not yet been achieved. Hence there was little chance to challenge the established theory. However, nobody believes that the Standard Model is the final word in high-energy physics. It is too arbitrary, too ugly a physical theory to be truly fundamental. In a sense, it is nothing but a patchwork of facts, sewn together by means of two insightful ideas, and tailored into the general shape of quantum field theories. One may think of it as a dress of sturdy material and comfortable fit, but still a dress that nobody would like to wear on a public occasion. Its workmanship is just too crude.

In the next and final chapter, I shall touch upon some ideas that try to impress a bit of style and principle into the Standard Model. I shall also summarize the understanding of matter achieved by modern physics and attempt to guess where the future will lead us.

10

OUTLOOK

UNANSWERED QUESTIONS
AND OPEN PROBLEMS

10.1 The Ancient Dilemma Revisited

Until then we believed in the old statement of Democritus "In the
beginning was the particle". . . . But perhaps this whole philosophy
was wrong. . . . In the beginning was a law of nature, mathematics,
symmetry? Perhaps we should approach Plato and indeed state that
"in the beginning was the symmetry."

—W. Heisenberg, rephrased from *Der Teil und das Ganze*

The oldest and most important question regarding the structure of
matter arises from the conflict of the atomic theory with the theory
of elements and concerns the issue whether matter consists of small,
indivisible entities or whether it can be divided into smaller pieces indefinitely. We followed the story of this debate through the centuries from
the earliest statements of these two basic intuitions until the understanding reached by twentieth-century physics. Modern physics provides a rather convoluted answer. We know definitely that matter does
not behave the same at all scales, but it is structured in many successive
layers. The material objects we perceive through our senses can be

divided up to the scale of a few millionths of a centimeter. At this scale, matter is organized in discrete objects, namely, molecules. Molecules are composed of smaller pieces of matter, which were mistakenly called atoms. Atoms are also composite: they consist of electrons and the nucleus. Electrons are no further subdivided according to our present knowledge, but the nucleus consists of protons and neutrons, which seem to be composed of other particles called quarks.

Our descriptions for the structure of matter seem to stop at quarks and electrons (or rather leptons). However, no one can preclude that these are composite objects too. Experiments at energies higher that the ones achieved so far may reveal an underlying structure. Matter is organized like a Russian doll: we open one and a new one appears inside. Only the new doll is very different from the one in which it is contained and very different from the one it contains. We simply cannot tell whether this procedure of opening the dolls will continue endlessly, or if there is some final level of description in terms of *really* elementary particles. These particles would then be the *true atoms*—the ultimate building blocks of matter. No matter how successful science may be, it cannot predict its own future—there exist questions to which there seem to be no final answer.

But even if we rested at the present level of knowledge and assumed that quarks and leptons are themselves the ultimate building blocks of matter, it would still be difficult to make the case that matter consists of discrete objects. The fundamental theory for the micro world, quantum field theory, relies on a rather elusive physical principle: the duality between particles and fields. Particles are genuinely discrete objects, but fields are supposed to be continuous up to the tiniest of scales, and both descriptions seem to hold simultaneously.

The description in terms of particles follows the spirit of the atomic theory of old times—matter consists of atoms that move in the Void. The field description, on the other hand, has inherited many of its insights from the theory of the elements. Indeed, one type of field differs from another through different *qualities*, namely, properties that cannot be interpreted in terms of motion and shape, such as the electric charge, charm, color, or baryonic number. These qualities are expressed mathematically in terms of conserved quantities, which according to

Noether's theorem correspond to the symmetries of the laws of motion. Somehow modern physics seems to have resurrected Plato's vision, presented in the *Timaeus*, that the qualitative differences between the elements of matter may be explained (or at least described) in terms of the different symmetries that characterize the organization of an underlying substratum.

Modern physics does perceive the field as material, but not in the way Descartes and the mechanistic tradition of nineteenth-century physics understood matter. The field introduces Aristotle's notion of potentiality and resurrects Leibniz's concept of force in the context of modern physics. The field-particle duality is therefore analogous not only to the duality between discrete and continuous, but also to the duality between concrete things and potentiality—matter and force. Modern quantum field theory tries to stride simultaneously on both edges of the divide. Still, the coexistence of these ideas is rather precarious. One will always be tempted to ask which of them is more fundamental—particle or field. The immediate answer is that particles seem to be the fundamental objects because (with the possible exception of the electromagnetic field) no field can be directly measured. The fields are then viewed as mathematical objects, which are only introduced in order to describe the interaction of particles in accordance with the principles of relativity and quantum theory.

However, counterarguments exist. They arise from the remarkable success of the gauge principle, which has proven indispensable for the description of the laws of motion in the micro world. The gauge symmetry is a symmetry of fields, not of particles. Moreover, it does not appear at the particle level—thanks to spontaneous symmetry breaking. It is natural to assume that the fundamental symmetry of matter—as the gauge symmetry is at present supposed to be—should appear at the level of the fundamental physical objects. One should therefore accept Faraday's intuition and accept fields, the carriers of symmetry, as the fundamental substratum of the material world. In this view, particles are nothing but a manifestation of fields, perhaps one among many.[1]

The perspective above is strengthened by the fact that the elementary particles of modern physics have one important difference from the atoms of the philosophers: *they are not eternal.* A d quark may disappear in a process, and other particles may appear in its place (e.g., a u quark together with leptons in beta decay). The laws of motion governing particle interactions do not preserve the particles' identities because these laws are expressed in terms of fields. They do preserve something, though—all these conserved quantities that correspond to the symmetries of fields: electric charge, leptonic and baryonic number, and so on. Hence one may say that the really fundamental entities are the field properties (i.e., the conserved quantities), which are transferred from one place to another in the only way that is compatible with the Poincaré symmetry of spacetime. In this view, particles are nothing but traveling "bundles of field qualities."

The theory of quantum chromodynamics is often invoked in support of the thesis above. The fact that fields are unobservable entities provides the key argument for the primacy of the particle concept. However, in QCD the fundamental particles (the quarks) are also unobservable. We have only observed composite particles, and we infer the properties of the quarks from the study of the conserved quantities that involve the composite particles. However, these conserved quantities refer to the quark fields and not to the quark particles. Indeed, we have no reason to believe that quarks really exist as particles, since they never appear isolated in nature.

The supporter of the particle primacy, however, will reply that our inability to observe fields has a different root from our inability to observe quarks. Fields (in particular Dirac fields) cannot be observed as a matter of principle because the fundamental structure of quantum field theories implies that the measurement of a field profile leads to the violation of important conservation laws (see the discussion in sec. 7.3). The fact that quarks do not appear isolated in nature is *not a matter of principle*, but a dynamical effect due to the specific form of the quarks' laws of motion. It is conceivable that experiments in very high energies may lead to a breakdown of QCD, and hence to an observation of free quarks.

The issue of the field or particle primacy cannot be settled, at least within the context of the present physical theories. Quantum theory does not care about the distinction between particles and fields: it is quite abstract and accommodates both descriptions simultaneously. One may construct the theory of quantum fields either through the language of particles or through the language of fields, employ both descriptions at the same time, and obtain identical physical predictions. The truth is that one description cannot do without the other. If we speak only of fields, we have no way of translating theory into experiment and experiment into theory because particles are what we directly measure. If we speak only of particles, we have no guide for the laws of motion that govern the interactions between them. This dilemma does not preclude one from taking sides, consciously or unconsciously. The younger generation of physicists, having entered research after the huge success of the standard model and of the gauge principle, has the almost instinctive tendency to view fields as fundamental. The older generation of physicists, the ones that had to struggle with the plethora of new particles and sought the order in that chaos, tends to focus on the particle aspect.

A common argument—originating from Niels Bohr—is that the duality between fields and particles is a fundamental feature of the world, one that forms the basis of the most successful physical theory ever constructed. It should therefore be taken at face value. No further elaboration is necessary, and if common sense seems to protest against it, so much the worse for common sense. The deepest nature of the world needs take no account of the human intuitions, which provide only a pale, imprecise, and unscientific reflection of reality. Field and particle are just words we employ to differentiate our experiences, so that we can reason about them. Hence, if our theories, which have been spectacularly verified by experiment, postulate a duality between those concepts, we should accept it as a fact: the paradox is in our minds, not in the world.

This point may, however, be contested. Quantum theory, like any scientific theory since Galileo's time, aims primarily to *describe* rather than explain of physical phenomena. It is very successful in that respect. However, it is not necessary that all features of a particular description reflect the structure of physical reality. Clearly, there will be some that do so (probably most of them) for otherwise the immense success of the theory would be extremely puzzling. However, some of these features may not correspond to anything real; they may be simply useful short-cuts that are effective for obtaining reliable results.

In other words, it is perfectly conceivable that specific features of our theories—the duality between particles and fields in particular—may be properties of our description of the world and not of the world itself. For this reason, the question about the primacy of the field or particle concept is hardly illegitimate. One or the other may be a more funda-mental object, but our present theories have not achieved the depth of understanding needed to make such a distinction. A better theory than quantum mechanics may be obtained in the future, which will be able to do so.

The discussion above hinges upon one of the major questions of modern physics: whether quantum mechanics is complete as a physical theory or not. This is a huge topic—starting from the conflict of Heisenberg's and Schrödinger's mechanics and from the legendary Bohr-Einstein debates—and it lies outside the scope of this book. The truth is that we have no reason to mistrust quantum theory based on experimental data; its predictions are remarkably successful. However, quantum mechanics says very little about what the world is really like; it draws no picture about the nature of physical reality that can be placed alongside the categories of common sense. One may choose to disregard common sense and trust quantum mechanics as a complete theory—most physicists do so. Nevertheless, a substantial minority believes that a better theory is possible. Common sense is too precious for our under-standing of the world: it cannot be disregarded simply because of the success of a single physical theory. In this perspective, the quantum the-ory is viewed as a very effective approximation to a deeper, more funda-mental theory, an approximation that will fail when it reaches beyond its limits. Whatever one's opinion about the merit of this stance may be, as things stand now, there is no other option but to accept quantum the-ory in its existing form and to follow its rules. For this reason, many questions about the theory's meaning have to be postponed until new discoveries allow us to phrase them in a way that can be meaningfully answered by recourse to experiment.

The interplay between particles and fields is also important in another context, apart from any considerations about the interpretation of quantum theory. The relation of the field description to that of particles rests explicitly on the assumption of the Poincaré symmetry, or in other words, on the selection of the special category of inertial reference frames. Ever since Einstein developed his general theory of relativity, we know that the concept of inertial frames is only an approximation that ignores the gravitational phenomena. In the presence of gravity, the relation between particles and fields is more complicated, and the question of the primacy of the field or the particle concept becomes very important because there is no guarantee that the two approaches will lead to the same physical theories.

At present, we lack a satisfactory theory that combines the features of quantum theory and general relativity, a theory of quantum gravity as it has come to be called. The construction of such a theory has been one of the most prominent and ambitious areas of research of the last two decades but still remains far from completion.

In spite of the absence of a full theory of quantum gravity, there has been substantial progress in elaborating on the relation between fields and particles in the absence of the Poincaré symmetry. The most spectacular result arose from the work of the British physicist Stephen Hawking in the study of the behavior of quantum fields around black holes.[2] (Black holes are regions in spacetime from which no light can escape.) Hawking's result was later made more explicit by the Canadian physicist William Unruh, who transferred Hawking's basic arguments into a less "exotic" physical situation.[3] It is Unruh's version that I describe here.

Consider an observer on an inertial frame of reference, who carries a detector that measures particles of a particular type. If the particle detector measures no particles, then one may say that the corresponding quantum field is in the vacuum state. Then consider the same particle detector on a reference frame that moves with constant acceleration with respect to the initial reference frame. It turns out that in this case the detector does measure particles, even though the state of the field has not been affected. Moreover, the energies of the measured particles are distributed as though the field's state is characterized by a constant temperature *whose value is proportional to the acceleration.*[4]

This result was completely unexpected, and its implications are truly remarkable. Two observers on inertial frames will always agree on the specification of the state of no particles. But this agreement will break down when one of the two observers starts accelerating. The vacuum of the field seems to be, therefore, an observer-dependent concept. In the absence of inertial frames (as in the real world, where the gravitational field is always present), there is no way to choose the vacuum uniquely. This strongly suggests that some of the basic ideas of traditional quantum field theory will have to be abandoned in a theory that incorporates the effects of the gravitational field.

The relation between acceleration and temperature is even more amazing. Temperature is a concept that refers to thermal phenomena. Ever since the development of kinetic theory in the nineteenth century (sec. 2.3), we know that temperature corresponds to the random motions of the molecules composing a physical body. In light of Hawking's and Unruh's results, temperature appears also to be related to acceleration between reference frames, a feature quite unrelated to its initial physical meaning. This result hints at a strange interplay among the quantum mechanical duality between field and particle, the relativistic concept of reference frames, and the theory of thermodynamics. This interplay may constitute a basic feature of the elusive quantum theory of gravity. It provides a glimpse at novel physical principles that may characterize a deeper theory of the world, the only glimpse that can be ascertained at this moment with a fair degree of confidence. We do not know yet what this theory will be, but if expectations are fulfilled, Hawking's discovery will be deemed comparable to that of Planck, which opened the first window to the world of the quantum.

10.2 The Singular Status of Quantum Field Theory

And they said, Go to, let us build us a city and a tower, whose top may
reach unto heaven; and let us make us a name, lest we be scattered
abroad upon the face of the whole earth.
—Genesis 11.4

The motions and interactions of microscopic particles are described by
quantum field theory. Hence any idea about the structure of matter
should be expressed in the language of this theory—the Standard Model
is simply a special case of a quantum field theory. In fact, one may argue
that quantum field theory is the most successful scientific theory ever
constructed. It is not only that it provides predictions in remarkable
agreement with the experiment, but also that it does so with minimal
assumptions. The Standard Model involves a relatively small number of
free parameters.

For a theory to make physical predictions, it is necessary to introduce
a number of parameters in its mathematical formulation, whose values
are not predicted by the theory itself. These parameters constitute the
mathematical input of the theory. Every physical magnitude that can be
calculated within this theory is a *function* of the input parameters, and
it corresponds to the theory's output. The input parameters are determined through experiment, and once this is done, the theory can be
immediately employed to make physical predictions. Thinking of a
physical theory in terms of input and output, we easily see that the best
theories are the ones that achieve maximal output with minimal input.
Hence, of two theories that describe the phenomena equally well, the
one involving fewer parameters is always to be preferred. Usually, when
a new theory supersedes an older one, it does so by predicting the value
of some of the latter's input parameters. In this sense, the ideal theory is
one that involves no input parameters at all.

In the days before quantum mechanics, the number of free parameters involved in the physical theories was enormous. We knew nothing
about the detailed processes responsible for the structure of the atoms.
Hence, every single number we obtained from experiment that corre-

sponded to a property of some material constituted a basic and unexplainable input parameter for the theory. Rutherford's model of the atom and the development of quantum mechanics transformed the situation dramatically. They made possible the prediction of all chemical properties of the elements (at least in principle) from the knowledge of the basic properties of the atomic constituents: essentially the mass and the charge of the electrons and nuclei. If we ignore the detailed properties of the nucleus (which is a reasonable approximation in the study of chemical phenomena), we can assume that the mass of a nucleus equals the sum of the masses of the protons and neutrons that compose it. One then simply needs to know four physical magnitudes (the masses of the electron, of the proton, and of the neutron, and the value of the electron's charge) in order to provide an adequate theoretical description of chemical phenomena, and to estimate the thousands of physical quantities measured in chemical experiments.[5]

However, matter's structuring does not stop at the atomic level: there are more things than chemistry in the micro world. Hundreds of new particles were discovered, each with the same right to be considered fundamental as the electron or the proton. Simplicity was therefore lost in high-energy physics, as one would have to take the properties of these particles as new input parameters in our fundamental theories. The quark model saved us from that disaster, and the development of gauge theories led to the construction of a quantum field theory that could account—in principle—for all observed properties of the subatomic world.

The Standard Model needs a remarkably small number of input parameters: only twenty-seven. This number may seem too large, if compared with the only four parameters needed for the description of chemical phenomena. Nonetheless, it is remarkably small if one takes into account the wealth of phenomena it purports to describe.

A basic principle of modern physics is that the behavior of a composite body is fully determined by the body's constituents. If we know the properties of quarks and leptons, we can predict the properties of nuclei and of atoms—at least in principle. In other words, the Standard Model contains hidden in its structure all laws of nuclear and atomic physics, of

TABLE 10.1
The parameters of the Standard Model

Description	Number of Parameters	Comments
Quark masses	6	
Lepton masses	6	Assumes that the neutrinos have nonzero mass.
Mediator masses in weak interactions	2	
Mass of the Higgs particle	1	In the simplest models, only one Higgs particle survives spontaneous symmetry breaking.
Mixing parameters of quarks	3	
CP-violating phase in quarks	1	
Mixing parameters of neutrinos	3	Arise only if the neutrinos have nonzero mass.
CP-violating phase in leptons	1	Arises only if the neutrinos have nonzero mass.
Coupling constants	3	One for each of the strong, weak, and electromagnetic interactions.
Theta parameter of QCD	1	A parameter that distinguishes between different choices for the QCD vacuum.
TOTAL	27	

chemistry, and of all theories that describe the bulk properties of matter. Hence, the twenty-seven parameters of the Standard Model can account for any macroscopic property of matter, whether this is chemical, thermal, or biological, if only we can find clever ways to extract this information from the theory's laws of motion. Twenty-seven is hardly a large number in this regard.

One may protest at such grandiose dreams for a physical theory, perhaps with good reason. There has been so far not even the outline of a proof that the macroscopic properties of matter arise from the principles of quantum field theory.[6] We cannot employ our theories to demonstrate conclusively that quarks lump together to form protons and neutrons, let alone prove that nuclei and atoms are the natural formations for matter. We cannot even reproduce the laws that govern the motion of macroscopic systems, like fluids, without *additional* assumptions, *extrinsic* to quantum field theory. A skeptic may simply point at this lack of proof and refuse to believe any claims about the range and scope of our fundamental theories.

However, the majority of physicists strongly believe that the laws of motion of the part determine the laws of motion of the whole. If this principle ever fails, one would have to revise the very concept of physical law. For this reason, the lack of explicit demonstration is taken only as an indication that the task at hand is too difficult. Perhaps it is not worth the effort to pursue it, at least at a time when one may delve deeper into the structure of the fundamental theories. "Quantum field theory also describes macroscopic phenomena," a supporter of this view would state, "and an explicit demonstration awaits only the sharpening of our present mathematical tools." One has to be careful on this issue, though. Quantum field theory is itself a quantum theory, and for this reason it is susceptible to all interpretational problems of quantum mechanics. This means that quantum field theory refers to measurement situations and assumes a split between the macroscopic world of the experimentalist and that of the microscopic system, the former being described with classical physics, the latter with quantum physics. How this can be possible, when macroscopic objects themselves consist of microscopic particles, remains a puzzle. Neither the Copenhagen interpretation nor any other approach has been so far able to provide a full and satisfactory solution.

The problem of the dichotomy between microscopic and macroscopic should put some water into our wine: one should keep some reservations about the claim of quantum field theory to contain the full description of matter in the macroscopic regime. One may at least entertain the possibility that the emergence of the macroscopic world of stones, rivers, planets, and of measurement devices involves some new physics, not covered by conventional quantum theory and its offspring.[7]

Quantum field theory is a singularity in the history of science. In spite of its remarkable success, the fact that it has not failed in any experimental test, and its ambitious claims to provide the full description of all material processes, most physicists believe that it provides only an approximate description to the world, and that it has to be replaced by a better theory. This is very unusual as theories in physics go: usually success is taken as the absolute criterion for a theory's validity. As long as the theory is successful, it is very rarely questioned, let alone deemed insufficient.

However, there is good reason for doubt. Quantum field theory constitutes by its very inception a marriage between the two great pillars of twentieth-century physics, quantum mechanics and the theory of relativity. A marriage it is, but not a fully consummated one. Quantum field theory incorporates in its description only the special theory of relativity, and it ignores all gravitational phenomena. Essentially, it assumes that matter (particles and/or fields) evolve within Minkowski spacetime, and that the spacetime is not affected by the matter it contains. The general theory of relativity, on the other hand, states that the spacetime is itself dynamical: its geometry changes as a result of its interaction with matter.

The description of matter as existing within an absolute independent "vessel" follows one of the oldest metaphors for matter. The only difference now is that the container is spacetime rather than space. Quantum field theory stands together with that metaphor. It will fail when the metaphor fails. And general relativity suggests—if it does not guarantee—that this will happen when the gravitational phenomena are fully incorporated into the quantum world order.

10.3 Grand Unified Theories, Supersymmetry, Strings, and All That

Could all the phaenomena of nature be deduced from only three or four general suppositions there might be a great reason to allow these suppositions to be true: but if for explaining every new Phaenomenon you make a new Hypothesis . . . your Philosophy will be nothing but a system of Hypotheses. And what certainty can there be in a Philosophy, which consists in as many Hypotheses as Phaenomena to be explained? . . . Tis much better to do a little with certainty and leave the rest for others that come after, than to explain all things by conjecture without making sure of any thing.

—Sir Isaac Newton, *Opticks*, discarded introduction to first edition

Quantum field theory provides an excellent account of the particles and their interactions through the electromagnetic, weak, and strong forces. However, this does not imply that improvement is not possible. The Standard Model involves twenty-seven parameters in its definition. Perhaps a brilliant idea will allow us to reduce this number by expressing some of these parameters as functions of the others. This possibility provides one of the major hopes and aims of present research. The Standard Model may be successful, but it is also inelegant, not to say downright ugly. Its structure is too arbitrary to be truly fundamental. A simpler theory should lie underneath, and the task would be to determine what this might be.

A new theory should help us resolve some of the major puzzles involved in the structure of the Standard Model. The fact that the latter provides an accurate description of processes up to the scale reached by experiment does not imply that it is entirely trouble free. It cannot provide an answer to many questions. For example, one may inquire why the strong forces are so much stronger than the weak ones. The only answer available at this moment is that this is the way things are: the value of the running coupling constant for the strong interactions is much larger than the one for the weak interactions—at least for the energy scales we trust our theory to be valid. This is a datum of experiment unexplainable by the present theories. However, the Standard Model predicts that the running coupling constant decreases much faster with energy in the strong interactions than it does in the weak ones. This implies that there must be an energy scale, at which the strength of the two forces is of the same order of magnitude. Perhaps in these energies the three different forces merge into a single one. If this is the case, the theory underlying the Standard Model will provide a unification of all forces, much like Maxwell's theory provided a unified description of the electric, magnetic and optical phenomena. This postulated theory is called grand unified theory (GUT).

Some rather naïve estimates suggest that the energy scale, at which the unification of forces takes place is of the order of 10,000,000,000,000,000,000 electron masses. This is about 10,000,000,000,000 times higher than the energies we currently access and about 100,000,000,000 times higher than the energies we estimate to achieve in our accelerators in the near future. Clearly, there is a long way to go until we reach these energies, and many things may happen in between: new forces and new particles may be discovered, particles thought of as elementary may turn out to be composite, even a breakdown of quantum theory is possible. Prudence would suggest that there is very little we can say about a grand unified theory at present.

Still, science does not need to be prudent, especially since it is costs little to be bold. The Standard Model is very successful, so why shouldn't a grand unified theory follow the same pattern, namely, that of a theory with a gauge symmetry and spontaneous symmetry breaking? Could the unified theory be defined by a larger gauge symmetry that incorporates the symmetries of the observed forces? Indeed, such models have been constructed. They typically predict the existence of a new type of force that appears at very high energies. This force violates one of the quantities that we believe at present to be conserved: the baryon number. In particular, these models predict that protons may decay into leptons and pions ($p \rightarrow e^+ + \pi^0$) even though these processes are extremely rare at low energies. One could then test these theories by looking for such decays. So far the result has been negative. In general, the predictions of grand unified theories either fail or cannot be made precise enough to allow for an experimental distinction from the Standard Model.

In any case, the GUTs we have so far constructed are nothing but elaborate hypotheses. They ignore the possibility that dramatic changes may occur in our understanding of fundamental particles during the 10,000,000,000,000 times we have to multiply our present experimental capabilities, before we reach what is commonly believed to be the

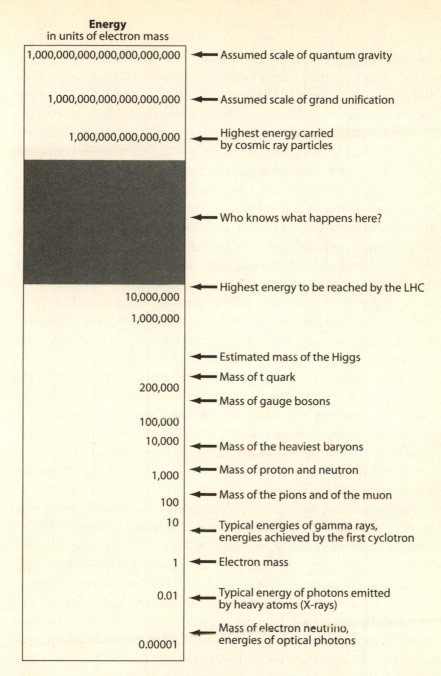

Figure 10.1 **The energy scales of particle physics.**

relevant energy scale for unification. Moreover, the GUTs answer none of the other important questions that are brought forward by the arbitrariness of the Standard Model: Why are there three families? Why are the particle masses so different? Why does symmetry breaking occur?

An alternative class of theories that has been extensively studied goes under the heading of supersymmetry. "Super" is an ambitious prefix and rather irrelevant for the description of the basic symmetry that characterizes these theories. The main postulate is that the number of degrees of freedom corresponding to fermions equals the number of degrees of freedom corresponding to bosons, and that the laws of motion must be invariant under a large class of transformations that transform bosonic fields into fermionic ones and vice versa. If supersymmetry is valid, then each fermion in nature should have its *supersymmetric partner*, namely, a boson of the same mass. This certainly contradicts our present observations, but perhaps this can be attributed to some form of symmetry breaking, which results in the fact that the particles we observe are the lightest of any supersymmetric pair.

Supersymmetry is a very powerful symmetry, and it places very strong constraints on the laws of motion that are satisfied by a theory's fields. Its adherents view this as a blessing because it considerably reduces the number of alternative theories that have to be examined. However, critics of supersymmetry find that it is too strong a symmetry, postulated for no good reason. They have jokingly dubbed it as *"an excellent answer begging for a question."*

Supersymmetry provides a resolution for some minor puzzles that appear when one tries to extend the Standard Model to energies higher than the ones accessible at present. This is in fact the main physical motivation for its introduction; there are, however, alternative theories that can do the same job. The good thing is that one expects to find at least some supersymmetric partners of the currently known particles in an energy scale that will be soon accessible. Hence, the idea of supersymmetry will be tested. If one such partner is found, it will provide the most spectacular verification of supersymmetry. This will constitute one of the rare (and highly celebrated) instants in the history of science, at which mathematical "necessity" predicted the existence of new physical objects. However, if no supersymmetric partner is detected, my guess is that supersymmetry will slowly recede from the limelight, and the historians of the future will view it as nothing but a late-twentieth-century scientific fashion.

A more ambitious program toward the unification of all forces is the theory of *superstrings*. Its basic tenet is that the fundamental objects of the world are not particles but strings, namely, one-dimensional objects that move in Minkowski spacetime. Strings may be open, in which case their endpoints move with the speed of light, or they may be closed, like hoops. Strings vibrate, and the "tunes" of their vibration are waves that correspond to physical particles. The key idea of string theory is that all physical interactions are due to the strings' "motions." Moreover, the theories that describe the strings have to satisfy supersymmetry because this alone can render their quantum description both consistent and nontrivial. The only free parameter in a supersymmetric string theory is a length scale, known as the string's *tension*. The adherents of string theory believe that all physical parameters of existing theories (the twenty-seven of the Standard Model plus any other that may be involved in other yet unknown interactions) are nothing but functions of the string's tension.

Superstring theory also aims to describe the gravitational interaction. One among the strings' vibrations corresponds to a particle with zero mass and spin equal to 2. Such a particle can act as a carrier of a force that mimics all the effects of gravity as predicted by general relativity. This description of gravity is rather remarkable because it does not abandon the idea of matter as something that moves within a background, unchanging vessel (usually Minkowski spacetime). In this sense, string theory is a rather conservative generalization of quantum field theory that does not overthrow its main tenet but has much higher ambitions for its domain of validity.

String theory has generated immense interest in the last twenty years, but so far it has failed to fulfill its promises. The theory is only well defined in a spacetime of ten dimensions, one for time and nine for space. Since this is definitely not what we observe, the only way out is to *postulate* (rather arbitrarily) that six of these dimensions are extremely small. In analogy, one may consider a thin sheet of paper as two-dimensional, even though it is strictly a three-dimensional object, because its thickness is very small in comparison to its other dimensions. There exists a very precise procedure of studying theories involving spacetimes with small dimensions. Unfortunately, the reduction of the full super-string theory to four dimensions is nonunique. One superstring theory has millions of different descriptions when it is restricted to four dimensions. It would take the lifetimes of a large number of scientists to work the details of only some of them. It is therefore very difficult to obtain any definite predictions.

However, the lure of a theory with only one free parameter is so strong as to seduce a very large part of the physics community. In recent years, more research papers have appeared in string theory than in any other branch of theoretical physics. Perhaps string theory is the philosopher's stone of the twenty-first century, *the theory of everything* being a new name for the alchemists' universal substance. Only time will tell whether superstring theory will also share the fate of the philosopher's stone or not.

10.4 Where Do We Go from Here?

Einstein: Why do you believe so strongly in your theory, when so many central issues remain unclear?
Heisenberg: Like you, I believe that the laws of nature have an objective character, which does not refer to the economy of thought alone. When one is led through Nature to mathematical forms of great simplicity and beauty . . . one cannot fail to believe that they are true, that they represent a real Way of nature.
— Heisenberg, *Der Teil und das Ganze*

Science, from Galileo onward, aims toward the description of the world, often claiming to have abandoned the search to identify the fundamental being of things, their reason for existence and their relation to a deeper structure of reality. Galileo suggested that this measure may perhaps be only temporary, but in this case, history upheld the old proverb that "nothing is more permanent than the temporary." However, the mere description of things happening in the world is of little interest by itself. Perhaps it would be more important to find the pattern, the rule that determines these happenings. To do so, one needs a way of expressing and relating all sorts of different phenomena, a language that can be both overwhelmingly abstract and discerningly specific: mathematics. Still, this is not enough by itself. A rigorous experimentation is also necessary, for otherwise it would be impossible to tie the mathematical theory to reality. The combination of both features led to the development of theories that not only describe the material world, but also make remarkably accurate predictions.

If the only aim of the physical sciences is to describe the structure and motion of matter in the simplest way possible, then every other feature traditionally attributed to matter lies outside the domain of scientific inquiry. The physical scientist should therefore not ask what the reason is for the existence of matter, if there is a purpose to the processes of nature, or how the material universe is related to that of thought and emotion. These issues may be irrelevant to the scientist's task, as they can be studied neither in experiments nor through the language of numbers. They are, therefore, unscientific and, for some, even meaningless. For this reason, many of the prescientific theories about matter were ignored or even aggressively attacked as hostile to the scientific worldview.

"Woe to the vanquished," some people will say. Science is overwhelmingly successful, and any competing worldview must inevitably bow to that success. However, for this attitude to be fully self-consistent, it has to make an extra step. If the description of things provides concrete knowledge much more than any attempt to find a meaning in them, it is meaningless to go on looking for any meaning beyond the level of description. "The whole modern conception of the world has been founded on the illusion that so-called laws of nature are explanations [rather than mere descriptions] of the world," Ludwig Wittgenstein wrote, expressing an attitude that for some time was popular in the philosophy of science.[8] We may say that a scientific theory explains a phenomenon, but what it does is only substitute a complex description with a simpler one. Even one of the most spectacular achievements of modern physics, the reduction of chemical phenomena to a combination of quantum mechanics and electromagnetism does not constitute an explanation. The underlying theories, however powerful and universal they may be, are still nothing but descriptions. They say nothing about what the world truly *is*; they only provide useful and successful rules for the description of physical phenomena. If one is tempted to say that quantum mechanics says something important about the deepest nature of the world, one should be quickly pointed to the fact that quantum mechanics does nothing more than provide efficient rules for the ordering of experimental results. In this view, electrons and protons, fields and particles, quarks and gluons are just linguistic conventions that translate the abstract mathematical structures of our theories into familiar images, which have nothing to do with the way things are in reality. Once you move beyond the efficient description of facts and start talking about the nature of reality and the true existence of things, you cross the line of honest scientific inquiry and stumble into the murky waters of metaphysical speculation. True science simply does not care about reality.

It is hard to avoid the conclusion that the attitude above becomes self-defeating when followed up to its ultimate logical consequences. It blatantly ignores the history of the scientific endeavor, not to mention the history of thought in previous eras. A theory about the world, even if its avowed aim is restricted to the mere description of things, is rarely philosophically neutral. Philosophical neutrality was far from Copernicus's mind when he started his investigations of the planetary motions. He would have saved himself great danger if he had stated that his heliocentric system was simply a mathematical description of things with no connection to the deeper structure of reality. Indeed, his editor took the precaution of making such a statement, but nobody in their right mind would take such a disclaimer seriously. Galileo, who is widely considered as the first to promote the descriptive character of the natural sciences, was persecuted because he could not bring himself to state that his system was nothing but a mere description of phenomena. The mechanics of Newton, married with a mechanistic philosophy, provided the basis of a worldview that spanned everything: from physics to social organization. These examples show we cannot mutilate scientific inquiry by accepting that scientific statements have no implications about the structure of reality. Even though speculative boldness often leads to grave errors, this is a small price to pay for the preservation of scientific creativity. As Hermann Weyl wrote, "science would perish without a supporting transcendental faith in truth and reality, and without the continuous interplay between its facts and constructions on one hand and the imagery of ideas on the other."[9]

Emphasizing the level of description is a good, indeed a necessary, discipline for the scientist. After all, a theory stands or falls according to its adequacy to describe the world we live in. However, science employs concepts and ideas that go beyond mere description. They are expressed in the language of mathematics, but they are not themselves mathematical. This is the reason that one and the same concept may appear in theories with distinct mathematical structures that provide very different (and even incompatible) descriptions of physical phenomena. Indeed, the most important concepts appearing in physical theories cannot be pinned down to a single mathematical or logical structure. They always present aspects that resist precise formalization, and for this reason they can adapt to different contexts and eras, always acting as a precious fertilizer to scientific creativity. In fact, many of these fundamental concepts have been bequeathed to science by earlier theories. The atoms and the Void, the fundamental elements, the intrinsic powers of matter, the idea of inertia, the image of the world as a machine are some examples of metaphors that persisted throughout the history of human thought and provided science with an armory of concepts that enables it to understand the world.

Modern physics describes matter through quantum field theory, which combines the basic insights of quantum mechanics and of the special theory of relativity. We saw earlier that quantum field theory is in excellent agreement with experiment, but its implications about the deeper nature of the world are difficult to grasp. In a sense, quantum field theory is the ultimate descriptive theory: it provides an excellent quantitative account of physical processes. However, it sheds little light on the nature of the things that participate in them. We do not know whether quantum field theory fundamentally describes fields or particles, or what the meaning of the wave function and the quantum phase is; we cannot agree whether this theory is valid only for measurements or describes physical objects in themselves; we cannot be sure whether it describes both the microscopic and macroscopic phenomena or whether the latter need a different description. The theory is remarkably successful in its predictions, but its foundations are still covered by a veil.

Quantum field theory—as a child of quantum mechanics—brings to its utmost the conception of a physical theory as a mere description of the regularities in experimental data. Galileo's dream has turned into a nightmare, it may seem. We have a successful map of the world, but we are not sure what its symbols really refer to, and how far they can be trusted. Perhaps, this is all we can expect from physical theories from now on: they will become increasingly more abstract, and there will be little we can say about their correspondence to reality. However, this attitude runs contrary to the gut feeling of any practicing physicist. Nobody working with quantum mechanics seriously entertains the thought that the theory does not describe real things. Still, when one comes to the point of discussing what these things are, the theory is not very helpful. Its structure cannot be easily reconciled with common sense. Perhaps we simply do not have the proper conceptual framework to realize the content of quantum theory, and new concepts need to be introduced for the apprehension of this new physics. Perhaps again the present theory is just an intermediate one that will be replaced by a deeper and better one, which will remove the veil cast upon the nature of physical reality. Any of the above may be true—we simply cannot tell at this moment.

The remarkable success of our theories about the material world has brought forward many new issues about the scope of scientific research. The highest energy we can achieve in the laboratory is many billions of times lower than the ones involved in grand unified theories. Still, the unification of forces is the avowed aim of modern physics, and to this purpose, we employ all our machinery of mathematical ingenuity and physical principle. We also see that gravity must enter into the picture, and for this reason we seek a theory of quantum gravity. This is even more ambitious because the scale where gravity and quantum theory will be united is naïvely estimated to be about 100 times higher than the scale for the unification of forces, which is itself 10,000,000,000,000 times higher than the one reached by present experiments. Already, many scientists speak about a theory of everything, and some believe that they have found it in superstring theory.

Given that we have a long way to go before we access these energy scales and will probably not reach them in the near future, the statement that we are in the position to construct a "theory of everything" may sound a bit premature. A skeptic may consider it as nothing but an advertising ploy to attract attention, or even view it as a kind of hubris like that of the nineteenth-century physicists who believed the mechanistic model and Newtonian physics to account for everything in the world.

Nonetheless, it remains a fact that we have set our eyes upon these issues. We may be far away in terms of experimental verification, but the theory of relativity describes gravity successfully, and it has identified its basic principles and symmetries. Perhaps it will be possible to provide a consistent theory that marries gravity with quantum theory, based solely on mathematical consistency and elegance. String theory is one such candidate, but there are also other theories supported by a smaller number of people. However, at present we possess no satisfactory and consistent theory of quantum gravity that can make testable physical predictions, and it may take some time until we have one.

This situation is all too novel in the history of science. We are secure in our knowledge of the microscopic processes within the domain of validity of the Standard Model. Our security is shaded only by the cryptic nature of quantum theory. Moreover, we can see from a very long distance where we have to go: the forces have to be unified and gravity has to be placed in the framework too. The latter task is particularly important: the future may uncover new forces and new particles, but gravity will always be the eventual goal. The reason is that the consideration of gravity takes us outside the paradigm governing our present quantum theories. General relativity does not describe matter as something that moves in a background "container" but views space and time as dynamical and coevolving with matter. The container and its content become so entangled that their separation is very difficult, if possible at all.

The construction of a quantum theory of gravity is the ultimate aim of modern physics, and it will probably remain the ultimate aim of any physics that relies on the same principles. Still, it is far away from any direct experimental reach. We need an accelerator with radius equal to the distance of Earth from the Moon in order to access these energies experimentally. We may never be able to achieve that, at least as far as we can project the development of modern technology into the future. This implies that our strategies and criteria for this physical theory must develop a different emphasis. Since exact matching between theory and experiment seems to be out of the question for now, the economy and self-consistency of a theory may start taking up a more prominent role. This is not to say that the quantum theories of gravity cannot be expected to make verifiable predictions. They could, for instance, provide an answer to the question why the dimensions of spacetime are 4 rather than 10—as string theory predicts, 11—as the so-called M-theory postulates, or 8¾—as may be suggested by another, yet unknown, relative of string theory that will be proposed in the future.

The novel features of this quest for quantum gravity may lead to a new perspective about what is involved in a scientific theory. It is not only that the focus and emphasis of our theories must change in this task, but that the ambitions become increasingly larger. Already people talk about theories of quantum cosmology, namely, the description of the Universe itself in terms of quantum theory.[10] Present quantum cosmological theories are rather primitive and very speculative, but if a theory of quantum gravity is constructed, then there is a good chance that quantum cosmology will become a mature science. Perhaps, then, the focus of scientific endeavor will move beyond the mere description of the physical phenomena and start examining issues like the reason and meaning of things, because it may prove simply impossible to shirk these considerations. Again, all this may be a big delusion, fueled by an intellectual arrogance that was borne out of the relative success of our present theories. The future may show that there are more things in the world that are dreamed of in our philosophies. We do not know and we cannot know. But since the effort will be made anyway, we can only "pray that the road ahead will be long, full of adventure, full of knowledge."[11]

N O T E S

Chapter 1. From Myth to Machine

1. The idea that the world comes out of water was stated much earlier in the creation myths of ancient peoples. The earliest recorded tradition of this sort goes back to the ancient Sumerians, during the third millennium A.D.

2. *Iliad*, 14:246.

3. The Indian version of the atomic theory was accepted with varying degrees of emphasis by different philosophical schools (mostly by the Nyaya-Vaisheshika school). Their concept of the atom was in many respects similar to that of their Greek counterpart, but there were important differences. The most important was that the Democritean atoms were supposed to explain the formation of the world in terms of purely mechanical motions without any intrinsic qualities—a feature inherited by early modern physics—while the Indian atoms also possessed intrinsic qualities (related to our sense experience) and arguably a kind of psychic dimension. See B. Pullmann, *The Atom in the History of Human Thought* (Oxford: Oxford University Press, 1998).

4. One can find analogous ideas about elements in Indian philosophical schools of about the same time. In fact, the Upanishads (eighth century B.C.) refer to the formation of the world by the gradual evolution of the four fundamental elements. See H. von Glasenapp, *Philosophie der Inder* (Stuttgart: A. Kroner, 1958).

Chinese cosmology also expresses similar ideas about the elemental forces of nature. All natural phenomena are guided by the rhythmic alternation of the fundamental forces of yin and yang. These have a huge number of connotations, usually associated with the female and male or the negative and positive principle of nature, respectively. Yin and yang compose the *wu-yun*, the five elements: wood, metal, fire, water, earth, often cited cyclically in that order (known as the Mutual Conquest Order), in which each element is able to conquer its predecessor (water, for example, conquers fire because it can extinguish it). Fire represents pure yang; wood, a mixture with more yang than yin; earth, balanced amounts of yang and yin; metal, a mixture with more yin than yang; and water, pure yin. Yin and yang together with the five elements have a strong numerological significance as well as astronomical connotations: the yin and yang represent the Moon and Sun, respectively, while the *wu-yun* represent the five planets known to the ancient world.

5. Aristotle, *De anima* 404b13.

6. Sextus, *Adv. Math.*, 7:135.

7. The Roman poet Lucretius in his philosophical poem "On the Nature of Things," the most eloquent exposition of Epicurus's philosophy that survives today, wrote: "While the first bodies are being carried downwards by their own weight in a straight line through the void, at times quite uncertain and uncertain places, they swerve a little from their course, just so much as you might call a change of motion. For if they were not apt to incline . . . no collision would take place . . . and nature would never have produced anything" (*De rerum natura*, 2:216, translated by W.H.D. Rouse). He continues: "If the first beginnings do not make a swerve to break the decrees of fate, that cause may not follow cause from infinity, whence comes this free will in living creatures all over the earth, whence I say is this will wrested from the fates by which we proceed whither pleasure leads each, swerving also our motions not at fixed times and fixed places, but just where our mind has taken us?" (*De rerum natura*, 2:251).

8. "The atoms move in the Void, because of their dissimilarities and the other differences [in shape, form, and size] and in their motion collide and become intertwined with each other, but in no circumstance does a new substance appear. . . . Some of the atoms are irregular triangles, some look like hooks, some convex, some concave, and others have innumerable differences. Democritus thinks that these bodies become connected to each other for some time, until some stronger external necessity shakes and disperses them." (An ancient account of Democritus ideas in Simplicius, *De Coel.*, 394, 33.)

9. The quotation is from the Emerald table, one of the oldest alchemical documents. It is attributed to Hermes the Trismegistus, a legendary, semidivine figure who is supposed to have been an Egyptian priest living about 2000 B.C. The translation here is by R. Steele and D. W. Singer, quoted in E. J. Holmyard, *Alchemy* (New York: Penguin, 1957).

10. A concise summary of the key views and practices of the early alchemists is given in the classic book by Marcelin Berthelot, *Les anciens Alchimistes Grecs et la naissance de l'alchimie* (Paris, 1884). Berthelot writes:

all physical bodies consist of a universal substance. To produce a specific body, for example gold, the most perfect of metals . . . we must use analogous bodies differing from it only with respect to a quality. After we remove their specific properties we reduce them to the primordial matter, the mercury of the philosophers. This is obtained from the usual mercury, after we subtract its fluidity (some kind of mobile fluid), which obstructs it from perfection. It will then become stable by removing . . . some gaseous state contained in the element and eventually, as some teach, remove some earthly element, like rust, which resists the refining of mercury. . . . They sought to remove from each metal all its specific properties. When one would form the primordial substance of all metals, namely the mercury of the philosophers, there remained its dying with Sulphur and Arsenic. . . . Gold was considered from the beginning as the most perfect of all transformations, [the closest to primordial matter,] the perfect body par excellence, not only because of its brightness, but because of its resistance to chemical reactions more than any other metal. . . . In the chemical phenomena, in the eternal recycling of transformations, which is a fundamental law of most physical processes . . . we understand why in the eyes of the alchemists this mysterious work had no beginning and end, and why the symbol of this phenomenon became the dragon in the form of a ring, symbol of nature remaining eternally the same beneath the flowing surface.

11. Plato, *Timaeus*, 54b.

12. Aristotle provides a concise description of the Pythagoreans' beliefs: "In numbers they see resemblance to things that exist and come into being . . . since the modifications and scales of musical harmonies are expressed through number; since all things seemed in their nature to be modelled in numbers; and numbers seemed to be the first things in the whole of nature; they assumed that the elements of numbers are the elements of beings and the whole heaven to be harmony and number." *Metaphysica* 1:985b 27.

13. Attributed to Philolaus of Croton, a disciple of Pythagoras. Cited in *Stobaeus Anthology*, 1:21, 7b.

14. Sir Francis Bacon, *The New Organon*, translated by James Spedding, Robert Leslie Ellis, and Douglas Denon Heath in *The Works*, (vol. 8) (Boston: Taggard and Thompson, 1863).

15. A rough translation of Aristotle, *Metaphysics*, A5.

16. Galileo, *The Assayer*, 1623 (Opere 6, 232).

17. K Muroi, *Historia Sci.* (2) 1 (3), 173 (1992).

18. Algebra grew out of arithmetic: in its earliest form it involved the determination of general rules for the manipulation of numbers and for the solution of equations. The Babylonians were the first to develop a substantial body of algebraic work. However, their treatment (like that of all mathematicians until the sixteenth century) was purely rhetorical, that is, the algebraic operations were described verbally and applied on a case-by-case basis. In Greece, algebraic problems were usually treated through geometry. However, there was a movement toward an independent study of arithmetic, which culminated in the work of Diophantus (second century A.D.), who introduced a primitive form of abstract notation. The Hindus later employed a more advanced abstract notation, and they are credited with the discovery of the positional notation for numbers we use today. The Arabic tradition of mathematics contributed significantly to algebra, mainly through the development of new procedures for the solution of problems. The most famous

mathematician was the Persian Mohammed ibn-Musa al-Khowarizmi (ninth century A.D.); the name "algebra" is derived from the title of one of his books. (The word "algorithm" also derives from his name.) Still, Arabic algebra was also rhetorical. After the introduction of Arabic numerals in western Europe, there was substantial development of algebra, especially after the Renaissance. The first to use letters to represent generic numbers was the French lawyer and amateur mathematician François Viète (1540–1603): his work is said to represent the birth of modern abstract mathematical symbolism. Symbolic formalism reached a mature stage in Descartes' work on analytic geometry.

19. In primitive form, the notion of the function can be found in ancient mathematics, for example, in the form of mathematical tables for reciprocals, square roots, and qubic roots that had been constructed by the Babylonians. Many mathematicians in the Middle Ages developed concepts that showed them to be familiar with the basic distinctions inherent in the definition of a function. Descartes and Newton effectively used the modern concept of function in their works without identifying it by name. The first person to use the name "function" was Gottfried Wilhelm Leibniz, who also introduced related mathematical terms such as "variable" and "parameter."

20. The idea of a graphical representation of functions appears in a primitive form in the work of the French philosopher Nicolas Oresme (fourteenth century A.D.). However, the development of analytic geometry needed more than that: an algebraic language for the description of functions was necessary. This was achieved by Descartes and Pierre de Fermat in the seventeenth century.

21. H. Bergson, *Creative Evolution* (New York: Dover, 1998). Bergson remarks in the same book that "ancient science supposes it understands the object of study, when it takes note of the moments of existence, while modern science studies its object at any moment of time."

22. Aristotle, *Physics*, 3, 200b12.

23. Copernicus's editor was fearful of persecution, and for this reason he had taken the precaution of adding a prologue in the book in which he downplayed Copernicus's discoveries by stating that the system he had presented claimed only mathematical consistency and not truth. Copernicus himself did not have the opportunity to read this disclaimer (or even his book) because he died shortly before the first copy arrived in his hometown.

24. Kepler did not shy away from trying to find an explanation of the laws of planetary motion. Being strongly influenced by the Platonic insistence on the mathematical structure of reality, he proposed that the distance relationships between the six planets known at that time could be understood in terms of the five Platonic solids. His 1596 book *Mysterium Cosmographicum* presented a model in which one Platonic solid fits between each pair of planetary spheres. This is a beautiful astronomical model that explains why there are only six planets. How can there be a seventh planet, when Euclid proved that there are only five Platonic solids? Kepler eventually accepted that his model did not work because it predicted the wrong values for the interplanetary distances.

25. R. Descartes, *The Principles of Philosophy*, translated by John Veitch (London: Orion House, 2001).

26. I. Newton, *The Principia*, def. 4, translated by A. Motte (New York: Prometheus, 1995).

27. The concept of momentum can be traced back to the works of the fourteenth-century philosopher Jean Buridan. Buridan tried to reshape the dominant Aristotelian theory of motion, according to which the action of an external agent is always necessary, in order to preserve a body's motion. For example, in the case of projectiles, the "force" was supposed to be exerted by the air surrounding the moving body. Buridan believed that only the existence of an internal motive "force" transmitted by the thrower to the projectile could explain the continuation of motion. This motive force he called "impetus." In his commentary on Aristotle's *Physics*, he explains: "It is because of this impetus that a stone moves on after the thrower has ceased moving it. But because of the resistance of the air (and also because of the gravity of the stone) which strives to move it in the opposite direction to the motion caused by the impetus, the latter will weaken all the time. Therefore the motion of the stone will be gradually slower, and finally the impetus is so diminished or destroyed that the gravity of the stone prevails." Buridan defined the impetus as the product of the body's weight times its velocity. Before Buridan, the idea of an internal motive force for motion is found in the works of the sixth-century Alexandrian philosopher John Philoponus, who had referred to it as an incorporeal energy. An important aspect of both Buridan's and Philoponus's ideas was that the motive force would allow motion to be preserved even in the void—hence no external agent would be needed to sustain the regular motion of the stars and all astronomical bodies. In some sense, their ideas are precursors to the concept of inertia: both, however, argued within the (erroneous) assumption that the state of rest is fundamentally different from that of motion.

28. Johannes Kepler, *Tertius interveniens*, (Frankfurt a. M.: Godtfried Tampachs, 1610). Quoted in Max Jammer, *Concepts of Force* (New York: Dover, 1999).

29. P. S. de Laplace, *Philosophical Essay on Probability* (Paris, 1814).

30. Quoted in Jammer, *Concepts of Force*.

31. From a letter of I. Newton to R. Bentley, 26.2.1692, reproduced in *Newton*, edited by B. Cohen and R. S. Westfall (New York: W. W. Norton, 1995).

32. Newton left some hints about his own speculations on the nature of forces in his *Opticks*:

It seems to me farther, that these Particles have not only a Vis inertiae, accompanied with such passive Laws of Motion as naturally result from that Force, but also that they are moved by certain active Principles, such as is that of Gravity, and that which causes Fermentation and the Cohesion of Bodies. These Principles I consider, not as occult Qualities, supposed to result from the specifick Forms of Things, but as general Laws of Nature, by which the Things themselves are form'd;

their Truth appearing to us by Phaenomena, though their causes be not yet discover'd. For these are manifest Qualities and their Causes only are occult. (*Opticks*, 2nd ed., bk. 3, part 1, query 31)

Chapter 2. Progress!

1. Boskovitch was the first to realize that material particles need not be impenetrable, but that their collisions may be caused by repulsive forces, which are exerted by their microscopic constituents. His theory was much beyond his time in the understanding of the role in the structure of matter. The following quotation from his work, *A Theory of Natural Philosophy* (Venice, 1763, translated by J. M. Child [Chicago: Open Court Publishing Co., 1922]) is particularly prescient.

Now the law of forces is of this kind; the forces are repulsive at very small distances, and become indefinitely greater and greater, as the distances are diminished indefinitely, in such a manner that they are capable of destroying any velocity, no matter how large it may be, with which one point may approach another, before ever the distance between them vanishes. When the distance between them is increased, they are diminished in such a way that at a certain distance, which is extremely small, the force becomes nothing. Then as the distance is still further increased, the forces are changed to attractive forces; these at first increase, then diminish, vanish, & become repulsive forces, which in the same way first increase, then diminish, vanish, & become once more attractive; and so on, in turn, for a very great number of distances, which are all still very minute: until, finally, when we get to comparatively great distances, they begin to be continually attractive & approximately inversely proportional to the squares of the distances.

Elsewhere he writes:

Now, because the repulsive force is indefinitely increased when the distances are indefinitely diminished, it is quite easy to see clearly that no part of matter can be contiguous to any other part; for the repulsive force would at once separate one from the other. Therefore, it necessarily follows that the primary elements of matter are perfectly simple, and that they are not composed of any parts contiguous to one another. This is an immediate and necessary deduction from the constitution of the forces, which are repulsive at very small distances & increase indefinitely.

2. G. W. Leibniz, *Essay on Dynamics* (1695), translation by J. Bennett at www.earlymoderntexts.com.

3. A. L. Lavoisier, *Elements of Chemistry* (New York: Dover, 1965).

4. The English chemist John Alexander Reina Newlands had noticed before Mendeleev (in 1865) that elements of similar type recurred at intervals of eight, which he likened to the octaves of music. His *law of octaves* was, however, ridiculed by his

contemporaries. Mendeleev developed the first periodic table almost simultaneously with the German Lothar Meyer in 1869. He went further than Meyer in employing the periodic table to correct mistakes in the values of several atomic masses and to predict the existence and properties of a few new elements in the empty cells of his table.

5. Lavoisier was not loath to the idea that the caloric consists of atoms, much smaller than those of ordinary matter, but he did not pursue it. He thought that the discussion about atoms was unwarranted by the experimental facts, and for this reason it should stay at the level of mere speculation.

6. The Swiss mathematician Daniel Bernoulli constructed the first model for a kinetic theory for gases in 1738.

7. Laplace had initiated an ambitious program of explaining the whole of physics in terms of imponderable fluids. These fluids—following Newton's ideas—were not viewed as fundamentally continuous but consisted of particles much smaller than the ones of ordinary matter. The task of physics would then be to find the forces and motions of these fundamental molecules. Laplace suggested that the postulate of a short-range attractive force between molecules would explain phenomena of such diversity as optical refraction, the solidity of solids, and chemical reactions. He believed that the precise understanding of such forces would lead physics to the same state of perfection as that reached in celestial mechanics by Newton's law of Universal Gravitation. Even though Laplace's program failed in the nineteenth century, his belief that the properties of matter are fully explained by the intermolecular forces proved remarkably successful in the twentieth century.

8. A large part of the debate between the dynamists and the supporters of Newton's theory focused on the issue of whether the living force or the momentum is a the true measure of the quantity of motion. This debate seems in retrospect rather futile, having to do only with the different a priori conditions each side demanded from the mathematical object that should represent the quantity of motion.

9. K. F. Mohr, *Zeitschr. Phys.*, 5, 419 (1837).

10. W. Thomson and P. G. Tait, *Treatise on Natural Philosophy* (Oxford, 1867; Prometheus Books, 2002).

11. In fact, thermal concepts are not fully reduced to the concepts of Newtonian mechanics because one is forced to make assumptions of a statistical nature. These assumptions are natural from the perspective of an external observer, who has to deal with an immensely large number of molecules. However, it is fundamentally unjustified in terms of first principles. A statistical description of molecules can be nothing else but an approximation, since the fundamental laws (Newton's) are not statistical but exact and deterministic. An analogous question still persists even after the advent of quantum theory. However, the theory works extremely well, and this qualifying "almost" for the reduction of thermal phenomena to mechanical ones is usually dropped in most presentations.

12. It is interesting to note that according to Newton, the forces of matter that cause the refraction of light are in essence gravitational. These are attractive; hence, the light

particles are supposed to move faster in matter than in the vacuum. The wave theory of light makes the opposite prediction, which as it turned out was the correct one.

13. The fundamental relation for waves is that their speed of propagation is proportional to the wavelength and inversely proportional to the period.

14. Coulomb announced his law for the electric force in 1785. We now know that John Robinson and Henri Cavendish had discovered this law before Coulomb, but their results were not published until well within the nineteenth century.

15. James Clerk Maxwell, *A Treatise of Electricity and Magnetism*, 3rd ed. (Oxford: Clarendon Press, 1891).

16. The electric charge was initially defined as a measure for the electric fluid, which was itself defined as the medium that gives birth to the electric forces. In other words, the electric charge is introduced to explain the electric phenomena but is defined in terms of the phenomena themselves. The definition is clearly circular, but this is not surprising. All definitions of new physical quantities are circular. One does not arrive at new physical quantities by logical deduction, but by a leap of intuition that "sees" what concept or metaphor is necessary to account for the phenomena. Logical deduction, by necessity, refers to a closed set of basic postulates. Any new idea of fundamental significance needs to sever some of the tight connections of a closely knit logical structure in order to find a space to ascertain itself. Too many worries about logical consistency at this stage may be simply disruptive of this process.

Although circular from a narrowly logical point of view, the definition of the charge is not an arbitrary theoretical construction. It must satisfy certain properties in order to account for experiments. For instance, we observe that the force exerted by two identical charged bodies (in the same position) on a third body is twice the force exerted by each of them separately. This, alone, states that the source of the electric phenomena is a quantity that adds up when electrically charged bodies are joined. Once we are persuaded of the reasonableness of the notion of the charge, we proceed to measure its numerical value. One may only measure an object that acts on its surroundings, and Newtonian mechanics states that one body acts on another by means of forces. Hence, any new physical quantity can be measured only by making reference to the forces it generates: but the law of force itself assumes the existence of the new physical quantity. Again, this argument sounds circular. The truth is that both definitions and experiments in physics involve a complicated mixture of theoretical thinking and experience that cannot be easily separated into its constituents.

17. The names *vitreous* and *resinous* for the two types of electric fluid are due to Charles Francois de Cisternay du Fay (1698–1739), after the substances in which they typically appear (vitriol and resin).

18. W. D. Niven, ed., *The Scientific Papers of James Clerk Maxwell* (Cambridge: Cambridge University Press, 1890), 358, 36.

19. The electric and the magnetic field are both vector fields—each of them needs three numbers at each point of space for a full specification.

20. Quoted in C. C. Gillispie, *The Edge of Objectivity* (Princeton: Princeton University Press, 1966).

21. In the language of the lines of force, this equation stated that the density of the electric charge equals the divergence of the electric field.

22. These equations were not all discovered by Maxwell. The first one is due to the German mathematician Carl Friedrich Gauss; the second is a natural consequence of Oersted's experiment; and the third had been a realization of Faraday. Only in the fourth did Maxwell add his contribution to an earlier work of André-Marie Ampere. Most important about Maxwell's theory was that he constructed a general conceptual framework that accommodated these equations.

23. Newton knew that light does not propagate instantaneously. He writes in the *Opticks* (book 2, part 3, proposition 11): "Light is propagated from luminous bodies in time, and spends about seven or eight minutes of an hour in passing from the sun to the Earth." This realization is attributed to the Danish astronomer Olaf Römer, who had studied in detail the eclipses of the moons of Jupiter. In Maxwell's time the best estimation (300,000 kilometers per second) came from experiments performed by Armand Fizeau in 1849.

Chapter 3. A New Arena Is Built

1. Newton writes:

The causes by which true and relative motions are distinguished, one from the other, are the forces impressed upon bodies to generate motion. True motion is neither generated nor altered, but by some force impressed upon the body moved; but relative motion may be generated or altered without any force impressed upon the body. For it is sufficient only to impress some force on other bodies with which the former is compared, that by their giving way, that relation may be changed, in which the relative rest or motion of this other body did consist. Again, true motion suffers always some change from any force impressed upon the moving body; but relative motion does not necessarily undergo any change by such forces. For if the same forces are likewise impressed on those other bodies, with which the comparison is made, that the relative position may be preserved, then that condition will be preserved in which the relative motion consists. And therefore any relative motion may be changed when the true motion remains unaltered, and the relative may be preserved when the true suffers some change. Thus, true motion by no means consists in such relations." (*The Principia*, scholium 11).

2. The essentials of this idea go back to Maxwell. See J. C. Maxwell, *Encyclopedia Britannica*, 9th ed., vol. 8, 1878. Reprinted in Niven, ed., *The Scientific Papers of James Clerk Maxwell*, 2:763.

3. The device designed by Michelson sent a light beam through a half-silvered mirror; there, the beam was split into two beams traveling at right angles to one another.

The beams were then reflected by mirrors, and they were recombined, producing (it was expected) an interference pattern. Michelson had carried out a first version of this experiment in 1881, but the precision had not been adequate for definite conclusions. In the experiment with Morley, the device was improved by allowing the beams to travel back and forth a few times, so that the distance they traveled would be larger, and the shift in the interference pattern more pronounced. To remove any external influence, the experiment was performed in the basement of a stone building and the device was built on a marble slab that floated in a pool of mercury. This allowed them to have the device rotated by a mechanical contraption. During each rotation, each of the two directions of the split beam would become twice parallel to Earth's velocity, and this would allow one to see a periodic behavior in the interference pattern.

4. A. A. Michelson and E. W. Morley, *Am. J. Sc.* 34, 333 (1887).

5. Inertial frames were initially conceived as the frames that move with constant speed along a straight line with respect to absolute space. In absence of absolute space, how can they be defined? The truth is that they cannot be defined; their existence is *postulated*. In effect, to identify an inertial frame we must study the form that the laws of motion take in it. If these laws are sufficiently simple and they have the same form in frames moving with constant velocity along a straight line with respect to the original frame, then we can call this frame inertial. Clearly, Einstein's postulate involves a definition that is circular from a logical point of view. However, it makes complete sense. First, for many phenomena it suffices to consider the surface of Earth as defining an inertial frame. This is not literally true, as Earth rotates around itself and it orbits around the Sun, but for any laboratory phenomena this assumption provides an excellent approximation. We would definitely not consider an accelerating elevator or a vehicle moving on a rollercoaster as defining inertial frames: it would be impossible to write any meaningful laws of motion there.

Second, when we try to write a law for a force from the study of experimental data, we had better come up with a simple expression. If the expression is not simple or is not derived by a simple principle, it is not worth taking the status of a law. One would be very suspicious of the law of Coulomb if the attraction or repulsion between electrically charged bodies was inversely proportional to their distance raised to a power of 1.5985785, while a power of 2 is perfectly reasonable. Hence, rounding up what we consider as fundamental force laws, we can ignore all errors due to the fact that we did not specify the inertial system with perfect accuracy. This commonsense approach may not be rigorous in terms of logic, but it is very clear and unambiguous about laboratory phenomena. Only at a cosmic scale does the specification of an inertial frame present us with grave difficulties. However, the general theory of relativity removes this problem because it considers all frames on an equal footing and does not single out the inertial ones.

6. When the person moves in a nonparallel direction to the tracks of the train, this rule needs to be generalized, but the result follows from simple addition or subtraction and a bit of geometry.

7. As an example of time dilation, we consider a rocket moving away from Earth. The observer on Earth perceives that the events in the rocket (from a heartbeat to a celebratory feast) last longer than the people in the rocket do. As motion is relative, the people in the rocket have the same impression about any events on Earth's surface. So what would happen if the astronaut and the observer on Earth were to compare their experiences? Which of them would be actually older? The answer is that if they are to make such a comparison, they have to meet. This means that the rocket will have to reverse course and return to Earth. To reverse course, the rocket must *accelerate* backwards. In that case, it will not define an inertial frame any more. We would have to employ different rules of transformation between the reference frames. General relativity provides these rules, and they lead to an unambiguous answer: at the time of their encounter, people on Earth will have aged more.

8. Lorentz discovered the transformations that carry his name in 1904. The same transformations had been derived much earlier (in 1887) by Woldemar Voigt (W. Voigt, *Goett. Nachr*, 1887, 41), and by J. Larmor in 1900 (J. Larmor, *Aether and Matter*, [Cambridge: Cambridge University Press, 1900]).

9. From a talk by Einstein at the University of Kyoto in 1922—see T. Ogawa, *Jap. St. Hist. Sci.* 18, 73 (1979).

10. H. Poincaré, *Science and Hypothesis* (New York: Dover, 1905).

11. Abraham Pais argues strongly about the influence of Poincaré's writings on Einstein—see A. Pais, *Subtle Is the Lord*, (Oxford: Oxford University Press, 1982), 133.

12. The word "measurement" should not be taken literally in the context of relativity. Einstein used this word in the first formulation of his theory because he wanted to stress that the notion of background time and background space are abstract constructions, which have nothing to do with our empirical perceptions of space and time. For this reason, he employed a concrete imagery of rods and clocks, and the word "measurement" was very adequate to bring forward the points he wanted to emphasize. The word "measurement" is also good for pedagogical purposes, when teaching the theory of relativity. However, it is not at all necessary for the conceptual framework of the theory, at least no more than it is in Newtonian mechanics. When we talk about time measurements, we need not assume the existence of an actual clock and of an observer that reads the clock. The word "clock" refers to any physical process, whose duration can be compared with the duration of phenomena in its vicinity: this can be the heartbeat of a person or a pulsation of a star. Phenomena like the time dilations are *not* subjective experiences of an observer, but actual behavior of the natural time parameters. Moreover, this behavior is due to the relative state of motion between physical bodies.

13. One should note that Plato's view is contrary to this statement, as he argues against it later in this dialogue.

14. We see in retrospect that the interpretation of energy as something active, "energetic," quite opposite to the concept of inertia, arose because for small speeds the measure of inertia is practically equal to the rest mass, which is constant for a given body.

15. Plato, *Timaeus*, 50c. Plato here refers to *hyle* (a concept usually translated as matter), which is the fundamental basic formless material from which every thing in the world is formed.

16. If massless particles ever moved with speed smaller than light, their energy would be zero because it is proportional to their rest mass. They would therefore stop existing.

17. In principle, *tachyons* are also possible, that is, particles that *always* move faster than light. One may obtain tachyons from the equations of special relativity, if one assumes that their rest mass is an imaginary number. So far, no tachyons have been unambiguously detected. If tachyons exist, they will have very counterintuitive properties: for example, their energy would decrease as the speed increases. The presence of tachyons in a physical theory is in general considered highly pathological, and for this reason most physicists dismiss them.

18. Maxwell's equations for the electromagnetic field are the same in all inertial frames of reference, but there is a slight complication involved. The electric field and the magnetic field depend on the reference frame. The magnetic field may vanish in one reference frame, and it may be nonzero in another. This is natural, after all, since the statement that moving charges produces the magnetic field invokes the concept of motion and implicitly that of a reference frame. However, other laws are not invariant under the Lorentz transformations, for example, Newton's law of Universal Gravitation (the elaboration of this point was to lead Einstein to his theory of general relativity).

Chapter 4. The Symmetry Beneath

1. There is a potential exception to this statement: the initial conditions of the whole Universe. This is the subject matter of modern (quantum) cosmology. It has been suggested in this field that there might be a simple law specifying the initial state of the Universe, and in one particular proposal by S. W. Hawking from Cambridge University and J. B. Hartle from the University of California, Santa Barbara, the same law would give both dynamics and initial condition (*Phys.Rev.* D28, 2960 [1983]). However, we are still very far away from an era in which we will be able to access such proposals in terms of observational data, so logical coherence and mathematical elegance can be the only judge of these ideas.

2. E. Noether, *Nachr. von der Gesell. Wiss. zu Göttingen* 235 (1918).

3. The relation between symmetries and conserved quantities is more intricate than what I have so far described. Conserved quantities refer to numbers we measure from experiments, while symmetries refer to changes of profiles of physical systems. They are very different concepts, as they refer to different things. However, the mathematical formalism of Newtonian mechanics together with the principle of energy conservation provides a relation between the two. This fact had been known ever since the nineteenth century from the work of the French mathematician Simeon-Denis Poisson. Poisson realized that there exists a precise and unambiguous rule that allows one to relate a

physical quantity to a rule of transformation for the profiles of a physical system. One may appropriately refer to this physical quantity as the *generator* of the transformation. Noether demonstrated that in the light of Poisson's theory, conserved quantities contain in themselves all information about the symmetry. Hence, symmetries to the laws of motion and conserved quantities are two different concepts that describe the same thing.

Poisson belonged to a large community of mathematical physicists who had rationalized the mathematical formulation of Newtonian mechanics. Joseph Lois Lagrange, Pierre–Simon de Laplace, Adrien–Marie Legendre, Carl Gustav Jacobi, and William Rowan Hamilton were the main figures in this effort. The rationalized mechanics described physical systems in terms of profiles, which were constructed from the knowledge of the positions and momenta of their constituent particles. Their description allowed the resolution of forces into a geometrical language. The geometry they employed, however, was not the physical geometry of space. It referred to the abstract mathematical space that hosts all possible profiles of a physical system. The fundamental concept in this rationalized mechanics was the energy, which turned out to be the generator of the profile's change in time according to Newton's laws.

4. In the most general case, the symmetry of space translation exists if the "empty space" exerts the same force at all places. In general relativity, where the spacetime can be different from Minkowski, there exist two classes of spacetimes that also have this property: the de Sitter and anti-de Sitter spacetime.

5. Noether's theorem states that the energy is conserved because of the symmetry of time translation. However, it does not specify that energy is positive; the positivity of energy is an additional and distinct assumption. From the transformation rules of Lorentz, we may prove that the positivity of energy does not depend on the choice of reference frame. If energy is positive in one coordinate system, it is positive in all. The stability of matter *only* requires that a state of minimum energy exists. This minimum energy could very well be negative for some systems. The only reason we are obliged to define energy in such a way that its minimum value is positive (or zero) is that the positivity of energy is a property that remains invariant under the Lorentz transformations.

In general relativity, energy positivity plays a crucial role—it guarantees the principle of causality, namely, that a physical system cannot travel to its own past. See S. W. Hawking and G. F. R. Ellis, *The Large-Scale Structure of Spacetime* (Cambridge: Cambridge University Press, 1973).

6. It is quite remarkable that a novel concept such as spin arises solely from the study of symmetries. We need only to refer to the Poincaré symmetry and to the requirement that our physical systems satisfy Newton's law in order to derive spin's existence mathematically. Spin could therefore have been discovered earlier than it actually was, if one took care to investigate all irreducible systems compatible with the Poincaré symmetry. But the mathematical theory necessary for this was developed much later, namely, in the 1960s and 1970s, when the mathematicians Bertram Kostant, Jean Marie Souriau, and

Anatol Kirillov independently tried to understand a surprising relation between the abstract notion of symmetry and the geometrical structures underlying Newton's equations. A detailed technical review of this theory is J. M. Souriau, *Structure of Dynamical Systems: A Symplectic View of Physics* (Boston: Birkhäuser, 1997). If this discovery had come earlier, it would have provided a nice example of mathematical consistency alone, without any empirical data, leading to the discovery of novel physical concepts.

In any case, spin was not introduced into physics through this route. It was first postulated during the 1920s in the context of the quantum theory of the atom, to explain certain phenomena in the emission of light from atoms (see sec. 7.1). For this reason, most physics textbooks state that spin is a quantum concept, which has no counterpart in Newtonian physics. Strictly speaking, this is not true. From a logical and structural perspective, spin is fully compatible with Newton's laws. The only difference in the quantum description is that the spin arrow's length may take only specific *discrete* values. Spin also appears outside the context of relativity. In prerelativistic physics, the analysis of the irreducible components associated with the symmetries of space and time separately would lead to the discovery of a physical quantity that has the properties of spin.

7. If a system is characterized by a specific symmetry, it is possible to exploit this symmetry in order to *freeze* one or more degrees of freedom. For instance, we may always define a reference frame, in which the particle is at rest. By doing so, we freeze the particle's velocity to zero. If we keep it frozen and employ the zero value in all calculations involving velocities, we have no right to refer to any transformations that change the value of velocity, namely, the boosts. The same principle is valid for all degrees of freedom.

An irreducible system associated with a specific type of symmetry is a physical system, *all the degrees of freedom of which may be frozen by exploiting its symmetry*. Equivalently, we may choose any profile of an irreducible system and obtain any other profile by acting upon it with symmetry transformations. Hence, any two profiles of an irreducible system may be related by a symmetry transformation.

We mentioned earlier that it is possible to freeze the momentum of a particle to zero by exploiting the boosts of the Poincaré symmetry. Moreover, we can always describe a particle as lying at the origin of our coordinate system. By doing this we use up our freedom to apply spacetime translations. But after we exhausted the boosts and the translations, we are still left with spatial rotations. In a description of an irreducible object, these degrees of freedom should also be employed. It is then necessary to assume that the particle has further structure, another degree of freedom *that can be frozen by means of a rotation*. The only way to achieve this is by introducing an new degree of freedom corresponding to an arrow with fixed origin and length. This is the spin. An arrow may always be made to point at a particular direction through a rotation. Hence, the choice of a specific direction involves freezing out all degrees of freedom related to rotations. This way, by using up all the freedom provided by the Poincaré symmetry, we can find a reference frame in which a particle is at its origin, does not move, and its arrow points toward a specific direction.

8. Different observers will not agree on the actual trajectory followed by a massless particle. Nonetheless, all observers will agree about the motion of the corresponding sheet.

Chapter 5. The Machine Breaks Down

1. The first discussion of black body radiation was by Gustav Robert Kirchhoff in 1860 (*Poggendorf's Ann. der Phys.* 109, 275 [1860]).

2. From Planck's letter to R. W. Wood (10.1.1931), in *Archive for the History of Quantum Physics,* Microfilm 66,5. A translation can be found in A Hermann, *The Genesis of Quantum Theory (1899–1913)* (Cambridge: MIT Press, 1971).

3. Ibid.

4. Autobiographical note in P. A. Schilpp, ed., *Albert Einstein: Philosopher-Scientist* (Chicago: Open Court, 1988).

5. A. Einstein, *Ann. Phys. Lpz.* 17, 132 (1905).

6. The name photon was coined by Gilbert Lewis, *Nature* 118, 874 (1926).

7. R. Millikan, *Phys. Rev.* 7, 355 (1916).

8. Thomson had actually measured the ratio of the particle's charge to its mass, and not the charge and the mass separately. The value of the electron's charge was identified later in a series of experiments by Millikan, who also established that all charges appearing in nature seem to be integer multiples of the electron charge.

9. J. J. Thomson, *Recollections and Reflections* (London: Bell, 1936).

10. These experiments were carried out in 1909 by Hans Geiger and Ernst Marsden, who were both Rutherford's collaborators (*Proc. Roy. Soc. A*, 82, 495 (1910)).

11. Rutherford proposed his model for the atom in *Philosophical Magazine* 21, 669 (1911). He was indebted to an earlier work of H. Nagaoka, who had proposed already in 1904 (*Philosophical Magazine* 445 [1904]) a model for the atom resembling the planet Saturn with its rings. A large number of electrons lie on a circle, in the center of which lies a heavy positively charged particle. The electrons in Nagaoka's model are in a state of equilibrium, and they only perform small oscillations around a point. In Rutherford's model, the electrons are free to move in any orbit around the positively charged particle and do not all occupy the same circle.

12. There was no way to reconcile the basic features of Bohr's model with Newtonian physics. As Planck explained in his Nobel lecture (1920), "the fact that the quite sharply defined frequency of an emitted photon should be different from the frequency of the emitting electron must seem to a theoretical physicist, brought up in the classical school, at first sight to be a monstrous and, for the purpose of a mental picture, a practically intolerable demand."

13. Letter from Rutherford to Bohr (3.20.1913), in N. Bohr, *Essays 1958–1962 on Atomic Physics and Human Knowledge* (New York: Vintage Books, 1963).

14. N. Bohr, "*On the Spectrum of Hydrogen*," in *The Theory of Spectra and Atomic Constitution* (Cambridge: Cambridge University Press, 1922).

15. M. Jammer, *The Conceptual Development of Quantum Mechanics* (New York: McGraw-Hill, 1966), 88.

16. Slightly modified from L. de Broglie, Nobel lecture, 1929.

17. L. de Broglie, *Philosophical Magazine* 47, 446 (1924).

18. De Broglie later realized that experiments that had been carried earlier by C. J. Davisson and C. H. Kunsman in New York (1919) verified his theory of matter waves.

19. Later disagreements of de Broglie with the dominant philosophy of quantum theory led him to propose the theory of the coexistence of a particle and of a wave that guides the particle's motion. This was an elaboration of his first idea about the duality between particles and waves. It is known as the "pilot wave" theory for quantum phenomena. This theory was strengthened and diffused by the British physicist David Bohm much later. It was proposed as an alternative to the description of quantum theory that came to be dominant; it does reproduce all its predictions—even though it cannot easily be generalized—but has the nasty feature of accepting the simultaneous existence of two fundamental entities in the buildup of the world. Furthermore, it violates the principle of locality (see sec. 7.6). To account for the phenomena, this theory has to accept the existence of signals faster than light, even though these signals never appear at a macroscopic level. For these reasons, and perhaps because of some prejudice toward any attempts to shake the foundations of such a successful theory as quantum mechanics, the theory of de Broglie and Bohm has few adherents and has not contributed significantly to the shaping of the worldview of modern physics.

20. P.A.M. Dirac, *Proc. Roy. Soc. London* 110, 561 (1926).

21. Auguste Comte, who lived in the beginning of the nineteenth century, is generally regarded as the father of positivism. Comte rejected the claim of science to describe objective reality, or any truth deeper than logical coherence or effectiveness. Positivism entered the sciences in the later half of the nineteenth century through the work of Pierre Duhem, Wilhelm Ostwald, and Ernst Mach. Positivism in that phase was critical of both the Newtonian mechanics and the atomic theory. Its advocates argued that the basic terms of these theories (atoms, forces, etc) were to a large extent linguistic conventions and should not be attributed to the objective character of reality.

Ostwald and Duhem were energists—successors of Mayer—and believed that sciences should be based on the concept of energy and that the phenomena of mechanics were only particular cases of energy exchange. Mach was perhaps the most influential positivist voice in the physics of the nineteenth century. He attacked the Newtonian concept of absolute space (but not absolute time). He also criticized the concept of inertia in Newtonian mechanics, and in these criticisms he came upon some basic ideas of the theory of relativity. The positivists of the nineteenth century mainly attacked the atomic theory. They thought that the existence of atoms was an assumption completely unwarranted by the empirical data. The atomic theory simply provided a convenient tool for the description of physical phenomena and reflected nothing of the deeper reality of the world. They were therefore highly critical of the development of statistical mechanics, which explained the macroscopic properties of matter in terms of the motion of its microscopic constituents. The positivists had attacked the wrong theory, though. The development of atomic theory in the first decade of the twentieth century demolished most of their arguments in physics. But the overall philosophical stance survived, namely, the condemnation of the uncritical application of theoretical constructs to objective reality and even the renunciation of science's ambition to state anything at all about the reality beyond a mere description of phenomena.

22. Einstein had been strongly influenced by Mach's thought, at least in the beginning of his research into gravity. He stated that Mach's book *The Science of Mechanics in Its Historical Evolution* had shaken in him for the first time the dogmatic faith of physicists in mechanics as the basis of their science. Mach's concept of inertia also played an important role in Einstein's development of the general theory of relativity. "It was not improbable," Einstein stated, "that Mach would have discovered the theory of relativity, if, at the time when his mind was still young and susceptible, the problem of the constancy of the speed of light had been discussed among physicists." However, general relativity introduced concepts that would have made Mach uncomfortable, and on some issues it ran contrary to Mach's ideas.

23. My description of the orbits in the hydrogen atom is rather simplified. To be precise, each orbit in the hydrogen atom is characterized by three integers (see sec. 6.1).

24. Heisenberg was strongly influenced by an earlier work of N. Bohr, H. A. Kramers, and J. C. Slater (*Zeit. Phys.* 24, 69 [1924]), who had provided an explicit renunciation of the principles of the classical theory. This work also suggested that for any atomic orbit there exists a virtual electromagnetic field in different potential configurations. These configurations correspond to the possible orbits to which an electron can jump. Heisenberg had collaborated with Kramers just a few months before he came up with his idea.

25. The theory of matrices and noncommutative objects had been developed in the mathematical literature sometime before Heisenberg. It originated in the work of Hamilton during the 1840s. However, in Heisenberg's days it was a rather esoteric branch of the mathematical science—Heisenberg himself did not know that the objects he had constructed were matrices. The application of the theory of matrices in quantum theory brought it into the limelight of physical research. Nowadays it is considered an indispensable tool for any physicist or mathematician.

26. The name "q-number" is from P.A.M. Dirac (*Proc. Roy. Soc. London* 110, 561 [1926]).

27. E. Schrödinger, *Annalen der Physik* 79, 361 (1926).

28. The configuration of a single particle is fully determined by its position, which is a point of physical space. For two particles, however, the possible configurations correspond to two points of physical space. Hence, the space of all possible configurations for two particles is the physical space taken *twice* and not the physical space itself, as would be necessary for the interpretation of the ψ-function to make sense.

29. W. Heisenberg, *Quantum Theory and Its Interpretations*, reprinted in *Quantum Theory and Measurement*, edited by J. A. Wheeler and W. H. Zurek (Princeton: Princeton University Press, 1983).

30. E. Schrödinger letter to Willy Wien (25.10.1926), quoted in W. Moore, *A Life of Erwin Schrödinger* (Cambridge: Cambridge University Press, 1994).

31. E. Schrödinger, *Annalen der Physik* 79, 734 (1926).

32. Quoted in Moore, *A Life of Erwin Schrödinger.*

33. W. Heisenberg, *Zeit. Phys.* 43, 172 (1927).

34. Starting with H. P. Robertson (Phys. Rev. 34, 163 [1929]), many proofs of uncertainty relations have been found, all of which are of a statistical nature. The statistical uncertainty relations refer to the spread of the measurement values in a large number of experiments and not to individual physical systems. For this reason, they are conceptually very different from the one originally derived by Heisenberg, which refers to measurement of individual quantum systems. Unlike the statistical uncertainty relations, for which a consensus has been reached concerning their interpretation, Heisenberg's original uncertainty relation remains a matter of controversy. The original proof involved the consideration of quantum particles interacting with an external measuring device and not isolated particles. How could one say that the uncertainty relation refers to the particles by themselves? Heisenberg dismissed the idea that one could talk about the values of physical observables without the consideration of a specific measurement scheme as metaphysical prejudice. However, this point of view was not and is not accepted by everyone. Different interpretations of quantum mechanics attribute different meaning to the uncertainty relation.

35. Roughly, the modulus square of the (complex) value of the wave function at a point of space gives the probability that the particle will be found at that point.

36. W. Heisenberg, *Physics and Philosophy* (New York: Prometheus Books, 1999).

37. Bohr's account of his discussions with Einstein is in P. A. Schilp, ed., *Albert Einstein: Philosopher-Scientist*, Library of Living Philosophers (Evanston, 1949). It has been reprinted in *Quantum Theory and Measurement*, edited by J. A. Wheeler and W. H. Zurek.

38. Rosenfeld provides a vivid picture of the debate: "At the sixth Solvay conference in 1930, Einstein thought he had found a counterexample to the uncertainty principle. It was quite a shock for Bohr . . . he did not see the solution at once. During the whole evening he was extremely unhappy, going from one to another and trying to persuade them that it couldn't be true, that it would be the end of physics if Einstein were right; but he couldn't produce any refutation. I shall never forget the vision of the two antagonists leaving the club: Einstein a tall majestic figure, walking quietly, with a somewhat ironical smile, and Bohr trotting near him, very excited. . . . The next morning came Bohr's triumph." After the sixth Solvay Conference Einstein changed his opinion about quantum theory, accepting it as a working compromise, but he remained critical of its claim to represent a final and complete picture of reality. He persisted in the hope of a better theory until the end of his life.

39. The term "quantum mechanics" first appeared as the title of an article by Born (M. Born, *Zeit. Phys.* 26, 379 [1924]). The term "mechanics of quanta," however, had been first used by Lorentz, who in an address at the Sorbonne in 1923 stated that "a mechanics of quanta, a mechanics of discontinuities has still to be made." See the discussion in Jammer, *The Conceptual Development of Quantum Mechanics.*

40. The quantum state is both similar to and different from the state or profile of classical systems we encountered previously. Both a quantum state and a profile encode the maximum information that we can have for a physical system, but in quantum theory, the uncertainty principle places a limit on how much information we may extract from a physical system. While the knowledge of the classical profile allows us to predict the values of all physical quantities, this is not true any more in quantum theory.

41. We should mention here a very important mathematical result: there exists a procedure by which the states characterized by a sharp value of energy can generate any possible state of the system. Moreover, most of the experimentally verified predictions of quantum theory—but not all—are obtained by the direct or indirect study of the energy q-number. This is a reason that allows many physicists to ignore the fine details of the meaning of quantum theory in their everyday work. It suffices that they know how to deal with the energy q-number and perhaps a couple more q-numbers that have a natural interpretation in terms of classical physics.

42. There are a few exceptions to this rule, as the so-called weak forces (see sec. 8.1) also play a role in a small class of atomic and chemical phenomena.

Chapter 6. So Familiar and Yet So Different

1. W. Pauli, *Zeit. für Phys.* 31, 765 (1925).

2. From a postcard from Heisenberg to Pauli, in W. Pauli, *Wissenschaftlicher Briefwechsel mit Bohr, Einstein, Heisenberg u.a., Band I: 1919–1929* (New York: Springer Verlag, 1979) (quoted from N. Strautmann, quant-ph/0403199).

3. O. Stern and W. Gerlach, *Zeit. Phys.* 8, 110 (1922).

4. G. E. Uhlenbeck and S. Goudsmit, *Die Naturwissenschaften* 13, 953 (1925); *Nature* 117, 555 (1926).

5. One may ask then why the electron spin had not been observed in electron beams and was only seen in the behavior of heavy atoms such as silver. The answer is that electrons have very small mass. As a consequence, their de Broglie wavelength is large, and any two trajectories arising out of a Stern-Gerlach experiment interfere so strongly (like in the two-slit experiment) as to wash out any distinction between them. On the other hand, the atoms of silver are quite heavy (about 100,000 times heavier than a single electron), so they have a much smaller de Broglie wavelength. For this reason, their beams exhibit much less interference, and the detected pattern is clear enough to distinguish the trajectory split due to spin.

6. Heisenberg, *Physics and Philosophy.*

7. There exist many schemes and ideas in modern physics that attempt to explain the world of classical physics as arising from quantum theory—a problem usually referred to as the quantum measurement problem. While classical behavior is indeed a limiting case of quantum behavior, it is not very robust—it can easily be drowned by the preponderance of quantum phenomena. The main issue is to find an explanation why this does not happen. At this point, all interpretations of quantum theory encounter serious problems—each of them has its own flaw. It is at the discretion of the theorist, whether he or she will see such flaws as lethal or as minor ones that will be resolved in the future.

8. Y. Aharonov and D. Bohm, *Phys. Rev.* 115, 485 (1959).

9. M. V. Berry, *Proc. Roy. Soc. Lond.* A392, 45 (1984).

10. The name helicity comes from "helix," which is Greek for spiral. This choice for the name is largely a historical accident and is not related to the spin of massless particles. It arose from the description of a certain property of light (known as polarization), which can be mathematically described by an image of spiraling motion.

11. When we view two or more particles as identical, we always make a tacit assumption that their motions are not constrained in a way that may allow us to distinguish them. For example, we can always distinguish two particles, even if they are of the same type, if we keep them separated. We can place two electrons in two parts of a box that are separated by means of an internal wall. We may then talk unambiguously about the electron in the left and the electron in the right part of the box. However, if we remove the wall and the electrons start moving freely within the box, there will be no way to trace their identities.

12. Einstein and Dirac significantly contributed in elaborating the implication of these statistics to the large-scale organization of matter. For this reason, the corresponding statistical laws are known as Bose-Einstein and Fermi-Dirac, respectively.

13. W. Pauli, *Phys. Rev.* 58, 716 (1942).

Chapter 7. Forging the Perfect Tool

1. The quantization of the electromagnetic field was first performed in one of the founding papers of quantum mechanics by Born, Heisenberg, and Jordan (*Zeit. Phys.* 35, 557 [1925]). Jordan and Wigner elaborated on the distinction between bosonic and fermionic particles in relation to quantum fields (*Zeit. Phys.* 47, 631 [1928]), while Heisenberg and Pauli were the first to develop the general theory for the quantization of fields (*Zeit. Phys.* 56, 1 [1929]; *Zeit. Phys.* 59, 168 [1930]).

2. Physical theories employ two conceptually distinct notions of time. The first highlights the relation of a physical system to the spacetime. It incorporates the notion of before and after, what physicists usually refer to as causal ordering. The second is related to the way the system changes as it evolves, and according to Noether's theorem it corresponds to energy. Even though it is usually not necessary to distinguish between the two notions, it is possible to rewrite all physical theories in a way that this distinction is

made manifest. See K. Savvidou, *J. Math. Phys.* 40, 5657 (1999), and, for a less technical account, C. J. Isham and K. Savvidou, "Time and Modern Physics," in *Time*, edited by K. Ridderbos (Cambridge: Cambridge University Press, 2002).

3. The protons (literally the "primary") had been initially postulated by Rutherford in his model of the atom, while the neutron was discovered in a famous experiment by J. Chadwick (*Proc. Roy. Soc. London* 81 [1932]).

4. Fock's method was applied to electrons in 1933 (C. R. Leningrad, 267 [1933]). One year later Furry and Oppenheimer (*Phys. Rev.* 45, 245 [1934]) established the equivalence of Dirac's hole theory to the description of electrons in terms of quantum fields. However, Pauli and Weisskopf proved in the same year (*Helv. Phys. Acta* 7, 709 [1934]) that Dirac's hole theory is not equivalent to the field theory description of bosonic particles. They constructed a field theory describing particles of zero spin and their antiparticles, in which the antiparticles could not be interpreted as holes in a negative energy sea.

5. It is mathematically possible to define classical Dirac fields. These fields behave like the quantum Dirac fields under the action of the Poincaré symmetry. However, the quantum Dirac field includes in its structure the Fermi connection, which has no analogue in classical mechanics. The correspondence with the classical Dirac field therefore does not go all the way.

6. One may phrase the argument about the inability to measure a Dirac field without referring to the electric charge. Even in Dirac fields that correspond to neutral particles such as the neutrinos, there exist conservation laws that are violated by the existence of states with sharp field values. Conversely, the conservation of charge forbids the measurement of quantum fields other than the Dirac field, if these fields correspond to charged particles.

7. Heisenberg associated potentia with the wave function of quantum mechanics rather than with the quantum fields. It can be argued, however, that he viewed the wave function and the fields as different mathematical manifestations of the same reality. D. C. Cassidy, *Uncertainty: The Life and Science of Werner Heisenberg* (New York: W. H. Freeman, 1992).

8. Feynman derived his graphical rules for quantum electrodynamics using a different method for quantization from the traditional one in terms of q-numbers (*Rev. Mod. Phys.* 20, 367 [1948]). Freeman Dyson then showed that Feynman's rules could be recovered from the ordinary rules of quantum theory (that involved q-numbers), in *Phys. Rev.* 75, 486 (1949).

9. During most of the 1930s and the early 1940s, the majority of physicists did not believe that the quantum field theory of interacting electrons and photons was adequate to explain physical phenomena. Most of the research diverged from the minimal assumptions involved in quantum electrodynamics. There was a strong tendency toward radical solutions, inherited by the revolutionary days of the birth of quantum theory. Moreover, the infinities involved in quantum field theory seemed too intimidating to allow quantum electrodynamics to become a physically meaningful theory.

The first indication that the infinities had some physical content came from precision measurements of a small energy difference (a "shift") between two specific energy levels of the hydrogen atom. This shift was named after Willis Lamb, who first detected it in 1947. It could be accounted neither by Dirac's theory of the electron nor by the part of quantum electrodynamics that was free from infinities. However, the full quantum electrodynamics predicted an infinite shift! Perhaps a deeper study of the infinities would allow one to isolate a finite number for the Lamb shift. Another measurement by Isidor Rabi (in 1947) of the magnetic moment of the electron—which was again slightly different from that predicted by Dirac's theory—further encouraged this idea. Schwinger (*Phys. Rev.* 74, 1439 [1948]), Tomonaga (*Prog. Theor. Phys.* 1, 27 [1946]), and Feynman (*Rev. Mod. Phys.* 20, 367 [1948]) developed three different formalisms by which to tackle this problem, essentially laying the foundations for a solid treatment of the renormalization procedure. It was then shown by Dyson that the three different approaches yield the same results. By performing a detailed study of the infinities in perturbation theory, Dyson showed that they could be consistently removed through the renormalization procedure.

We should also remark that the idea of renormalization had been suggested already in the 1930s the work of Oppenheimer (*Phys. Rev.* 35, 461 [1930]) and Weisskopf (*Zeit. Phys.* 89, 27 [1934]), and it was being extensively discussed in the years before the war.

10. To absorb the infinities in a redefinition of the physical parameters, we must have some mathematical control over the form and structure of the infinities. For this purpose, one introduces an additional parameter into the theory, which has the aim to *quantify the infinity*. This procedure is called *regularization*. The new parameter changes the physical content of the theory to such an extent as to render the terms of infinite magnitude appearing in the physical quantities finite. This is too drastic a step, and it is done at the *price* of some cherished physical principle: the resulting theory violates either the Poincaré symmetry or causality or lacks a spacetime character. For this reason, the theory with the inclusion of the new parameter is unphysical. We introduce this parameter only for purposes of mathematical control, and we remove it after renormalization. No physical quantity depends on it.

11. The conventional value of the electric charge and mass is defined with respect to a reference process characterized by the lowest possible value of energy.

12. The contribution of the infinite number of additional parameters is often negligible for the physical predictions at low energy. The nonrenormalizable field theory provides then an adequate description—an *effective theory*—for low-energy phenomena. An effective theory may be viewed as an approximation to a more fundamental theory that can be adequately defined only at high energies.

13. Renormalizability refers to the study of quantum field theory through perturbation methods, hence to a particular mathematical technique. It is conceivable that a theory that is nonrenormalizable, and hence intractable by means of perturbation theory, may be perfectly well defined when treated with alternative methods. This possibility

has been verified in some simple models, e.g., B. Rosenstein, B. J. Warr, and S. H. Park, *Phys. Rev. Lett.* 62, 1433 (1989).

14. S. Tomonaga, *Nobel Lectures (Physics): 1963–1970* (Amsterdam: Elsevier, 1972).

15. P. A. M. Dirac, "The Inadequacies of Quantum Field Theory," in *Reminiscences about a Great Physicist: P. A. M. Dirac*, edited by B. Kursunoglu and E. Wigner (Cambridge: Cambridge University Press, 1987).

16. A comment of Chen-Ning Yang in his textbook on *Particle Physics* (Princeton: Princeton University Press, 1961) reflects this change of perspective: "To those of us who were educated after light and reason had struck the final formulation of quantum mechanics, the subtle problems and the adventurous atmosphere of these pre-quantum mechanics days, at once full of promise and despair, seem to take on an almost eerie quality. We could only wonder what it was like when to reach correct conclusions through reasonings that were manifestly inconsistent constituted the art of the profession."

17. Newton's laws of motion provide an axiomatic system for classical mechanics. They describe a large class of theories, and this description is made specific once we identify the laws of the forces. Similarly, quantum theory can be written in an axiomatic framework: Paul Dirac was the first to write one, and there were many to follow his example. The difference in quantum field theory is that the only available technique (perturbation theory) leads to meaningless results, and the cure (renormalization) is not very sure-footed from the mathematical point of view. In other theories, one may obtain successful predictions from specific models, without caring too much about an exact specification of axioms, so that a skeptic might depreciate the axiomatic treatment as an indulgence to an overrigorous way of thinking that is not very relevant to the common practice of physicists. However, in quantum field theory, the axiomatic treatment is indispensable because there is no other way to verify that important concepts and ideas can be properly defined.

18. Wightman's scheme is fully developed in R. Streater and A. Wightman, *PCT, Spin, Statistics and All That* (Princeton: Princeton University Press, 1955). An alternative and potentially more general axiomatic scheme was developed by Rudolph Haag, *Local Quantum Physics* (Berlin: Springer, 1996).

19. The principle of local causality is not applied the same way to all fields. It involves a distinction between fields corresponding to particles of the Fermi and the Bose type, the assumption that there exists no third alternative and is implemented through a different mathematical procedure for each type of field. Hence, it involves a tacit recognition of the relation between fields and particles. For this reason, even if axiomatic quantum field theory is phrased solely in terms of fields, the physical justification of its basic axioms is impossible without the subtle introduction of the particle concept.

It should also be noted that the mathematical implementation of the locality principle for fermion fields does not conform to the description of the principle given in the main text in a straightforward way. The condition that the fermion fields have to satisfy is mathematically natural and fully compatible with the fact that measurements at spacelike separated regions should not affect each other. However, it also involves further restrictions, which cannot be explained without an analysis of the corresponding formalism. This complication is partly related to the fact that fermion fields cannot be directly measured.

20. The principle of locality is rather more general than the principle of local causality, the latter being a special case of the former in the context of quantum field theory. Recent years have witnessed an increasingly elaborate discussion of the locality principle, mainly motivated by a widespread perception that it is violated by the phenomenon of quantum entanglement. As a result many different versions of the locality principle have been studied, each subtly different from the other. In my presentation, I have chosen to ignore these fine distinctions, emphasizing the key idea underlying this principle.

21. To be precise, it is possible to prove a version of spin-statistics theorem if one includes additional assumptions into the quantum theory of particles. However, these assumptions are unjustified from first principles. For this reason, the existing consensus is that the introduction of quantum fields is essential for the proof of the spin-statistics theorem.

22. Another important consequence of the basic axioms of quantum field theory is that it is impossible to avoid the infinities in perturbation theory. This statement goes by the name of *Haag's theorem*. It can be justified as follows. We consider a specific quantum field theory characterized by a number of physical parameters (coupling constants, particle masses, etc). To change the value of one of these parameters, we would have to drastically modify the structure of the field, and we would have to do this at all space and in the smallest of scales. The energy we would expend for this purpose would be necessarily infinite. There exists, therefore, an infinite energy gap between two different values of the physical parameters, however close one might be to the other. In particular, this energy gap is present when one attempts to go from a theory of free particles (zero coupling constant) to a theory of interacting particles (a nonzero value of coupling constant). It is for this reason that infinities appear when we use perturbation theory: if we start with the free theory, we cannot go to the interacting one without burdening our equations with expressions that involve this infinite energy gap.

23. The system of axioms sketched here needs to be strengthened by some extra assumptions in order to enable us to derive certain physically interesting properties. If, for example, we study the general properties of processes that involve particle scattering, we need to add an assumption about the existence of states that correspond to isolated particles.

24. A. Angelopoulos et al., *Phys. Lett.* B444, 43 (1998).

Chapter 8. Pieces of a Puzzle

1. W. Pauli, letter to a physicists' gathering at Tübingen (12.4.1930). Reprinted in *Wolfgang Pauli, Collected Scientific Papers*, edited by R. Kronig and V. Weisskopf, vol. 2, p. 1313 (New York: Interscience, 1964).

2. E. Fermi, *Zeit. Phys.* 88, 161 (1934).

3. F. Reines and L. C. Cowan, *Phys. Rev.* 92, 830 (1953).

4. To be precise, the way the argument is presented, it would also be possible for the neutron's L-charge to be a noninteger number. To avoid this, we should use more complex arguments that involve a larger number of processes. For this reason, we restrict the possible values of the L-charge to integers.

5. Isospin may be may be generalized for N particles that behave identically under the strong interactions, in complete mathematical analogy to spin. In that case they form an isospin N-tuplet, which is characterized by isospin value equal to $(N-1)/2$—see fig. 6.7 for spin. The elements of the N-tuplet are distinguished by means of the n numbers $N(N-1)/2, -(N-2)/2, \ldots, (N-3)/2, (N-1)/2$. So for N = 4, the isospin value is 3/2 and the values of the projection are $-3/2, -1/2, 1/2, 3/2$.

6. H. Yukawa, *Proc. Phys.-Math. Soc. Japan* (3) 17, 48 (1935).

7. G. Danby, J-M. Gaillard, K. Goulianos, L. M. Lederman, N. Mistry, M. Schwartz, and J. Steinberger, *Phys. Rev. Lett.* 9, 36 (1962).

8. The mu particles are characterized by a new conserved quantity, known as muon number. This is in all respects similar to the leptonic number, but it refers exclusively to the muon. Its introduction was motivated by the fact that processes of the form $\mu^- \rightarrow e^- + \gamma$ and $\mu^- \rightarrow e^+ + e^- + e^+$ have not been observed, in contrast to the processes $\mu^- \rightarrow e^- + \nu_\mu + \nu^*$. This suggests that one has to distinguish between two types of lepton number, both of which are simultaneously conserved. The first is the electron lepton number, which was the one that had been originally defined. Muons and the associated neutrino have zero value of this number. The second conserved quantity is the muon lepton number, of which only muons and their neutrinos had nonzero values.

9. The charged pions were detected in 1947 (C.M.G. Lattes, H. Muirhead, G.P.S. Occhialini, and C. F. Powell, *Nature* 159, 694 [1947]). Like the muon before them, they were identified from the study of the products of collisions from cosmic rays in the atmosphere. The cosmic rays are particles of extremely high energies that arise out of cataclysmic cosmic processes.

10. C. N. Yang, Nobel lecture, 1964.

11. T. D. Lee and C. N. Yang, *Phys. Rev.* 104, 254 (1956).

12. Wu in her experiment (C. S. Wu, E. Ambler, R. W. Hayward, D. D. Hoppes, and R. P. Hudson, *Phys. Rev.* 105, 1413 [1957]) studied the beta decay of Cobaltium nuclei. These nuclei consist of 60 nucleons and are characterized by a nonzero value of spin. When a specimen of Cobaltium is placed in a strong magnetic field, the spin arrows of the nuclei tend to align with the direction of the magnetic field. If the temperature is sufficiently low, almost all nuclei will have their spins aligned in the same direction. The Cobaltium nuclei may emit electrons (beta decay) by means of the weak interactions. The presence of the magnetic field and the alignment of the spins that follows as consequence imply that the system is characterized by a preferred direction in space. If, however, the P-transformation is a symmetry of the system, the presence of the preferred

direction does not affect the beta decay, and the intensity of the emitted electrons is the same, either parallel or antiparallel to spin. Wu and her group implemented the P-transformation physically by inverting the direction of the magnetic field. The nuclear spin then followed the field's direction. One simply then compared the properties of the beta particles for the different directions of the field. The results were unambiguous: the P-transformation is not a symmetry of the weak interactions.

13. The name *kaon* for these particles is a nickname that arose because they were labeled as K_0 and $\bar{K}_0$ when they were first discovered. The kaons in question have no electric charge (they are therefore named neutral) and no spin; they are strange particles, and one is the antiparticle of the other.

14. C. G. Jung, "Archetypes of the Collective Unconscious," in *Collected Works*, vol. 9, pt. 1, edited by G. Adler and R.F.C. Hull (Princeton: Princeton University Press, 1981).

15. M. Gell-Mann, *Phys. Rev.* 125, 1067 (1962).

16. M Gell-Mann, *Phys. Lett.* 8, 214 (1964).

17. Gell-Mann explains the choice of the name quark:

In 1963, when I assigned the name "quark" to the fundamental constituents of the nucleon, I had the sound first, without the spelling, which could have been "kwork." Then, in one of my occasional perusals of *Finnegans Wake*, by James Joyce, I came across the word "quark" in the phrase "Three quarks for Muster Mark." Since "quark" (meaning, for one thing, the cry of a gull) was clearly intended to rhyme with "Mark," as well as "bark" and other such words, I had to find an excuse to pronounce it as "kwork." But the book represents the dreams of a publican named Humphrey Chimp den Ear wicker. . . . From time to time, phrases occur in the book that is partially determined by calls for drinks at the bar. I argued, therefore, that perhaps one of the multiple sources of the cry "Three quarks for Muster Mark" might be "Three quarts for Mister Mark," in which case the pronunciation "kwork" would not be totally unjustified. In any case, the number three fitted perfectly the way quarks occur in nature. (M. Gell-Mann, *The Quark and the Jaguar* [New York: W. H. Freeman, 1994])

18. G. Zweig, CERN preprint 8409/Th. 412 (1964).

19. J. D. Bjorken and S. L. Glashow, *Phys. Lett.* 11, 255 (1964); S. L. Glashow, J. Iliopoulos, and L. Maiani, *Phys. Rev.* D2, 1285 (1970).

20. The latter τ comes from the Greek word "triton" (third), as the tau was the first representative of the third lepton family.

21. Theoretical arguments together with present experimental results suggest that there exist only three families of elementary particles. However, these arguments are based upon an extrapolation of known theories into energies higher than any we have yet accessed, and for this reason they may be spoiled by any surprising new discoveries. There is, however, conclusive evidence that a fourth particle family will be very different from the first three—its neutrino, for example, will have to be ultraheavy.

Chapter 9. Reaching the Limits

1. Noether's contribution to the gauge theory arose from a second version of her theorem—known as Noether's second theorem—which reveals a different manifestation of symmetries in physical theories.

2. M. Atiyah, *Biographical Memoirs 82: Hermann Weyl* (Washington, DC: National Academies Press, 2002).

3. C. N. Yang and R. L. Mills, *Phys. Rev.* 96, 191 (1954).

4. In high-energy experiments, we detect particles with the same spin and small differences in their mass (e.g., the neutron and the proton, the three pions). This implies that the corresponding symmetry in the field is approximate. Still, even an approximate symmetry may be very useful for mathematical calculations, especially if we attribute the failure of symmetry to processes other than the ones on which we focus. For example, the mass difference between proton and neutron can be ignored in the study of strong interactions, and it is attributed to the much weaker electromagnetic force.

5. The first ideas of spontaneous symmetry breaking in physics with reference to the nonuniqueness of the vacuum are found in Heisenberg's quantum mechanical study of ferromagnetism (*Zeit. Phys.* 49, 616 [1928]). These results were subsequently strengthened by Lev Landau (*Phys. Z. Sovjetunion* 11, 26 [1937] and 545 [1937]). Heisenberg was again the first to introduce the idea of spontaneous symmetry breaking in quantum field theory, at the 1958 Annual International Conference on High Energy Physics (CERN, 119). His work was very influential, even though it was off the mark in its details. Y. Nambu gave a concrete account of the relation between spontaneous symmetry breaking in condensed matter and in particle physics (*Proceedings of the 1960 Annual International Conference on High Energy Physics*, Rochester, 858). He was followed by J. Goldstone (*Nuov. Cim.* 19, 965 [1961]). The incorporation of spontaneous symmetry breaking into gauge theory followed in the work of M. Baker and S. L. Glashow (*Phys. Rev.* 128, 2462 [1962]) and P. W. Anderson (*Phys. Rev.* 130, 439 [1963]).

6. Higgs was the first to study the effects of the spontaneous symmetry breaking on the mass of the mediating fields in gauge theories that have the same symmetry as electromagnetism (*Phys. Lett.* 12, 132 [1964]; *Phys. Rev.* 145, 1156 [1966]). The study of the most general case is by T.W.B. Kibble (*Phys. Rev.* 155, 1554 [1967]).

7. The paradox of the Buridan's ass is a satire of a theory of the French philosopher Jean Buridan, who lived in the fourteenth century A.D. He was an advocate of moral determinism, stating essentially that a human faced with alternative courses of action must always choose rationally the greater good. Buridan allowed that the choice could be delayed in order to assess more fully the possible outcomes. He envisioned an example of a dog faced with the choice of two equally desirable and accessible meals. Having no reason to prefer one to the other, its choice must necessarily be random. The satirist substituted the dog with an ass and suggested that Buridan should remain consistent with

his determinism. The poor animal will then starve to death because it will be perpetually unable to make a choice.

8. S. Weinberg, *Phys. Rev. Lett.* 19, 1264 (1967); A. Salam, in *Elementary Particle Theory: Proceedings of the Novel Conference VIII* (Stockholm: Almvist and Wiksell, 1968). Apart from spontaneous symmetry breaking, the models in these papers were by S. L. Glashow (*Nucl. Phys.* 22, 569 [1961]).

9. N. Cabibbo, *Phys. Rev. Lett.* 10, 531 (1963).

10. This rule may seem somewhat arbitrary; it is in fact the result of a rather complicated method of counting the possible rotations of the quantum phases in relation to the gauge symmetries of the theory. It depends strongly on the fact that we have three particle families. If the families were only two, this change in the quantum phase would not have taken place. If, on the other hand, there were four particle families, the quantum phases would have to change in four different places during the mixing. For n families, the number of such changes is $(n-1)(n-2)/2$ (M. Kobayashi and M. Maskawa, *Prog. Theor. Phys.* 49, 652 [1973]).

11. Slightly rephrased from M. Goldberger, *The Quantum Theory of Fields: 12th Solvay Conference* (New York: Interscience, 1961).

12. M. Gell-Mann, *Physics* 1, 63 (1964).

13. M. Veltman, *Nucl. Phys.* B7, 637 (1968); G. 't Hooft, *Nucl. Phys.* B33, 173 (1971); G. 't Hooft, *Nucl. Phys.* B35, 167 (1971); G. 't Hooft and M. Veltman, *Nucl. Phys.* B44, 189 (1972).

14. Color was proposed in the mid-1960s to solve some of the problems characterizing Gell-Mann's original quark model (O. W. Greenberg, *Phys. Rev. Lett.* 13, 598 [1964]; M. Han and Y. Nambu, *Phys. Rev.* 139B, 1006 [1965]).

15. The fact that asymptotic freedom is present in quantum chromodynamics was discovered independently by David Politzer (*Phys. Rev. Lett.* 30, 1346 [1973]) and by David Gross with Frank Wilczek (*Phys.Rev.* D8, 3633 [1973]; *Phys.Rev.* D9, 980 [1974]).

16. The most important consistency problem of QCD is known as the strong CP problem. There are strong arguments that the full quantum treatment of QCD should lead to an additional source of CP violation. The reason is that the complexity of the gauge symmetry makes the choice of vacuum in the theory nonunique. Different choices of vacuum are labeled by a parameter known as the theta parameter of QCD. There is no theoretical argument that allows one to predict the value of theta. It is a free parameter of the theory, which can be determined only by experiment. The mathematical terms that depend on theta provide a source for CP violation in the strong interactions. However, no such violation has ever been observed. This implies that the value of the theta parameter is extremely small, almost zero. Since theta is a free parameter, the value zero is not forbidden. However, it is very unlikely because the value of theta arises as a sum of contributions from different fields. Therefore, a zero value of theta would involve a near miraculous cancelation of different mathematical terms.

There are several proposals for the resolution of the problem above. The most popular one involves the postulate of a new particle, the *axion*, which is related to an additional gauge symmetry of the system. In this view, spontaneous symmetry breaking results in the absorption of the theta parameter within the structure of the vacuum that corresponds to the axion.

Chapter 10. Outlook

1. A description of quantum field theory in a language that emphasizes the primacy of the particle concept is contained in one of the most influential modern textbooks in quantum field theory, S. Weinberg, *The Quantum Theory of Fields: I. Foundations* (Cambridge: Cambridge University Press, 1995). The opposing view—that particles are nothing but the manifestation of fields in our measuring devices—is developed, for example, in an article by H. D. Zeh (*Phys. Lett.* A309, 329 [2003]).

2. S. W. Hawking, *Comm. Math. Phys.* 43, 199 (1975).

3. W. G. Unruh, *Phys. Rev.* D14, 870 (1976).

4. Unruh's and Hawking' s results were obtained initially in the context of quantum field theory in a curved spacetime. This description is an approximation in the sense that it ignores the quantum nature of spacetime and studies only the effects of the classical "gravitational fields" of general relativity on matter that satisfies the laws of quantum mechanics. In this method, fields are treated as primary objects and particles as secondary, derived ones. However, Hawking's and Unruh's results persist even if one starts from the opposing assumption and considers particles as fundamental.

An experimental verification of the Unruh effect is extremely difficult: to obtain a temperature of 0.1 degrees that can be barely distinguished with present methods, one needs to achieve accelerations 200 times larger than the one Earth is exerting on bodies on its surface.

5. The combination of a fast computer with huge memory, very efficient programmers, and a lot of theoretical work may allow the theoretical prediction and description of any chemical phenomenon solely in terms of the basic physical parameters. In practice, however, computers have only a finite speed and memory, and for this reason the properties of only a limited class of atomic and molecular systems have been determined from first principles.

6. One of the most important results in this direction is the proof of the stability of matter in bulk in the context of nonrelativistic quantum mechanics. This means that the energy per particle in a system of fermions interacting through electromagnetic forces cannot take arbitrarily large negative values. The proof involves certain simplifying assumptions (i.e., treating the electromagnetic field as nondynamical and subject to classical physics), which are nonetheless reasonable as first approximations. The result above was obtained by F. J. Dyson and A. Lenard (*J. Math. Phys.* 8, 423 [1967]; *J. Math. Phys.* 9, 698 [1968]) and then strengthened by E. H. Lieb and W. Thirring (*Phys. Rev.*

Lett. 35, 687 [1975]). Pauli's exclusion principle plays a crucial role in these proofs. Matter would not have been stable if the fundamental particles were bosons. Analogous proofs exist about the stability of atoms and stars (see the review in E. H. Lieb, *Rev. Mod. Phys.* 48, 553 [1976]). However, a definite proof should involve the full theory of quantum electrodynamics, something that is extremely difficult and has not yet been achieved. There are some results on the stability of matter in the bulk and on the stability of atoms that involve quantum field theory, but in a rather restricted context. (See a review in E. H. Lieb, Proceedings of the Werner Heisenberg Centennial, Munich, Dec. 2001, math-ph/0209034.)

7. There exist certain models for quantum theory, in which a random "force" is exerted upon the wave functions of a physical system, with the result that all quantum behavior is lost in macroscopic systems (see, for example, G. C. Ghirardi, A. Rimini, and T. Weber, *Phys. Rev.* D34, 470 [1986]). These forces are negligible in the micro world, but they strongly affect the behavior of larger systems. Models of this type involve variations of standard quantum mechanics, which can neither be proved nor disproved by the present state of experiments. Their main benefit is that they provide a solution of the so-called measurement problem, which has plagued quantum mechanics ever since its inception. (In my opinion, this is the best resolution available at the moment, even though it appears rather inelegant and arbitrary.) The main drawback of such models is that they cannot be easily reconciled with relativity.

8. L. Wittgenstein, *Tractatus Logicophilosophicus* (London: Routledge, 1997), 6.371.

9. H. Weyl, *Philosophy of Mathematics and Natural Science* (Princeton: Princeton University Press, 1949).

10. Present-day cosmology is based on general relativity, which is a classical deterministic theory. Quantum effects are usually appended to this classical description, but the spacetime is not treated with the language of quantum theory. Quantum cosmology ιinvolves a description of the Universe in which the spacetime itself is subject to the rules of quantum theory. It was a very active topic of research in the 1980s, but related research declined in later years. There are two reasons for this: the first is that we possess at present no realistic description of quantum gravity; the second is that we are not sure how to understand quantum theory when it refers to an individual system such as the Universe, in which case the interpretation in terms of measurements is problematic.

11. K. Cavafy, *Ithaca* (1911).

GLOSSARY

Absolute (time and space): The basic assumption of Newtonian mechanics was that time and space are absolute and predetermined structures, not dependent in their flow and extension, respectively, on any material processes. The existence of absolute space and time implies that motion is also absolute, namely, there exists a reference frame of perfect rest.

Analytic geometry: The study of geometry with numbers and functions through the introduction of coordinate systems.

Boost: See Lorentz transformations.

Boson: A particle with spin equal to an integer multiple of Planck's constant. Bosons do not satisfy the exclusion principle. Photons, gauge particles, and the Higgs particles are all bosons.

Color: A "charge" special to quarks, which takes three possible values and appears in the context of quantum chromodynamics, the theory of the strong interactions.

Commutative: The multiplication of any two numbers is commutative in the sense that the product does not depend on which term is the first. The multiplication of matrices, however, is not commutative: the product of AB of two matrices A and B is not equal to the product BA.

Complex numbers: Numbers that can be written as the sum of ordinary numbers with numbers that arise from the multiplication of ordinary numbers with the square root of the number -1. Complex numbers can be represented by points on a plane.

Copenhagen interpretation: The dominant interpretation of quantum theory, which was developed by Bohr, Heisenberg, and Born. It affirms that the rules of quantum theory refer to the concrete physical context of measurement and not to the properties of the material things in themselves.

Degree of freedom: A degree of freedom is a physical quantity that needs to be specified in order to determine the classical profile or the quantum state of a physical system.

Determinism: The concept that the complete knowledge of the present state of the world allows a full determination of the future. In classical physics, this determination arises from the laws of motion.

Dirac field: A type of field that refers to particles with spin $\frac{1}{2}$. Its mathematical structure was determined through Dirac's theory of the electron.

Dynamical: A physical magnitude is dynamical (in the context of mechanics) if its definition involves further concepts than geometrical ones. Mass, force, and energy are dynamical magnitudes.

Entanglement: The quantum phenomenon according to which a composite system seems to have more "potential properties"—possible quantum states—than the sum of "potential properties" of its constituents.

Exclusion principle: The statement that at most one particle of a given type can fill each energy level of a physical system.

Fermion: Particle with spin equal to a half-integer multiple of Planck's constant. Fermions satisfy the exclusion principle. Electrons, neutrinos, and quarks are all fermions.

Force: In Newton's theory, the force is the cause of the acceleration of physical bodies. In philosophy the concept of force referred to a principle of nature, which was the intrinsic cause of motion in things. In high-energy physics, the word "force" is used heuristically to describe the three types of microscopic interaction: strong, weak, and electromagnetic forces.

Function: In the simplest possible case, a function is defined as a set of rules that assigns to every number a specific mathematical object (usually another number).

Gauge transformation: A transformation that corresponds to the change of the quantum phase (or of a similar object) by a different amount at each spacetime point.

Generator of a transformation: A physical quantity that may can employed in order to construct a specific transformation of profiles of a given physical system.

Gluons: The eight types of particle with spin 1 and zero mass that mediate the strong interactions according to quantum chromodynamics.

Helicity: A physical quantity for massless particles with spin: it takes value $+1$ if spin is parallel to the direction of motion and -1 if it is antiparallel. A specific particle cannot have its helicity changed in any way.

Higgs fields: Fields corresponding to particles of spin zero, which are introduced in quantum field theories, in order to effect the spontaneous symmetry breaking of the gauge symmetry.

Irreducible system (associated with a specific symmetry): One of the simplest possible physical systems that is characterized by this symmetry. An irreducible system cannot be mathematically decomposed into two smaller systems characterized by this symmetry.

Isospin: A physical quantity that is mathematically similar to the quantum mechanical spin and is conserved in the strong interactions. Particles come in isospin doublets (proton and neutron) or triplets (the three π particles), etc.

Kinematical: In the context of mechanics, a physical quantity is called kinematical if it can be fully defined in terms of the geometry of space and time. The position and velocity of a particle are, for, example kinematical quantities.

Kinetic theory: A theory that describes the phenomena of heat in gases as arising from the motions of their constituent molecules. Temperature arises in the kinetic theory as a measure of the mechanical energy carried by the molecules.

Locality principle: A basic principle of modern physics that any interaction between physical bodies must be mediated between them by means of some material object and cannot happen instantaneously. In the context of relativity, the locality principle implies that no signal can travel faster than light.

Lorentz transformations: The set of transformations between inertial frames that move with respect to each other. The key feature of these transformations is that they preserve the speed of light. One usually includes spatial rotations in the definition of the Lorentz transformations. The pure Lorentz transformations, namely, the ones that transform between reference frames that move with respect to each other, are referred to as **boosts**.

Massless: A particle is massless if its rest mass equals zero. In this case, the particle moves with the speed of light. All particles with nonzero rest mass are called **massive**.

Matrix: A table whose entries are numbers. Matrices can be added, subtracted, and multiplied like numbers. Their multiplication is, however, noncommutative.

Minkowski spacetime: The mathematical space obtained from the union of Newton's absolute space and time in a way that respects the Lorentz transformations between inertial frames.

Noether's theorem: The statement that the existence of a symmetry to the laws of motion for a physical system implies the existence of conserved quantities.

Observable: According to the Copenhagen interpretation, the physical quantities of quantum theory refer to measurements, and for this reason they should be designated as observables (see also **Q-numbers**).

Perturbation theory: A mathematical technique that is used in quantum field theory. It describes the effects of interactions between particles by adding them as small corrections to the description of the motion of free (noninteracting) particles.

Photoelectric effect: A phenomenon of interaction between light and atoms. Its explanation played a crucial role in the development of quantum theory.

Positivity of energy: The statement that the energy of a physical system cannot take arbitrarily large negative values (in fact it can only take positive ones), for otherwise matter would not be stable.

Planck's constant: The fundamental constant of nature that appears in all laws that describe quantum phenomena.

Poincaré symmetry: The basic symmetry of Minkowski spacetime, which includes the transformations of Lorentz and the translations in space and time.

Renormalization: The absorption of the infinities of perturbation theory into a redefinition of the basic physical parameters of a quantum field theory.

Running coupling constant: A function that determines how the strength of an interaction varies with energy.

Q-numbers: The mathematical objects that represent physical magnitudes in quantum mechanics. Q-numbers are noncommutative and essentially correspond to matrices.

Quantum phase: A mathematical object appearing in quantum mechanics that acts like a register keeping track of a system's past history. It cannot be measured in individual systems, but only through statistics of a large number of systems.

Separation (timelike-spacelike-null): Two events are timelike separated if they can be connected by a signal that moves with speed lower than that of light. If they can only be connected by a signal that moves with exactly the speed of light, they are null separated. Otherwise, they are spacelike separated.

Spin-statistics theorem: The statement that identical particles with spin a half-integer multiple of Planck's constant satisfy the exclusion principle (they are connected by Fermi connections), and ones with spin an integer's multiple of Planck's constant do not satisfy the exclusion principle (they are connected by Bose connections).

Spontaneous symmetry breaking: The nonmanifestation of a physical system's symmetry at the level of observation because of the nonuniqueness of its vacuum state.

Statistical mechanics: The theory that uses probabilistic arguments to describe the thermal properties of macroscopic objects in terms of the laws of motion of their constituents.

Vacuum: The state of minimum energy for a quantum field. It remains invariant under the Poincaré symmetry; in most quantum field theories, it is unique.

Ψ-function (or wave function): The function introduced by Schrödinger, which incorporates the effects of the wave behavior of ordinary matter. According to the interpretation of Born, the Ψ-function describes the probabilities relevant for the description of a physical system.

GUIDE FOR
FURTHER READING

This note is not intended to provide a comprehensive bibliography (let alone an exhaustive one) for the subject. Its only aim is to list some books that I found useful or enjoyable, and which in my opinion provide a good starting point for any reader who would like to explore in more detail the topics covered in the book.

For a history of the theories about matter in antiquity, Plato's *Timaeus* still makes pleasant reading and Aristotle's *Physics* may be hard to follow, but it is very informative about the ancients' way of thinking. There are many good translations in print. I also recommend R. Sorabji's *Matter, Space and Motion: Theories in Antiquity and Their Sequel* (Ithaca: Cornell University Press, 1992). For the era of the Scientific Revolution, C. C. Gillispie's *Edge of Objectivity* (Princeton: Princeton University Press, 1966) is classic; the same holds for T. S. Kuhn's *Structure of Scientific Revolutions* (Chicago: University of Chicago Press, 1996). Galileo's *Dialogue Concerning the Two Chief World Systems* is still enjoyable, while a summary of Newton's thought can be found in *Newton*, edited by B. Cohen and R. S. Westfall, (New York: W. W. Norton, 1995).

Three nice books that deal with the development of important physical concepts from antiquity until the twentieth century are M. Jammer's *Concepts of Mass* (New York: Dover, 1993) and *Concepts of Force* (New York: Dover, 1999), and B. Pullmann's *The Atom in the History of Human Thought* (Oxford: Oxford University Press, 2001). For the development of physics in the nineteenth century, see R. D. Purrington's *Physics in the Nineteenth Century* (New Brunswick, NJ: Rutgers University Press, 1997) and P. M. Harman's *Energy, Force and Matter* (Cambridge: Cambridge University Press, 1982).

For the theories of relativity, a large number of books at all levels exist. My favorite popular accounts are R. Geroch's *General Relativity from A to B* (Chicago: University of Chicago Press, 1981) and D. Bodanis's *E=mc²: A Biography of the World's Most Famous Equation* (New York: Berkley Books, 2001). For the development of Einstein's thought, see the scientific biography by A. Pais, *Subtle Is the Lord: The Science and Life of Albert*

Einstein (Oxford: Oxford University Press, 1982). If someone would like to enter into the mathematical niceties of the theory, a book I am very fond of is the rather old-fashioned account of P.A.M. Dirac, *General Theory of Relativity* (Princeton: Princeton University Press, 1996).

For quantum mechanics, I cannot recommend Heisenberg's *Physics and Philosophy* (New York: Prometheus Books, 1999) highly enough. M. Jammer's *The Conceptual Development of Quantum Theory* (New York: Mc-Graw-Hill, 1966) is full of detail but seemingly out of print. For the birth of modern atomic theory, S. Weinberg's *The Discovery of Subatomic Particles* (Cambridge: Cambridge University Press, 2003) is very informative. For a technical introduction to the theory, my preferences lie with L. E. Ballentine's *Quantum Mechanics: A Modern Development* (River Edge, NJ: World Scientific, 1998); for an introduction with an emphasis on the major conceptual issues, see C. J. Isham's *Lectures on Quantum Theory* (River Edge, NJ: World Scientific, 1995).

The closest to a general history of the twentieth-century developments in the understanding of matter I have come across is A. Pais's *Inward Bound: Of Matter and Forces in the Physical World* (Oxford: Oxford University Press, 1988). For a nontechnical account of modern particle physics, see Y. Ne'eman and Y. Kirsh, *The Particle Hunters* (Cambridge: Cambridge University Press, 1986). Some popular accounts from physicists who participated in these developments, like R. P. Feynman's *Q. E. D.: The Strange Theory of Light and Matter* (London: Penguin Books, 1990), G. 't Hooft's *In Search of the Ultimate Building Blocks* (Cambridge: Cambridge University Press, 1996), and M. G. Veltman's, *Facts and Mysteries in Elementary Particle Physics* (River Edge, NJ: World Scientific, 2003), are very interesting in their perspective. For a technical introduction to quantum field theory, a good point of departure is A. Zee's *Quantum Field Theory in a Nutshell* (Princeton: Princeton University Press, 2003).

One relevant book that resists classification is H. Weyl's beautiful essay on *Symmetry* (Princeton: Princeton University Press, 1952). Finally, I cannot recommend highly enough a recent work that contains the full perspective of modern physics within a single volume. It is R. Penrose's *The Road to Reality: A Complete Guide to the Laws of the Universe* (London: Jonathan Cape 2004).

INDEX

absolute space, 49–50

absolute time, 49–50, 107, 116

accelerators, 285

Aharonov, Yakir, 210

alchemy, 20–21, 57, 368n10; and philosopher's stone, 20

alpha particles, 155–157, 272–275

analytic geometry, 35, 42

angular momentum, 138; discrete values of, 161–162

antiparticles, 233–237

Aristotle, 9, 17–18, 26, 342

asymptotic freedom, 334, 394n15

atomic number, 67–69; explanation of, 158

atoms, 4; Dalton's, 65–66, Democritus's, 14, 368n8; Epicurus's, 16, 368n7; in Indian philosophy, 367n2; jelly model of, 156; Nagaoka's model of, 381n11; quantum description of, 199–202; Rutherford-Bohr model of, 157–163; spectra of, 160–161; stability of, 159, 272; and the Void, 16, 62

axiomatic method, 31; in quantum field theory, 255–260, 389n17

axion, 394n16

Bacon, Francis, 25

Bacon, Roger, 26

baryon number, 282, 284, 289, 324

baryons, 284, 289, 291, 296–297, 331

B-charge. *See* baryon number

Bequerel, Henri, 155

Bergson, Henri, 35

Berry, Michael, 211

beta particles, 155, 272, 274, 276–277, 298

Black, Joseph, 63

black body radiation, 147–151

Bohm, David, 210, 382n19

Bohr, Niels, 159–163, 170, 173, 177–178, 181, 191–194, 207, 344, 384n38

Boltzmann, Ludwig, 74, 147

Born, Max, 169, 177–179, 188–190

Bose, Satyenda Nath, 222

Boskovitch, Roger Joseph, 60, 239, 372n1

bosons, 222

Buridan, Jean, 316, 371n27, 393n7

caloric, 69–70

Carnot, Sadi, 71

Cartan, Elie, 332n

cathode rays, 154–155, 158

Cavendish, Henry, 63, 374n14

charge conjugation. *See* C-symmetry

charm, 300

classical physics, 162; in relation to quantum mechanics, 170, 206–208

complementarity principle, 192–193, 207, 229, 259

Compton, Arthur, 153

Comte, Auguste, 382n21

connection, 213–215; of Bose type, 221–223, 226–227; of Fermi type, 221–223, 236

INDEX